R.W. Ridge, A.M.C. Emons (Eds.)

Root Hairs

Cell and Molecular Biology

With 77 Figures, Including 1 in Color

Springer

Robert William Ridge
Department of Biology
Division of Natural Sciences
International Christian University
3-10-2 Osawa, Mitaka, Tokyo 181-8585
Japan

Anne Mie C. Emons
Department of Plant Sciences
Wageningen University
Arboretumlaan 4, NL-6703 BD Wageningen
The Netherlands

Cover photo credit:
Scanning images showing frozen hydrated root hairs on a primary root of *Vicia sativa* spp. *nigra* (front cover image, courtesy of Norbert de Ruijter and Adriaan van Aelst, Wageningen University) and *Arabidopsis thaliana* Columbia (back cover image, courtesy of John Esseling and Adriaan van Aelst, Wageningen University).
The roots were grown in air, frozen in liquid nitrogen, freeze-dried for 2 min at -90°C, 10 nm Pt sputter-coated, and observed in a cryo field-emission scanning electron microscope at 5 kV at -180°C.

ISBN-13: 978-4-431-68372-8

Library of Congress Cataloging-in-Publication Data

Root hairs : cell and molecular biology / R.W.Ridge, A.M.C.Emons, (eds.).
p.cm.
Includes bibliographical references and index.
ISBN-13: 978-4-431-68372-8 e-ISBN-13: 978-4-431-68370-4
DOI: 10.1007/978-4-431-68370-4
1. Roots (Botany) I. Ridge, R.W. (Robert William), 1951- II. Emons, A.M. (Anne Mie C.), 1942-

QK644 .R653 2000
575.5'4—dc21

00-036556

Printed on acid-free paper

Softcover reprint of the hardcover 1st edition 2000

Typesetting: Camera-ready by editors/authors

SPIN: 10691887

Springer
Tokyo
Berlin
Heidelberg
New York
Barcelona
Hong Kong
London
Milan
Paris
Singapore

Springer
Tokyo
Berlin
Heidelberg
New York
Barcelona
Hong Kong
London
Milan
Paris
Singapore

Springer
Tokyo
Berlin
Heidelberg
New York
Barcelona
Hong Kong
London
Milan
Paris
Singapore

Springer
Tokyo
Berlin
Heidelberg
New York
Barcelona
Hong Kong
London
Milan
Paris
Singapore

Contributors

T.V. Bhuvaneswari
Department of Biology, Faculty of Science, University of Tromsø, N-9037 Tromsø, Norway
Fax: +47 77 6333. Email: bhuvana@ibg.uit.no

Tatiana Bibikova
Plant Molecular Science Group, Division of Biochemistry and Molecular Biology, Institute of Biomedical and Life Sciences, Bower Building, University of Glasgow, Glasgow G12 8QQ, UK
Fax: +44 141330 4620. Email: tbibi001@udcf.gla.ac.uk

Ton Bisseling
Laboratory of Molecular Biology, Department of Plant Sciences, Wageningen University, Dreijenlaan 3, 6703 HA Wageningen, The Netherlands
Fax: +31 317 483584. Email: Ton.Bisseling@mac.mb.wau.nl

Nawal M. Boukli
L.B.M.P.S., Université de Genève, 1, ch. De l'Impératrice, 1292 Chambésy, Genève, Switzerland
Fax: +41 22 732 0734. Email: boukli@sc2a.unige.ch

William J. Broughton
L.B.M.P.S., Université de Genève, 1, ch. De l'Impératrice, 1292 Chambésy, Genève, Switzerland
Fax: +41 22 732 0734. Email: brougthw@sc2a.unige.ch

Alison Cavell
IACR-Long Ashton Research Station, Department of Agricultural Sciences, University of Bristol, Long Ashton, Bristol BS41 9AF, UK
Fax: +44 1275 394281. Email: alison.cavell@bbsrc.ac.uk

Norbert de Ruijter
Department of Plant Sciences, Laboratory of Plant Cytology and Morphology, Wageningen University, Arboretumlaan 4, 6703 BD, Wageningen, The Netherlands
Fax: +31 317 485005. Email: Norbert.deRuijter@algem.pcm.wau.nl

Anne Mie C. Emons
Laboratory of Experimental Plant Morphology and Cell Biology, Department of Plant Sciences, Wageningen University, Arboretumlaan 4, NL-6703 BD Wageningen, The Netherlands
Fax: +31 317 485005. Email: annemie.emons@algem.pcm.wau.nl

John Esseling
Laboratory of Experimental Plant Morphology and Cell Biology, Department of Plant Sciences, Wageningen University, Arboretumlaan 4, NL-6703 BD Wageningen, The Netherlands
Fax: +31 317 485005. Email: John.Esseling@algem.pcm.wau.nl

Hubert H. Felle
Botanisches Institut, Justus-Liebig-Universität, Senckenbergstr. 17, D-35390 Giessen, Germany
Fax: +49 641 99 35119. Email: Hubert.Felle@bio.uni-giessen.de

Charlotta A. Fredriksson
Biology Department, International Christian University, 3-10-2 Osawa, Mitaka-shi, Tokyo 181-8585, Japan
Fax: +81 422 33 1449. Email: fredriksson@ icu.ac.jp

Theodorus W.J. Gadella Jr.
MicroSpectroscopy Center Wageningen, Department Biomolecular Sciences, Agricultural University Wageningen, Dreijenlaan 3, 6703 HA Wageningen, The Netherlands
Fax: +31 317 483584. Email: Dorus.Gadella@laser.bc.wau.nl

Moira E. Galway
Department of Biology, St. Francis Xavier University, Antigonish, Nova Scotia, B2G 2W5, Canada
Fax: +1 902 867 2389. Email mgalway@stfx.ca

Joachim Goedhart
MicroSpectroscopy Center Wageningen, Department Biomolecular Sciences, Agricultural University Wageningen, Dreijenlaan 3, 6703 HA Wageningen, The Netherlands
Fax: +31 317 483584. Email: Goedhart@gcg.tran.wau.nl

Simon Gilroy
Biology Department, The Pennsylvania State University, 208 Mueller Lab, University Park, PA, 16802, USA
Fax: +1 814 865 9131. Email: sxg12@psu.edu

Claire Grierson
IACR-Long Ashton Research Station, Department of Agricultural Sciences, University of Bristol, Long Ashton, Bristol BS41 9AF, UK
Fax: +44 1275 394281. Email: Claire.Grierson@bbsrc.ac.uk

Frédérique C. Guinel
Department of Biology, Wilfrid Laurier University, Waterloo, Ontario N2L 3C5, Canada
Fax: +1 519-746-0677. Email: fguinel@wlu.ca

Almut Herrmann
Botanisches Institut, Justus-Liebig-Universität, Senckenbergstr. 17, D-35390 Giessen, Germany
Fax: +49 641 99 35119. Email: Almut.Herrmann@bio.uni-giessen.de

Ann M. Hirsch
Department of Molecular, Cell and Developmental Biology and Molecular Biology Institute, 405 Hilgard Avenue, University of California, Los Angeles, CA 90095-1606 USA
Fax: +1 310-206-5413. Email: ahirsch@ucla.edu

Helen R. Irving
Department of Pharmaceutical Biology & Pharmacology, Monash University, 381 Royal Parade, Parkville, VIC 3052, Australia
Fax: +61 39903 9638. Email: helen.irving@vcp.monash.edu.au

Mariko Izumo
Biology Department, International Christian University, 3-10-2 Osawa, Mitaka, Tokyo 181-8585, Japan. Now at: Dept Biology, Vanderbilt University, Box 1812-B, Nashville, TN 37235, USA
Fax: +1 615 343 0336. Email: izumom@ctrvax.vanderbilt.edu

Andrea Jahraus
Laboratory of Molecular Biology, Department of Plant Sciences, Wageningen University, Dreijenlaan 3, 6703 HA Wageningen, The Netherlands
Fax: +31 317 483 584. Email: Andrea.Jahraus@mac.mb.wau.nl

Masayuki Katsumi
Biology Department, International Christian University, 3-10-2 Osawa, Mitaka, Tokyo 181-8585, Japan
Fax: +81 422 33 1449. Email: katsumi@icu.ac.jp

Marilyn N. Kelly
Department of Pharmaceutical Biology & Pharmacology, Monash University, 381 Royal Parade, Parkville, VIC 3052, Australia
Fax: +61 39903 9638. Email: marilyn.kelly@vcp.monash.edu.au

Tijs Ketelaar
Laboratory of Experimental Plant Morphology and Cell Biology, Department of Plant Sciences, Wageningen University, Arboretumlaan 4, NL-6703 BD Wageningen, The Netherlands
Fax: +31 317 485005. Email: Tijs.Ketelaar@Guest.PCM.WAU.NL

Roger R. Lew
York University, Biology Department, 4700 Keele Street, Toronto, Ontario M3J 1P3, Canada
Fax: +1 416 736 5698. Email: planters@yorku.ca

R. Larry Peterson
Department of Botany, University of Guelph, Guelph, Ontario N1G 2W1, Canada
Fax: +1 519 767 1991. Email: lpeterso@uoguelph.ca

Robert W. Ridge
Biology Department, International Christian University, 3-10-2 Osawa, Mitaka, Tokyo 181-8585, Japan
Fax: +81 422 33 1449. Email: ridge@icu.ac.jp

John Schiefelbein
Department of Biology, 830 North University Avenue, University of Michigan, Ann Arbor, MI 48109, USA
Fax: +1 734 764 0884. Email: schiefel@umich.edu

Bjørn Solheim
Department of Biology, Faculty of Science, University of Tromsø, N-9037 Tromsø, Norway
Fax: +47 77 6333. Email: bsolheim@ibg.uit.no

Kevin J. Stevens
Department of Botany, University of Guelph, Guelph, Ontario N1G 2W1, Canada
Fax: +1 519 767 1991. Email: kjsteven@uoguelph.ca

Antonius C.J. Timmers
Laboratoire de Biologie Moléculaire des Relations Plantes-Microorganismes, Chemin de Borde-Rouge, BP27, 31326 Castanet-Tolosan, France
Fax: +33 561 285061. Email: timmers@toulouse.inra.fr

Jan A. Traas
Laboratoire de Biologie Cellulaire, Institut National de la Recherche Agronomique, Route de Saint Cyr, 78026 Versaille Cedex, France
Fax: +33 01 30 83 34 58. Email: traas@versailles.inra.fr

Chapter 1
Root Hair Ultrastructure and Tip Growth

Moira E. Galway

Department of Biology, St. Francis Xavier University, Antigonish, Nova Scotia, B2G 2W5, Canada

1 The Ultrastructural Approach to Tip Growth

The major focus of this review is the small but dynamic volume of cytoplasm that fills the tips of all growing root hairs and in which tip growth takes place. Electron microscopy is one way to investigate this space and to obtain "snapshots" of tip-growing cells in action. Combined with other methods, electron microscopy provides a powerful means to investigate the mechanism of tip growth in root hairs that has yet to be fully exploited. For example, immunoelectron microscopy has revealed the distribution of cell wall matrix components in the secretory vesicles of *Vicia* root hairs (Sherrier and VandenBosch 1994); quantitative analysis of electron micrographs has been used to determine secretory vesicle production rates (Picton and Steer 1981), and to identify sites of endocytosis and exocytosis in pollen tubes (Derksen et al. 1995b). Electron microscopy can be used to explore the effects of experimentally perturbing tip growth with drugs such as cytochalasin (Ridge 1990) or to determine how tip growth is altered in mutant root hairs (Galway et al. 1997; 1999). Thus electron microscopy remains an essential tool for cell biologists, and it allows the findings of biochemists and molecular biologists to be placed "in the context of a living, functioning cell" (Steer 1991).

2 The Ultrastructure of Root Hairs and Other Tip-Growing Cells

All tip-growing cells exhibit a characteristic polarized organization of the cytoplasm in which the organelles and vesicles involved in the synthesis and secretion of new cell wall and plasma membrane are concentrated at the growing tips of the cells (reviewed by Battey and Blackbourn 1993; Harold 1990; Schnepf 1986; Sievers and Schnepf 1981). All tip-growing cells must carry out certain

---

Key words. *Arabidopsis thaliana*, cryofixation, electron microscopy, endocytosis, endoplasmic reticulum, exocytosis, fixation, membranes, morphogenesis, mutants, secretion, tip growth, ultrastructure, vacuoles, vesicles processes essential for polarized growth, such as directed vesicle transport and membrane turnover. The components required to support these processes such as the endomembrane system, motor proteins, coated vesicles, and vesicle docking and fusion proteins are found in similar forms in plants, animals and fungi (Battey and Blackbourn 1993; Hawes and Satiat-Jeunemaitre, Low and Chandra 1994; Thiel and Battey 1998). Nevertheless, tip-growing cells like root hairs, pollen tubes, hyphae and moss protonemata have different origins and functions, resulting in considerable structural and physiological diversity among these cells. Pollen tubes and root hairs exhibit major structural differences, for example in cell wall composition and organization, as well as more subtle differences in the control of tip growth (Bibikova et al. 1997). Thus it is not surprising that mutations in genes which cause dramatic changes in root hair morphology do not always affect pollen tube structure or function (Schiefelbein et al. 1993; Wen and Schnable 1994). Such diversity between different types of tip-growing cells has led to the recognition that the components of different tip-growing cells may be assembled and controlled in different ways (Battey and Blackbourn 1993; Harold 1997; Mahló 1998).

In this review, major features of root hair ultrastructure will be presented in light of our current understanding of tip growth, and comparisons will be made with other tip-growing cells, particularly pollen tubes. Areas which require more research will be indicated. Recent major investigations of root hair ultrastructure that are referred to include: Emons (1987; *Equisetum hyemale*, *Limnobium stoloniferum*), Ridge (1988, 1995; *Vicia hirsuta*), Galway et al. (1997, 1999; *Arabidopsis thaliana*). In all of these studies, optimum preservation of the hairs for electron microscopy was obtained by using freeze substitution. Comprehensive reviews of root hair structure and function can be found in Peterson and Farquhar (1996), Ridge (1996) and Miller et al. (1997), and recent reviews of tip growth include: Harold (1997), Kropf et al. (1998) and Yang (1998).

3 Impediments to Understanding Root Hair Ultrastructure

At present our knowledge of root hair fine structure is limited in three major ways. First, there have been fewer studies of root hairs than studies of pollen tubes or fungal hyphae. Second, our knowledge is derived from root hairs of seedlings grown in soil-free conditions such as humidified air, liquid or agar, because soil-free roots are easier to grow, observe, and prepare for electron microscopy. However, root hair lifespan and morphology differ in soil compared to soil-free media. In soil-free media, root hairs are long-lived, straight, tubular cells, whereas in soil hair lifespan is shorter and hair shape more varied due to contact with soil particles (Farr 1928; Haberlandt 1914; for recent discussion see McCully 1995) or due to interactions with soil microorganisms (see Chapters 14, 17 and 18, this volume).

The third major limitation to our understanding root hair ultrastructure has been the difficulty in preserving these cells for electron microscopy. They are poorly preserved by chemical fixation because their large vacuoles make them sensitive to osmotic changes during fixation (Cooper 1982; Ridge 1988). Only short, young hairs were successfully fixed in earlier studies (e.g., Bonnett and Newcomb 1966; Newcomb and Bonnett 1965; Robertson and Lyttleton 1982). This problem has been overcome in the last decade by the use of cryofixation, in the form of plunge-freezing followed by freeze substitution. Further information on cryofixation methods suitable for root hairs can be found in Galway et al. 1995, and in references therein.

Chemically fixed and cryofixed root hairs exhibit a similar complement of organelles and a similar polarized organization of the cytoplasm. However, membranes and organelles are much better preserved by cryofixation (Emons 1985; 1987; Ridge 1988; 1995; Sherrier and VandenBosch 1994). Chemical fixation should be avoided not only because it can change the morphology of membranes and organelles, it can also induce cytoplasmic movements that redistribute organelles within cells (Doris and Steer 1996; Heath et al. 1985; Kaminskyj et al. 1992; Mersey and McCully 1978; Wilson et al. 1990).

4 Root Hair Organization in Relation to Tip Growth

New roots hairs form continuously on the root epidermis behind the tips of growing roots. Root hairs initially appear as small round protrusions on the surface of specialized root epidermal cells known as trichoblasts (Cutter 1978; Farr 1928; Peterson and Farquhar 1996; Ridge 1996). Each protrusion, called a papilla by Haberlandt (1914), expands into a tip-growing tubular cell, the tip of which is filled with cytoplasm, while the rest of the hair is vacuolated and contains only thin layers of actively streaming cytoplasm (Fig. 1a,b; Galway et al. 1997; Miller et al. 1997; Peterson and Farquhar 1996; Ridge 1996). Under soil-free growing conditions, root hairs continue to elongate, growing away from the root surface, until they reach a length that varies with the species and growing conditions. Each hair then stops elongating, the tip cytoplasm rapidly disperses, and the tip becomes as vacuolated as the rest of the hair (Galway et al. 1997; Meekes 1985; Miller et al. 1997).

At present little is known about the ultrastructural organization of trichoblasts during root hair outgrowth. Most research has focused on the ultrastructure of root hairs, especially hair tips. Two distinct regions of cytoplasm are apparent in the tips of growing root hairs, an apical region and a subapical region (Bonnett and Newcomb 1966; Emons 1987; Galway et al. 1997; Ridge 1988; Sherrier and VandenBosch 1994; Volkmann 1984). The apical region is the site of tip growth where the plasma membrane expands and new cell wall material is secreted. Thus the cytoplasm is filled with large numbers of vesicles interspersed with ribosomes (Fig. 1c; Emons 1987; Galway et al. 1997; 1999; Ridge 1988; 1995). In one instance, root hair secretory vesicles have been shown to contain xyloglucan and polygalacturonic acid destined for the cell wall (Sherrier and VandenBosch 1994). No rough or smooth (tubular) endoplasmic reticulum (ER) has been reported to be present in the apical regions of growing root hairs or pollen tubes with the

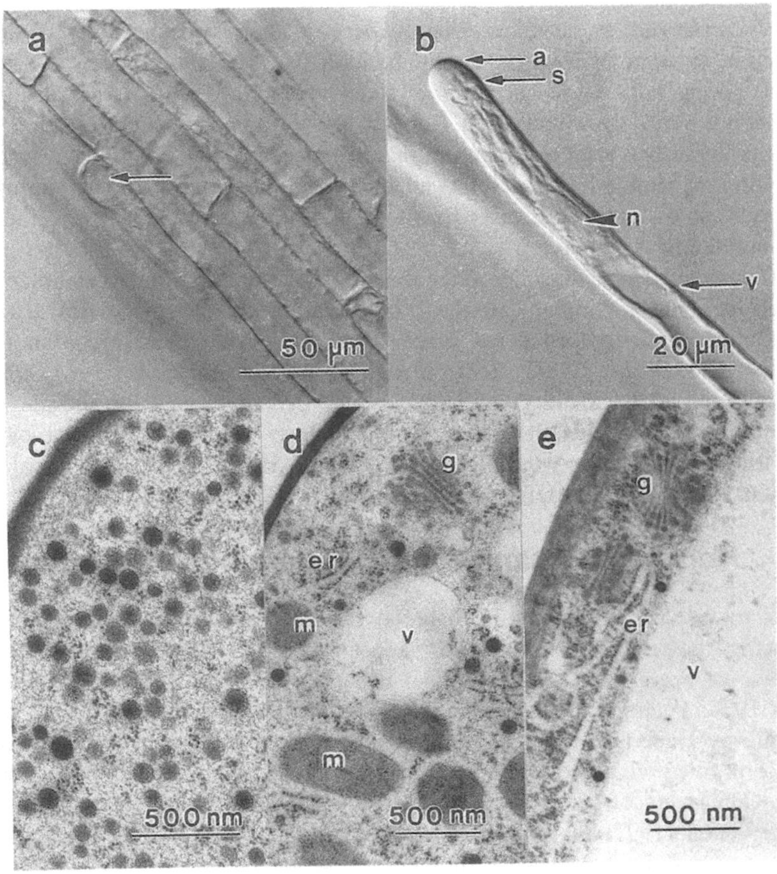

Fig. 1. *a-e*. Survey of *Arabidopsis* root hairs (Columbia wild type) by light (*a, b*) and electron microscopy (*c-e*). *a,b* Differential interference contrast microscopy of a young root hair (*arrow*) emerging from the apical end of an epidermal cell (*a*) and an older root hair (*b*) in which the nucleus (*arrowhead*) is positioned between the cytoplasmic tip region and the vacuolated basal region (*arrow labelled v*); the apical and subapical regions are indicated by *arrows* labelled *a* and *s* respectively. *c-e* Transverse sections of root hairs oriented with cell walls to the upper left, depict (*c*) the apical region filled with secretory vesicles and ribosomes, (*d*) the subapical region containing mitochondria (*m*), Golgi bodies (*g*), endoplasmic reticulum (*er*) and a small vacuole (*v*) and (*e*) the vacuolated region, containing a thin layer of cortical cytoplasm with Golgi bodies (*g*) and endoplasmic reticulum (*er*) located between the central vacuole (*v*) and the cell wall

exception of lily pollen (Derksen et al. 1995a, Lancelle and Hepler 1992 and references therein).

The subapical region is filled with ER and Golgi bodies, which presumably support tip growth through the biosynthesis of proteins and polysaccharides destined for the growing tip, as well as mitochondria, supporting energy metabolism, plus clathrin-coated pits and vesicles, actin microfilaments, cortical microtubules and the occasional plastid (Fig. 1d; Bonnett and Newcomb 1966; Galway et al. 1997; Sherrier and VandenBosch 1994; Ridge 1988; Volkmann 1984). The same types of organelles present in this region are also found in the thin layers of cortical cytoplasm located in the basal vacuolated region of hairs (Fig. 1e). Cytoplasmic streaming circulates the cortical cytoplasm and organelles to and from the cytoplasmic tip region (Miller et al. 1997; Peterson and Farquhar 1996; Ridge 1996).

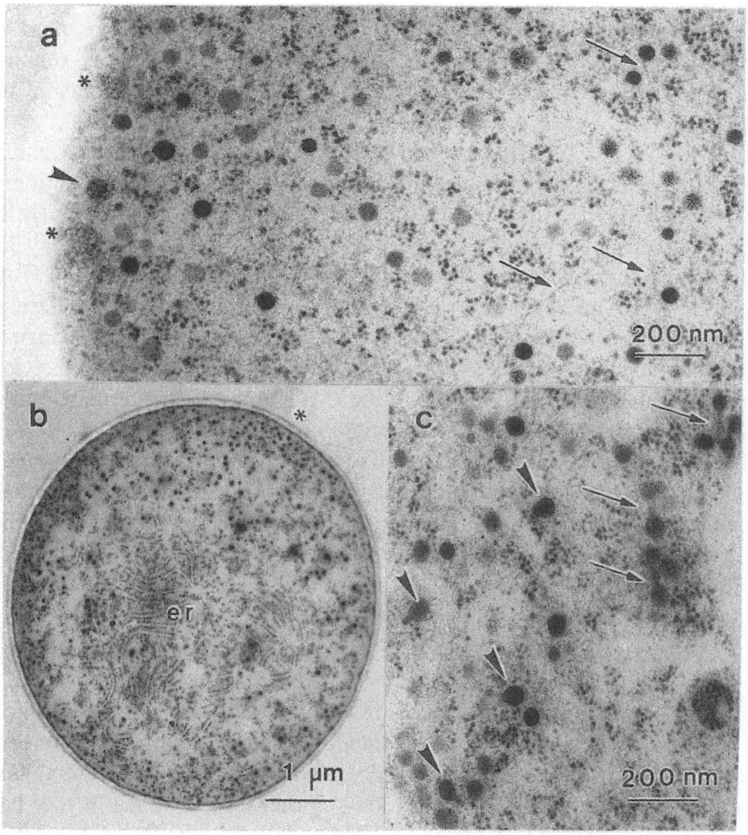

Fig. 2. *a-c.* Vesicle distribution and morphology in *Arabidopsis* root hairs. *a* Microfilaments (*arrows*) lie among the ribosomes and secretory vesicles in a transverse section from the apical region of a *rhd3* mutant hair; clathrin-coated pits (*asterisks*) have formed on the plasma membrane, and a coated vesicle (*arrowhead*) is present in the cytoplasm; note the poor staining of the cell wall, which is common in freeze-substituted hairs. *b* Transverse section from the upper part of the subapical region of an *rhd3* hair; endoplasmic reticulum (*er*) occupies the center of this hair, surrounded by areas of ribosome-free cytoplasm; secretory vesicles are located in the cortex and are more numerous in the upper half of the hair (*asterisk*). *c* Ovoid and stalked secretory vesicles (*arrowheads*) some with coats visible, in the subapical region of a wild type hair; note groups of secretory vesicles which may be attached (*arrows*). Bars *a* 200 nm, *b* 1 µm, *c* 200 nm

Together the vesicle-rich apical region and the organelle-rich subapical region make up the volume of cytoplasm visible by light microscopy in the tips of all growing root hairs. In *Arabidopsis* hairs the nucleus is usually positioned at the base of the cytoplasmic tip region (Fig. 1b; Galway et al. 1997; 1999). Nuclear position and the control of nuclear position by elements of the cytoskeleton in root hairs has been studied by immunofluorescence microscopy and will not be discussed further here, but the topic has been reviewed by Peterson and Farquhar (1996) and Ridge (1996).

The distribution of microtubules and actin microfilaments has been examined in root hairs, but the functions of these cytoskeletal elements require further clarification. Both microtubules, lying in the cortical cytoplasm, and actin microfilaments, which are distributed throughout the cytoplasm, are oriented parallel to the axis of hair elongation (Miller et al. 1997; Peterson and Farquhar 1996; Ridge 1996). Microtubules are believed to be involved in positioning of the nucleus, and may be involved in the control of root hair diameter (Miller et al. 1997; Peterson and Farquhar 1996; Ridge 1996). More recently, they have been implicated in the regulation and stability of tip growth in root hairs (Bibikova et al. 1999). Actin microfilaments are thought to mediate cytoplasmic streaming and vesicle transport (Miller et al. 1997; Peterson and Farquhar 1996; Ridge 1996; for other potential functions see Kropf et al. 1998). It is not clear whether intact actin microfilaments usually penetrate into the extreme apex of root hairs and other tip-growing cells, or what function they might have there (Kropf et al. 1998; Miller et al. 1997). In *Arabidopsis*, short actin-like filaments were detected among secretory vesicles in the apical regions of some wild type and mutant root hairs (Fig. 2a). Since the cytoskeleton of root hairs is covered in Chapters 2 and 3 in this volume, it will not be discussed further here, except where it relates to other aspects of hair cell fine structure.

5 Vacuoles, Root Hair Expansion, and Morphogenesis

Light microscopy highlights the continual streaming of the cytoplasm in living root hairs (Galway et al. 1999). In *Arabidopsis* hairs, the tip cytoplasm extends an average of 20 μm below the hair apex (Galway et al. 1999 and unpublished data). This is comparable to the 20-80 μm long cytoplasmic tip zone in root hairs of the cress *Lepidium sativum* (Volkmann 1984). The tip cytoplasm represents only a small part of a root hair, as the average length of *Arabidopsis* hairs grown in water is 300 μm, and in air 800 μm (Galway et al. 1999). Electron microscopy indicates that there is a gradual transition between tip cytoplasm and apical cytoplasm. Some vacuoles appear in the subapical region and these become larger towards the base of this region (Fig. 1d). The presence of these vacuoles raises questions such as: Do vacuoles form de novo in root hairs? Do small vacuoles exist independently of the central vacuole? Are some of these vacuoles actually extensions into the tip of the central vacuole? Turgor pressure is believed to drive root hair expansion, and a major function of vacuoles in growing cells like root hairs is to maintain turgor pressure by expanding and accumulating solutes (Kropf et al. 1998; Peterson and Farquhar 1996; Sievers and Schnepf 1981; Wink 1993).

Vacuole expansion must occur rapidly to keep pace with root hair growth, so root hairs may prove to be particularly suitable for investigating the development of plant vacuoles.

Ultrastructural analysis of root hair mutants with altered hair morphology is a new and promising way to study the function of vacuoles and other cell components in tip growth. Two *Arabidopsis* root hair mutants, *rhd3* and *rhd4*, have been investigated to date (Galway et al. 1997; Galway et al. 1999). Mutations in the *RHD3* gene of *Arabidopsis*, which encodes a putative GTP-binding protein, reduces cell expansion throughout the seedlings, including the root hairs which are shorter and contain smaller vacuoles than wild type hairs (Galway et al. 1997; Wang et al. 1997). Reduced expansion of vacuoles seems to be the major cytological defect, and it can account for the general inhibition of cell expansion throughout the mutant seedlings. The mutant root hairs are also wavy (Galway et al. 1997; Schiefelbein and Somerville 1990). Straight tubular cells grow by a graded, uniform expansion of the growing apex, so that expansion is maximal at the apex and diminishes rapidly with distance below the apex (Harold 1990; Kropf et al. 1998; Sievers and Schnepf 1981). The waving growth of *rhd3* hairs is caused by differential expansion of the hair apices, so that the apices bulge in different directions during growth (Galway et al. 1997). These autonomous tip deformations are similar to those that normally occur when root hairs contact physical barriers including soil particles (Bibikova et al. 1997; Farr 1928; Haberlandt 1914). Wavy growth like that in *rhd3* hairs can also be induced by the drug-induced disassembly or stabilization of microtubules in wild type *Arabidopsis* root hairs (Bibikova et al. 1999). Ultrastructural analysis of *rhd3* hairs revealed asymmetrical distributions of secretory vesicles in the apical and subapical regions (Fig. 2b; Galway et al. 1997). The waving growth of the hairs is probably generated by a corresponding asymmetry in exocytotic events (Galway et al. 1997). But what relationship, if any, exists between the abnormal distribution of secretory vesicles and the reduced expansion of vacuoles in the root hairs? This question cannot be answered until we know more about the function of the RHD3 protein and the regulation of tip growth in *Arabidopsis* (Wang et al. 1997).

Mutations in the *RHD4* gene of *Arabidopsis* (which has not been cloned) result in short, bulging root hairs of variable diameter (Galway et al. 1999; Schiefelbein and Somerville 1990). Root hair diameter varies with hair growth rate: during periods of slow growth the *rhd4* hairs bulge, and during periods of more rapid growth their diameter shrinks. During slow growth, the volume of the tip cytoplasm diminishes and the central vacuole extends closer to the hair tip (Galway et al. 1999). Electron microscopy revealed a normal distribution of organelles, vesicles and cytoskeleton in these hairs, except for localized cell wall thickenings which are discussed below (Galway et al. 1999). The straight growth of hairs indicates that apical expansion and the pattern of exocytosis is normal. The relationship between hair growth rate, expansion, cell wall extension, and vacuole expansion and the maintenance of turgor pressure requires further investigation in this interesting mutant.

6 Is There a Polarized Distribution of Endoplasmic Reticulum in Root Hairs?

The tips of root hairs and pollen tubes maintain a tip-focused calcium ion concentration gradient which is required for tip growth, perhaps by triggering exocytosis (Bibikova et al. 1997; Miller et al. 1997 and references therein). It has been suggested that certain organelles located below the vesicle-rich apical regions, such as the ER, mitochondria or vacuoles may act as internal calcium ion sinks or storage sites and aid in regulating calcium ion concentrations in the cells (Derksen et al. 1995a,b; Hepler et al. 1990; Miller et al. 1997). In normal and mutant *Arabidopsis* root hairs, ER is well positioned to play this role, because ER (mainly rough ER) is the most abundant type of organelle immediately below the vesicle-rich apical region (Fig. 2b; Galway et al. 1997). Rough ER was also identified as the first organelle to appear below the apical regions in root hairs of the crucifer, *Raphanus sativum* (Bonnett and Newcomb 1966). A similar distribution of ER has been noted in several studies of fungal hyphae (Harold 1990), and smooth tubular ER is positioned adjacent to the vesicle-rich region in tobacco pollen tubes (Derksen et al. 1995b). However more information on patterns of organelle distribution in a variety of root hairs is needed to determine if there is a significant polarity in ER distribution in root hairs. Recently, confocal laser scanning microscopy has been used very effectively to determine the overall form, distribution and movement of ER in *Arabidopsis* root cells and root hairs, including the presence of ER at the tips of the hairs (Ridge et al. 1999).

7 Secretory Vesicle Pleiomorphy: Origin and Possible Significance

In *Arabidopsis thaliana* and *Vicia hirsuta* root hairs, secretory vesicle shape varies as first reported by Ridge (1988): there are spherical secretory vesicles and pyriformis (pear-shaped) secretory vesicles, so-called because they have small stalks or tails. Pyriformis vesicles have not yet been reported in other types of tip-growing cells (Ridge 1996). In cryofixed *Arabidopsis* hairs, secretory vesicle shapes include stalked-spherical, teardrop-shaped, and ovoid (Fig. 2c; Galway et al. 1997). Since spherical and pyriformis secretory vesicles are similar in size and intensity of staining and can occur in close proximity (Fig. 2c and 3a), it was concluded that pyriformis vesicles are probably pleiomorphic forms of "normal" spherical secretory vesicles (Galway et al. 1997). Non-clathrin coats cover secretory vesicles of all shapes; in well-stained vesicles the coats appear as a short fuzzy ring or halo around the vesicle (Fig. 2a,c; Galway et al. 1997; Ridge 1988). These coats likely include proteins required for secretory vesicle transport, membrane docking and membrane fusion, and perhaps integral proteins destined for insertion into the plasma membrane (e.g., calcium ion channels) but the biochemical analysis of plant vesicles is still at an early stage (Battey and Blackbourn 1993; Derksen et al. 1995a, Hawes and Satiat-Jeunemaitre 1996; Thiel and Battey 1998; Yang 1998).

In *Arabidopsis*, pyriformis vesicles were not observed in all cryofixed root hairs. These vesicles may be labile and perhaps are only preserved by optimal freezing conditions (Ridge 1988). Since pyriformis vesicles haven't been observed in chemically fixed cells, chemical fixation may alter vesicle morphology (Ridge 1988).

Could pyriformis vesicles be spherical vesicles deformed by interactions with other structures (Galway et al. 1997)? Some groups of pyriformis vesicles appear

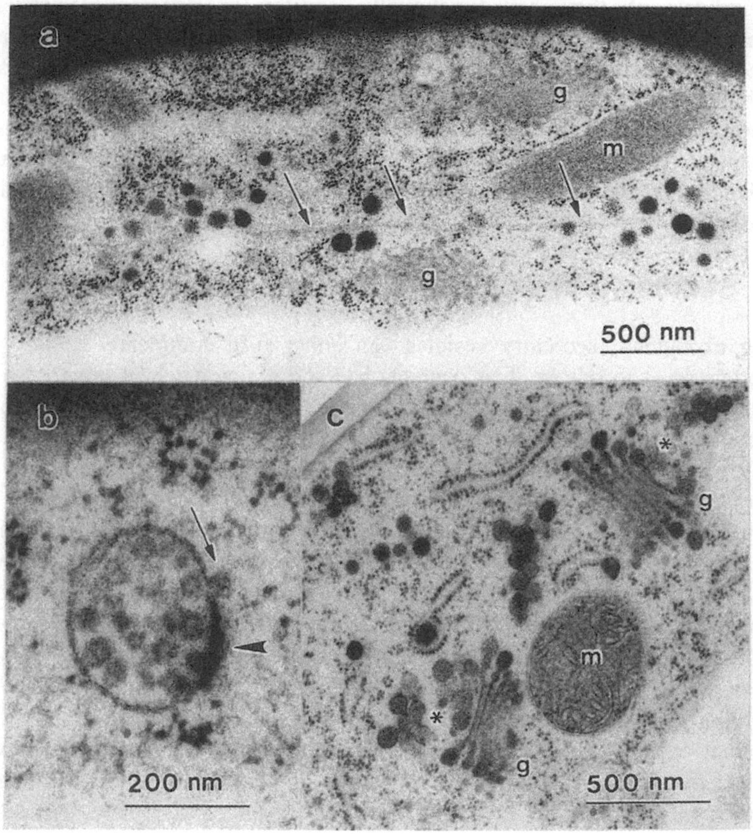

Fig. 3. *a-c*. Vesicles, Golgi bodies, and multivesicular bodies in *Arabidopsis* root hairs. *a* In a wild type hair, spherical and pyriformis vesicles are aligned along a microfilament bundle (*arrows*) located in cortical cytoplasm between the cell wall (*top*) and central vacuole (*below*); two Golgi bodies (*g*) and a mitochondrion (*m*) are also present. *b* A multivesicular body in the vacuolated region of a wild type hair; it contains many vesicles and a forming or fusing vesicle is attached to its surface (*arrow*), just above a darkly-stained plaque (*arrowhead*); similar plaques were frequently observed on other multivesicular bodies in the hairs. *c* Typical Golgi bodies (*g*) with five or six cisternae in the subapical region of an *rhd3* hair exhibit a distinct cis-trans polarity; in each Golgi body trans cisternae are thinner and lie adjacent to a cluster of vesicles interpreted as part of the trans-Golgi network (*asterisks*); a mitochondrion (*m*) and other secretory vesicles are also present. Bars *a* 500 nm, *b* 200 nm, *c* 500 nm

to be connected (Fig. 2c), and both spherical and pyriformis secretory vesicles are frequently observed in close association with microfilaments in the subapical regions of *Vicia* and *Arabidopsis* hairs (Fig. 3a; Galway et al. 1997; Ridge 1988). However, no connections were detected between pyriformis vesicles in the apical regions of *Vicia* root hair tips and any other structure including actin (Ridge 1988). Several approaches could help solve the mystery of pyriformis vesicle function and origin. First, immunolabelling vesicles to determine their contents may indicate if they are functionally similar or different (Sherrier and VandenBosch 1994). Second, if pyriformis vesicles are deformed by interactions with actin microfilaments or other elements of the cytoskeleton, then treatment of root hairs with specific cytoskeleton inhibitors should reduce or eliminate pyriformis vesicles. Ridge (1990) did treat *Vicia* root hairs with the anti-actin drug cytochalasin D to study the drugs effects on exocytosis and the ultrastructural organization of the hairs, but vesicle morphology was not specifically addressed (Ridge 1990).

8 Secretory Vesicles and Exocytosis

During exocytosis, secretory vesicles can either fuse completely to the plasma membrane so as to release their contents and add a new patch of membrane to the plasma membrane, or else they can fuse, release their contents and detach again (recycle) without increasing plasma membrane area (Thiel and Battey 1998). In rapidly growing polarized cells such as root hairs, full fusion of secretory vesicles is expected to be the dominant form of exocytosis (Thiel and Battey 1998). Under certain experimental conditions, cell expansion can be blocked in root hairs and other tip-growing cells, while exocytosis continues (reviewed by Battey and Blackbourn 1993; Sievers and Schnepf 1981; see also Ridge 1990). In such cases a thickened cell wall is deposited at the non-growing tip. A similar process is thought to occur at *rhd4* mutant hair tips during episodes of reduced growth, since electron microscopy revealed localized thickenings of the primary cell walls in these hairs (Galway et al. 1999).

Actual fusion of secretory vesicles to the plasma membrane has not been observed in cryofixed *Arabidopsis* root hairs or reported in other root hairs, and has been rarely observed in pollen tubes (Derksen et al. 1995a, Lancelle and Hepler 1992). One reason is the rapidity of vesicle fusion, estimated to take about 0.1 s (Derksen et al. 1995a, Thiel and Battey 1998). A second reason, evident in *Arabidopsis* hairs, is the steep curvature of the hairs in the apical region where secretory vesicles are most dense and maximum vesicle-membrane fusion is expected to occur. The plasma membrane-cell wall interface is not distinct in cross sections from this region (Fig. 2a). Methods that reveal large areas of the surface of the plasma membrane in the apical region should provide more opportunities to view vesicle fusion. For example, freeze fracture of the tips of *Lepidium* root hairs revealed numerous blisters in the plasma membrane thought to be caused by secretory vesicle fusion (Volkmann 1984). However it is technically difficult to obtain fractures in the apical regions of root hairs (Miller et al. 1997; see also Chapter 4, this volume).

9 Clathrin-Coated Vesicles and Membrane Recycling

Vesicles form by budding from the plasma membrane and from endomembranes in plant and animal cells. Their protein coats aid in their formation and separation from the source membrane. In both plant and animal cells, clathrin-coated pits and buds form at two sites: on the plasma membrane and at the trans faces of Golgi bodies (Low and Chandra 1994). Clathrin coats disassemble soon after vesicle formation, so clathrin-coated vesicles are seldom observed far from their source (Low and Chandra 1994). Here I focus on the distribution and function of clathrin-coated pits, buds and vesicles associated with the plasma membrane in root hairs. The function of clathrin-coated vesicles is to remove and recycle plasma membrane as well as to endocytose extracellular material such as fluid or receptor-bound ligands, although it is not yet clear how common the latter process is or how important it is to plant cell metabolism (Low and Chandra 1994). Clathrin coats appear as hexagonal/pentagonal protein lattices in surface view, and as spiky coats in cross section (Low and Chandra 1994). The coats are well-preserved by chemical fixation, so coated vesicles were readily identified in early ultrastructural studies of root hairs (Bonnett and Newcomb 1966).

In *Arabidopsis* root hairs, clathrin-coated pits and vesicles are most abundant in the apical and subapical regions (Fig. 2a; Galway et al. 1997; 1999). This agrees with other reports that coated pits and vesicles associated with the plasma membrane are most abundant in or near the growing tips of root hairs (Emons and Traas 1986; *Equisetum*, Ridge 1988; *Vicia*, Robertson and Lyttleton 1982; *Trifolium*), and pollen tubes (reviewed by Derksen et al. 1995a). Morphometric analysis indicates that clathrin-coated pits reach a peak of abundance just below the growing tips of *Nicotiana tabacum* pollen tubes (Derksen et al. 1995b). In contrast, a survey of coated pit densities in various plant cells found the lowest densities to be in fully-grown root hairs (*Equisetum*, Emons and Traas 1986).

Calculations of secretory vesicle production rates in pollen tubes indicate that if complete fusion of secretory vesicles occurs, more membrane is added to the plasma membrane than is required for tip growth (reviewed by Derksen et al. 1995b, Steer 1988). Thus membrane retrieval is probably the most important function for coated vesicles in tip-growing cells. Coated pits might also retrieve specialized tip-restricted membrane proteins such as calcium ion channels, which need to be removed (or alternatively inactivated, although there are no data to support this possibility) once they are displaced from the tip onto the flanks of the cell (Derksen et al. 1995a,b). If the major function of coated pits is retrieving tip-restricted proteins, then they should be absent from the growing tip and concentrated on the flanks just below the growing tip, as in tobacco pollen tubes (Derksen et al. 1995b). On the other hand, if they are functioning in simple membrane retrieval, they need not necessarily be excluded from the growing tip. In life, the distribution and concentration of coated pits may vary according to hair growth rate and the rate of secretion. Hair growth rate varies with age (Emons and Traas 1986), but it has also been shown to fluctuate continuously in elongating *Arabidopsis* root hairs (Galway et al. 1999; Wymer et al. 1997). Such variation could explain why coated pits and vesicles have been detected in the apical regions of some root hairs (Galway et al. 1997; 1999), and not others (Ridge 1988).

10 The Fate of Clathrin-Coated Vesicles in Root Hairs

The fate of clathrin-coated vesicles in root hairs, pollen tubes or other tip-growing cells has not been explored at the ultrastructural level. However investigations on the fate of endocytosed markers in other plant cells, especially protoplasts, have shown that coated vesicles deliver these markers to a membrane-bound structure called the partially coated reticulum (PCR), an organelle of disputed origin, since some authors say it is identical to the trans-Golgi network or TGN (Low and Chandra 1994). From this sorting/recycling compartment, markers are either delivered to the Golgi complex for recycling by exocytosis or delivered to multivesicular bodies (MVBs) for sorting and delivery to vacuoles for degradation. The PCR and MVBs are therefore related to the endosomes of animal cells (Low and Chandra 1994).

In actively secreting cells such as root hairs recycling of coated vesicles and their contents via the PCR and the Golgi bodies is likely to be the more active pathway (Low and Chandra 1994). Nevertheless, MVBs are common in wild type and mutant *Arabidopsis* root hairs (Fig. 3b). They are located in the subapical regions, and in the cortical cytoplasm in the vacuolated regions. Due to the poor staining of membranes in freeze-substituted hairs of *Arabidopsis* they are often difficult to see. MVBs or MVB-like structures also occur in pollen tubes and fungal hyphae (for example, Roberson and Fuller 1988; hyphae, de Win et al. 1996; pollen). In *Arabidopsis* hairs, darkly stained clusters of vesicles and tubular-vesicular structures that lie near the trans faces of Golgi bodies are interpreted as TGN (Fig. 3c). Other groups of vesicles not associated with Golgi bodies may be in transit to the tip (Fig. 2c, 3c). It has not been determined if there is another distinct type of membrane-bound compartment that could function as PCR in these cells.

11 Conclusions and Future Directions

For cell and molecular biologists the rapid and highly localized growth of root hairs is advantageous for investigating cell morphogenesis and tip growth, as well as specific growth processes that are common to all cells such as exocytosis, membrane recyling and vacuole expansion. As tubular extensions of root epidermal cells, root hairs are also more similar to the average plant cell than pollen grains and pollen tubes. However, a better understanding of root hair ultrastructure is required in order to exploit these cells for studies of growth and morphogenesis. Cell and molecular research ultimately depends on microscopy to visualize cells, organelles and other cell components, or to localize molecules and macromolecules within cells. Although confocal microscopy is now being used increasingly for all of these purposes, due to superior optics (compared to traditional light microscopy) and reduced specimen preparation time (compared to electron microscopy), electron microscopy is likely to remain the method of choice for high resolution subcellular localization of proteins and other molecules in plant cells.

ACKNOWLEDGMENTS

I would like to thank J.W. Schiefelbein (University of Michigan) with whom I began work on the ultrastructure and morphogenesis of *Arabidopsis* root hairs as a postdoctoral fellow, and J.W. Heckman, Jr. (Michigan State University) for his skill and advice with cryofixation. My research on root hairs is supported by a research grant from the Natural Sciences and Engineering Research Council of Canada.

References

Battey NH, Blackbourn HD (1993) The control of exocytosis in plant cells. New Phytol 125:307-338

Bibikova TN, Blancaflor EB, Gilroy S (1999) Microtubules regulate tip growth and orientation in root hairs of *Arabidopsis thaliana.* Plant J 17:657-665

Bibikova TN, Zhigilei A, Gilroy S (1997) Root hair growth in *Arabidopsis thaliana* is directed by calcium and an endogenous polarity. Planta 203:495-505

Bonnett HT Jr, Newcomb EH (1966) Coated vesicles and other cytoplasmic components of growing root hairs of radish. Protoplasma 62:59-75

Cooper KM (1982) Callose-deposit formation in radish root hairs. In: Brown RM (Ed) Cellulose and other natural polymer systems. Plenum Press, New York London, pp 167-184

Cutter EG (1978) Plant anatomy. Part 1: Cells and tissues, 2nd ed. Edward Arnold, London, pp 97-106.

de Win AHN, Knuiman B, Pierson ES, Geurts H, Kengen HMP, Derksen J (1996) Development and cellular organization of *Pinus sylvestris* pollen tubes. Sex Plant Reprod 9:93-101

Derksen J, Rutten T, van Amstel T, de Win A, Doris F, Steer M (1995a) Regulation of pollen tube growth. Acta Bot Neerl 44:93-119

Derksen J, Rutten T, Lichtscheidl IK, de Win AHN, Pierson ES, Rongen G (1995b) Quantitative analysis of the distribution of organelles in tobacco pollen tubes: implications for exocytosis and endocytosis. Protoplasma 188:267-276

Doris FP, Steer MW (1996) Effects of fixatives and permeabilisation buffers on pollen tubes: implications for localisation of actin microfilaments using phalloidin staining. Protoplasma 195:25-36

Emons AMC (1985) Plasma-membrane rosettes in root hairs of *Equisetum hyemale.* Planta 163:350-359

Emons AMC (1987) The cytoskeleton and secretory vesicles in root hairs of *Equisetum* and *Limnobium* and cytoplasmic streaming in root hairs of *Equisetum.* Ann Bot 60:625-632

Emons AMC, Traas JA (1986) Coated pits and coated vesicles on the plasma membrane of plant cells. Eur J Cell Biol 41:57-64

Farr CH (1928) Root hairs and growth. Quart Rev Biol 3:343-376

Galway ME, Heckman JW Jr, Hyde GJ, Fowke LC (1995) Advances in high-pressure and plunge-freeze fixation. In: Galbraith DW, Bourque DP, Bohnert HJ (Eds) Methods in cell biology 49. Methods in plant cell biology, Part A. Academic Press, San Diego, pp 3-19

Galway ME, Heckman JW Jr, Schiefelbein JW (1997) Growth and ultrastructure of *Arabidopsis* root hairs: the *rhd3* mutation alters vacuole enlargement and tip growth. Planta 201:209-218

Galway ME, Lane DC, Schiefelbein JW (1999) Defective control of growth rate and cell diameter in tip-growing root hairs of the *rhd4* mutant of *Arabidopsis thaliana*. Can J Bot 77:494-507

Haberlandt GFJ (1914). Physiological plant anatomy. Drummond M (Trans) Physiologische Pflanzenanatomie, 4th ed. Macmillan Co, London, pp 221-223.

Harold FM (1990) To shape a cell: an inquiry into the causes of morphogenesis of microorganisms. Microbiol Rev 54:381-431

Harold FM (1997) How hyphae grow: morphogenesis explained? Protoplasma 197:137-147

Hawes C, Satiat-Jeunemaitre B (1996) Stacks of questions: how does the plant Golgi work? Trends Plant Sci 1:395-400

Heath B, Rethoret K, Arsenault AL, Ottensmeyer FP (1985) Improved preservation of the form and contents of wall vesicles and the Golgi apparatus in freeze substituted hyphae of *Saprolegnia*. Protoplasma 128:81-93

Hepler PK, Palevitz BA, Lancelle SA, McCauley MM, Lichtscheidl I (1990) Cortical endoplasmic reticulum in plants. J Cell Sci 96:355-373

Kaminskyj SGW, Jackson SL, Heath IB (1992) Fixation induces differential polarized translocations of organelles in hyphae of *Saprolegnia ferax*. J Microsc 167:153-168

Kropf DL, Bisgrove SR, Hable WE (1998) Cytoskeletal control of polar growth in plant cells. Curr Op Cell Biol 10:117-122

Lancelle SA, Hepler PK (1992) Ultrastructure of freeze-substituted pollen tubes of *Lilium longiflorum*. Protoplasma 167:215-230

Low PS, Chandra S (1994) Endocytosis in plants. Ann Rev Plant Physiol Plant Mol Biol 45:609-631

Mahló R (1998) Expanding tip-growth theory. Trends Plant Sci 3:40-41

McCully ME (1995) How do real roots work? Some new views of root structure. Plant Physiol 109:1-6

Meekes HTHM (1985) Ultrastructure, differentiation and cell wall texture of trichoblasts and root hairs of *Ceratopteris thalictroides* (L) Brongn (Parkeriaceae). Aquat Bot 21:347-362

Mersey BG, McCully ME (1978) Monitoring the course of fixation in plant cells. J Microsc 114:49-76

Miller DD, de Ruijter NCA, Emons AMC (1997) From signal to form: aspects of the cytoskeleton-plasma membrane-cell wall continuum in root hair tips. J Exp Bot 48:1881-1896

Newcomb EH, Bonnett HT Jr (1965) Cytoplasmic microtubule and wall microfibril orientation in root hairs of radish. J Cell Biol 27:575-589

Peterson RL, Farquhar ML (1996) Root hairs: specialized tubular cells extending root surfaces. Bot Rev 62:1-40

Picton JM, Steer MW (1981) Determination of secretory vesicle production rates by dictyosomes in pollen tubes of *Tradescantia* using cytochalasin D. J Cell Sci 49:261-272

Ridge RW (1988) Freeze-substitution improves the ultrastructural preservation of legume root hairs. Bot Mag (Tokyo) 101:427-441

Ridge RW (1990) Cytochalasin-D causes abnormal wall-ingrowths and organelle-crowding in legume root hairs. Bot Mag (Tokyo) 103:87-96

Ridge RW (1995) Micro-vesicles, pyriform vesicles and macro-vesicles associated with the plasma membrane in the root hairs of *Vicia hirsuta* after freeze-substitution. J Plant Res 108:363-368

Ridge RW (1996) Root hairs: cell biology and development. In: Waisel Y, Eshel A, Kafkafi U (Eds) Plant roots: The hidden half. Marcel Dekker, New York, pp 127-147

Ridge RW, Uozumi Y, Plazinski J, Hurley U and Williamson RE (1999) Developmental transitions and dynamics of the cortical ER of *Arabidopsis* cells seen with green fluorescent protein. Plant Cell Physiol. 40(12):1253-1261

Roberson RW, Fuller MS (1988) Ultrastructural aspects of the hyphal tip of *Sclerotium rolfsii* preserved by freeze substitution. Protoplasma 146:143-149

Robertson JG, Lyttleton P (1982) Coated and smooth vesicles in the biogenesis of cell wall, plasma membranes, infection threads and peribacteroid membranes in root hairs and nodules of white clover. J Cell Sci 58:63-78

Schiefelbein J, Galway M, Masucci J, Ford S (1993) Pollen tube and root-hair tip growth is disrupted in a mutant of *Arabidopsis thaliana*. Plant Physiol 103:979-985

Schiefelbein JW, Somerville C (1990) Genetic control of root hair development in *Arabidopsis thaliana*. Plant Cell 2:235-243

Schnepf E (1986) Cellular polarity. Ann Rev Plant Physiol 37:23-47

Sherrier DJ, VandenBosch KA (1994) Secretion of cell wall polysaccharides in *Vicia* root hairs. Plant J 5:185-195

Sievers A, Schnepf E (1981) Morphogenesis and polarity of tubular cells with tip growth. In: Kiermayer 0 (Ed) Cytomorphogenesis in plants. Cell Biology Monographs 8. Springer-Verlag, Vienna New York, pp 265-299

Steer MW (1988) Plasma membrane turnover in plant cells. J Exp Bot 39:987-996

Steer MW (1991) Quantitative morphological analysis. In: Hall JL, Hawes C (Eds) Electron microscopy of plant cells. Academic Press, London San Diego, pp 85-104

Thiel G, Battey N (1998) Exocytosis in plants. Plant Mol Biol 38:111-125

Volkmann D (1984) The plasma membrane of growing root hairs is composed of zones of local differentiation. Planta 162:392-403

Wang H, Lockwood SK, Hoeltzel MF, Schiefelbein JW (1997) The *ROOT HAIR DEFECTIVE3* gene encodes an evolutionarily conserved protein with GTP-binding motifs and is required for regulated cell enlargement in *Arabidopsis*. Genes Devel 11:799-811

Wen T-J, Schnable PS (1994) Analyses of mutants of three genes that influence root hair development in *Zea mays* (Gramineae) suggest that root hairs are dispensable. Am J Bot 81:833-842

Wilson TP, Canny MJ, McCully ME, Lefkovitch LP (1990) Breakdown of cytoplasmic vacuoles: a model of endoplasmic membrane rearrangement. Protoplasma 155:144-152

Wink M (1993) The plant vacuole: a multifunctional compartment. J Exp Bot 44(suppl):231-246

Wymer CL, Bibikova TN, Gilroy S (1997) Cytoplasmic free calcium distributions during the development of root hairs of *Arabidopsis thaliana*. Plant J 12:427-439

Yang Z (1998) Signaling tip growth in plants. Curr Opp Plant Biol 1:525-530

Chapter 2
The Role of Microtubules in Root Hair Growth and Cellulose Microfibril Deposition

Tijs Ketelaar and Anne Mie C. Emons

Laboratory of Plant Cytology and Morphology, Department of Plant Sciences, Wageningen University, Arboretumlaan 4, 6703 BD Wageningen, The Netherlands

1 Introduction: Cortical Microtubules in Plant Cells

Cortical microtubules (CMT) were first observed in plant cells (Ledbetter and Porter 1963) by transmission electron microscopy. In intercalary growing cells, CMT have always been found oriented with a right angle to the elongation axis of the cell (reviewed in Cyr 1994). Moreover, if CMT are depolymerized by addition of pharmalogical agents, cells expand isotropically (colchicine: Green 1962; dinitroaniline herbicides: Upadhyaya and Noodén 1977; Baskin et al. 1994). Stabilization of CMT by taxol also leads to more isotropic growth (Baskin et al. 1994). Furthermore, growth regulators affect the orientation of the CMT (e.g., van Spronsen et al. 1995). Most researchers agree that the orientation of the CMT determines the direction of cell expansion.

One hypothesis as to how the CMT could determine the orientation of cell expansion, is orienting the deposition orientation of the cellulose microfibrils (CMF) of the cell wall (Heath 1974). The CMT could perform this process by channeling the cellulose synthases through the plasma membrane (Giddings and Staehelin 1991). This is a plausible hypothesis, since in all growing cells the CMT and the nascent CMF are found in parallel to each other, both directed transverse to the orientation of cell elongation (reviewed in Cyr 1994). One can imagine that abundant transverse hoops of stiff cellulose microfibrils would limit cell elongation to the direction transverse to the hoops (Gertel and Green 1977). However, there is no agreement among researchers that CMT determines the orientation of the CMF by default (Emons et al. 1992).

Some of the reasons for the lack of agreement are that the pharmacological agents that depolymerize microtubules have many different effects on the orientation of the CMF (review: Robinson and Quader 1982; Emons et al. 1992), such as random deposition (Schnepf et al. 1975), deposition is swirls (Hepler and Fosket

Key words. cytoskeleton, cellulose microfibril, microtubule, root hair

1971), ordered deposition without regular changes in the orientation that occurs without treatment (Green et al. 1970) or no change in deposition (Emons et al. 1990; Schmid and Meindl 1992). Furthermore, a co-alignment of CMT and CMF is often not found in secondary walls which have an even deposition along the plasma membrane (review: Emons et al. 1992). Since in the absence of CMT ordered deposition occurs, one expects an ordering mechanism independent of CMT. A drawback in many studies is that researchers did not check whether the microtubules were really depolymerized, or whether growth really continued; and in many studies appropriate methods for the visualization of CMF were not used. Uranyl acetate and lead citrate post staining after glutaraldehyde osmium tetroxide fixation does not reliably stain the CMF. Stripes may be seen on the images, which have no relationship whatsoever with the CMF (Emons 1988). Appropriate visualization methods for CMF are (Emons 1988): freeze fracturing, cleaving and shadowing with or without wall matrix extraction (Chapter 4 this volume), staining of sections of matrix-extracted material, and specific polysaccharide staining of fixed cells.

Experiments with drugs that inhibit CMF formation indicate that the CMF provide spatial cues for CMT organization (Fisher and Cyr 1998). Thus, CMT determine the orientation of cell elongation, and this process may be mediated by CMF. However, such correlations do not prove that CMT regulate the orientation of CMF deposition by default. As mentioned above, in growing cells the orientation of the CMF and the orientation of the CMT both correlate with the direction of growth. Therefore, these cells are not suited to discriminate between the function of the CMT in cell growth, whether or not mediated by CMF, and CMF deposition orientation itself.

Because root hairs are tip-growing cells, the wall deposited at the tip is a primary cell wall. A secondary cell wall is deposited in the tube of the hair, against this primary cell wall. The new wall deposited in the hair tube is a secondary wall by definition, since in the tube cell expansion does not take place. Furthermore, since this secondary wall is deposited sequentially from root hair base to root hair tip, all stages of wall development are represented in a single root hair. Since the deposited CMF in this non-growing tube do not change position, they form a record of the deposition history. This makes root hairs, especially those with non-uniform CMF deposition angle, ideal cells to study the function of the CMT in CMF deposition orientation, independently from their function in cell growth.

In this chapter, we will review our knowledge of interphase microtubules in root hairs. We will discuss the role of the CMT in tip growth and show that there is no role for CMT in CMF orientation during deposition. Therefore, other possible functions will be discussed as well as the role of the occurrence and possible role of endoplasmic CMT.

2 Cortical Microtubules in Root Hairs

2.1 Growing Root Hair Tip

Newcomb and Bonnett (1965) and Seagull and Heath (1980) reported an absence of CMT in the tip of radish root hairs. However, they used chemical fixation, which is now known to cause fixation artifacts. Lloyd and Wells (1985) used various chemical fixation protocols and showed that the CMT in tips of radish root hairs are sensitive to fixation conditions. With improved chemical fixation they could detect CMT at the tip of growing root hairs (Lloyd and Wells 1985). Currently, the most reliable fixation method is freeze fixation, where the first fixation step is physical. In freeze substitution studies of *Equisetum hyemale* root hairs (Emons and Derksen 1986; Emons 1987), Emons (1989) has shown that CMT are present at the plasma membrane of the tips of growing hairs. CMT have a random orientation in the hemisphere (Fig. 1a). The microfibrils in the hemisphere are also randomly oriented (Emons 1989). Thus, in the root hair tip, like in other expanding cells, the nascent CMF lie parallel to the CMT, perpendicular to the direction of expansion. Expansion in a semi-hemisphere is in all orientations. This is the only study so far in which the microtubule cytoskeleton was investigated in depth in tips of growing root hairs with freeze substitution electron microscopy. In published micrographs of another study using freeze substitution electron microscopy, but not primarily directed to the investigation of CMT, CMT can be observed in areas where tip growth takes place (Galway et al. 1997). We conclude that in the growing area of root hairs, just like in other growing cells, the orientation of both CMT and CMF is perpendicular to the direction of growth, and may determine the orientation of expansion.

2.2 Root Hair Tube

2.2.1 Techniques for Visualization of CMT, Orientation of CMT

The CMT in the root hair tube of a variety of species have been analyzed with many different techniques. Radish (*Raphanus sativus*) root hairs were one of the first cell types in which the CMT were studied (Newcomb and Bonnett 1965). Like in all early ultrastructural studies, the cells were chemically fixed, followed by resin embedding and post staining. High concentrations of glutaraldehyde (up to 6%) were used, in combination with osmium tetraoxide. In grazing sections of the cortical cytoplasm of the root hair tube, microtubules were observed in the electron microscope in axial orientations. These observations were confirmed by Seagull and Heath (1980).

In *Raphanus sativus* (Newcomb and Bonnett 1965; Seagull and Heath 1980) and *Equisetum hyemale* (Emons 1982; Emons and Wolters-Arts 1983) net-axially oriented CMT were observed in chemically fixed electron microscope sections. Two different orientations of CMT in root hairs have been reported with immunofluorescence after chemical fixation. Net-axial CMT were found in *Equisetum hyemale*, *Limnobium stoloniferum*, *Lepidium sativum*, *Ceratopteris*

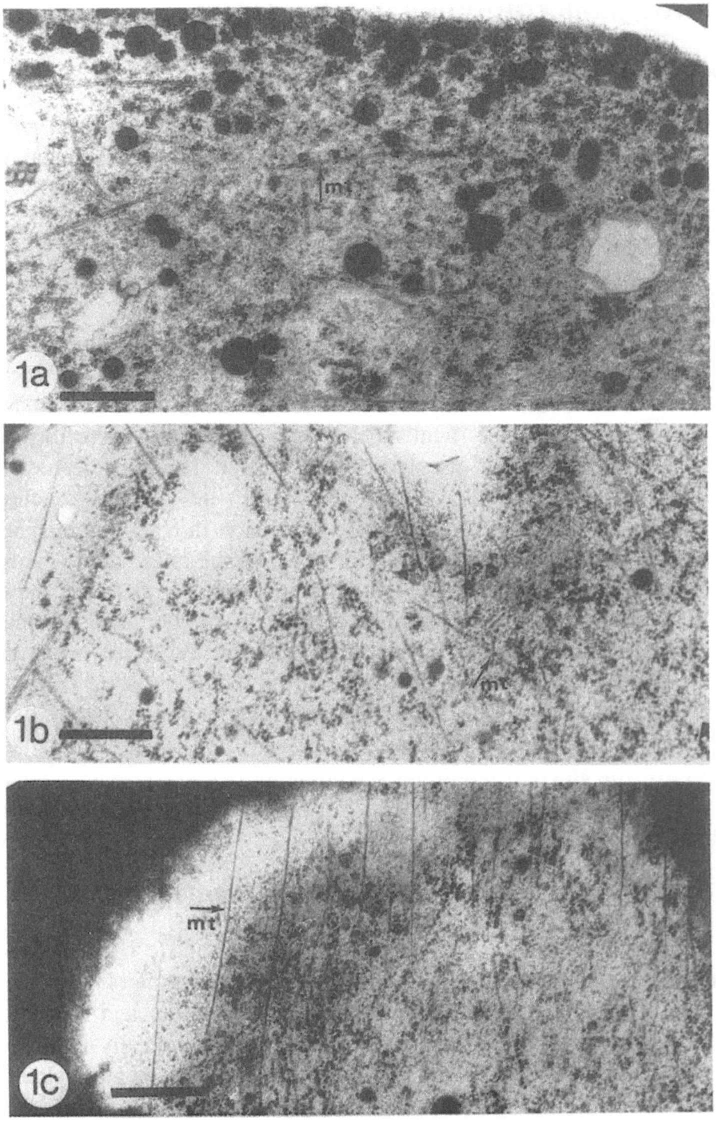

Fig. 1. The orientation of cortical microtubules (*mt*) at different distances from the growing tip of an *Equisetum hyemale* root hair. *A* Random orientation in the dome; *B* Subapically a change in orientation to net-axial; *C* Net-axially oriented cortical microtubules in the tube. Bars 500nm

thalictroides (Traas et al. 1985), *Raphanus sativus* (Lloyd and Wells 1985; Traas et al. 1985) and helical CMT were found in *Allium cepa* (Lloyd 1983; Lloyd and Wells 1985; Traas et al. 1985), and *Urtica dioica* (Traas et al. 1985; van Amstel and Derksen 1993). *Equisetum hyemale* and *Raphanus sativus* root hairs were studied with dry cleaving, which confirmed their net-axial orientation (Traas et al. 1985).

The only freeze substitution study on the CMT cytoskeleton carried out previously has been done on *Equisetum hyemale* (Emons 1989). This paper confirmed the net-axial orientation seen before with chemical fixation. However, CMT were more straight and diverted less from the axial orientation in comparison to the orientations found in chemically fixed cells. Light microscopic freeze fixation studies, which have given good results for the actin cytoskeleton (Miller et al. 1999; see also Chapter 3 this volume), are promising for CMT, but remain to be done.

2.2.2 Relationship of CMT and CMF in Root Hair Tubes?

As mentioned above, secondary cell wall formation takes place in the non-expanding tube of the root hair. Therefore, the role of the CMT in CMF deposition orientation can be studied in this part of the cell independently of the role of the CMT in cell expansion. In most terrestrial plant species studied, the CMF (Emons and van Maaren 1987) in the root hair tube run net-axially, as do the CMT (Traas et al. 1985) (Table 1). This observed correlation, however, does not mean that the one orients the other. Root hairs with helicoidal walls (Pluymaekers 1982) would be the perfect cells to study the relation between CMT and nascent CMF. Helicoidal cell walls consist of one microfibril thick lamellae. The microfibrils in every subsequent lamella make a constant angle with the CMF in the previous lamella. During growth at the tip, all stages of secondary wall development are present in the hair tube (Fig. 2) (Emons and Wolters-Arts 1983). Therefore, lamellae having microfibrils in different orientations are present along the plasma membrane (Emons 1989). If microtubules would orient the microfibrils during deposition, the microtubules should lie in those orientations as well. However, the microtubules are net-axially oriented in the whole hair tube (Fig. 3) (electron microscopy: Emons 1982; Emons and Wolters-Arts 1983; immunofluorescence: Traas et al. 1985; freeze substitution electron microscopy Emons 1989). Figure 3a is an immunofluorescence micrograph of a growing root hair of *Equisetum hyemale* with net-axial microtubules. CMT in the tube of other root hairs with helicoidal wall texture (*Hydrocharis morsus ranae*: Emons and van Maaren 1987; *Ceratopteris thalictroides*: Meekes 1985; *Limnobium stoloniferum*: Traas et al. 1985) are net-axial as well. Although in growing cells or cells parts like root hair tips, CMT may have a role in orienting the CMF, we conclude that CMT do not function in determining the orientation of deposition of cellulose microfibrils by default.

2.2.3 Function of the Cortical Microtubules in the Root Hair Tube

In expanding cells or cell parts such as the tips of growing root hairs, the CMT may be linked to the plasma membrane by Microtubule Associated Proteins (MAP). MAP have been found in plant cells (Chang Jie and Sonobe 1993; Bokros et al. 1995; Durso et al. 1996; Rutten et al. 1997). CMT have a very specific halo

Table 1. Orientations of cortical microtubules (CMT) and texture of the recently deposited cellulose microfibrils (CMF) in the cell wall

Species	Microfibril deposition (references)	Microtubule orientation (references)
Equisetum hyemale		
Growing tip	Random (1,2,11)	Random (1,2,11)
Tip-300μm	Random (1,2,11)	Axial (1,2,11)
300-3000μm	Helicoidal (1,2,11)	Axial (1,2,11)
Full-grown	Helical (1,2,11)	Axial (1,2,11)
Old	Axial (13)	Axial (11)
Ceratopteris thalictroides		
Growing tip	Random (3)	?
Tip-base	Helicoidal (3,12)	Axial (3,12)
Limnobium stoloniferum		
Growing tip	Random (3,4)	Random (3,4)
Tip-base	Helicoidal (3,4)	Axial (3,4)
Raphanus sativus		
Lepidium sativum		
Growing tip-25μm	Random (3,5)	?
25μm-base	Axial (3,5)	Axial (3,5)
Urtica dioica		
Growing tip	Random (3)	
Tip-base	Helical (3,6)	Net-axial (3,6)
Allium cepa		
Growing tip	Random (7,9)	Helical or absent (9)
Tip-base	Helical (7) or Helicoidal (9)	Helical (7,10)
Hydrocharis morsus ranae		
Tip-base	Helicoidal (8)	Axial (8)

References: (1) Emons (1982); (2) Emons and Wolters-Arts (1983); (3) Sassen et al. (1981); (4) Pluymaekers (1982); (5) Newcomb and Bonnett (1965); (6) Van Amstel and Derksen (1993); (7) Lloyd and Wells (1985); (8) Emons and Van Maaren (1987); (9) Traas et al. (1985); (10) Lloyd (1983); (11) Emons (1989); (12) Meekes (1985); (13) Emons unpublished 1983

around them, seen well in electron micrographs. This halo and the distance of the core of the microtubule to the plasma membrane is not different in transverse sections of the tube of root hairs of *Equisetum hyemale* (Emons unpublished) and other plant cells. Therefore, the electron microscope studies do not provide evidence for a difference in function between the CMT in growing cells and in full-grown cells.

The function of the CMT in the tube of root hairs remains unclear. A few observations point to a possible role. Emons et al. (1990) have shown that when

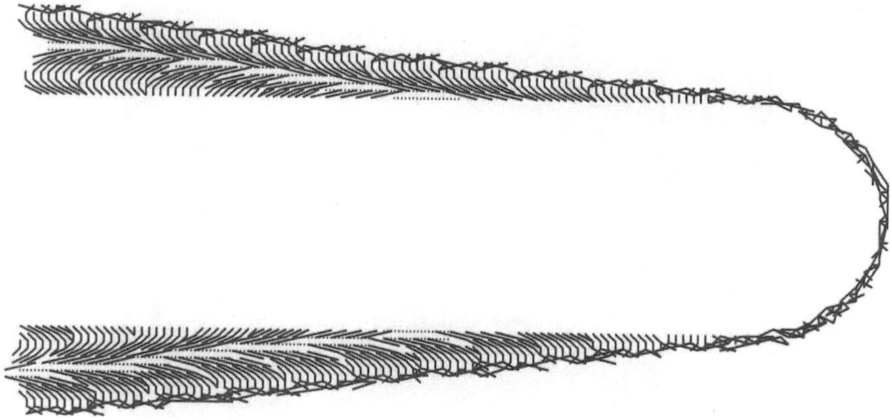

Fig. 2. Drawing displaying the different stages of helicoidal cell wall development which are present in the tube of a growing *Equisetum hyemale* root hair, showing that in a growing root hair with a helicoidal wall deposition the cortical microtubules lie in different orientations along the plasmamembrane at the same time. The root hair diameter and the cell wall width are not proportional

the CMT are depolymerized by colchicine at such a concentration that root hair growth proceeds (10 mM), the new tube has a larger diameter than the existing tube. Figure 3b shows a root hair of *Equisetum hyemale* treated with colchicine. The part of the hair that had been growing in the colchicine solution has no CMT and is wider. Bibikova et al. (1999) conducted a pharmacological study to decipher the role of CMT in root hairs of *Arabidopsis thaliana*. Both CMT depolymerization by the drug oryzalin and CMT stabilization by the drug taxol cause a wavy root hair phenotype, and taxol is able to induce branching, indicating that CMT are involved in the determination of growth direction of root hairs. Tominaga et al. (1997) show that cytoplasmic streaming does not recover in the proper orientation after actin filament depolymerization by cytochalasin B, when microtubules are not present. They conclude that functioning microtubules are required for recovery of cytoplasmic streaming after actin depolymerization. This indicates that the CMT may be involved in directing newly formed actin filaments. Obviously more work is needed in this field.

3 Endoplasmic Microtubules in Root Hairs

The observation of endoplasmic microtubules in tips of root hairs of *Vicia sativa* by Bakhuizen (1988), together with the absence of CMT is an indication that this observation could be an artifact of the chemical fixation of the root hairs. Bakhuizen suggests that the microtubules radiating from the nuclear envelope

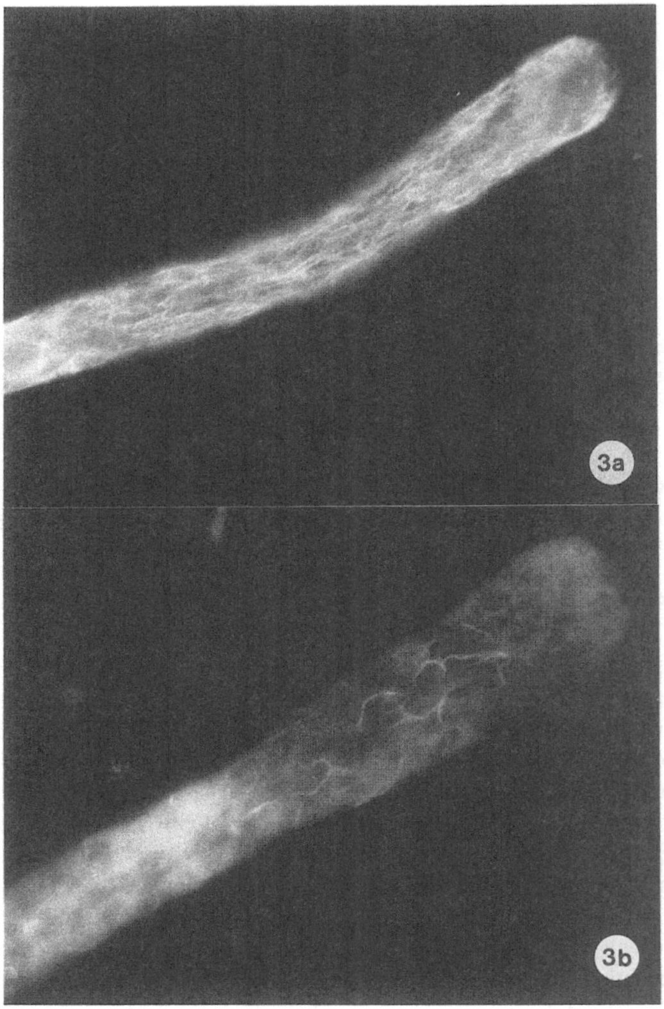

Fig. 3. *a* Immunofluorescence micrograph of a growing root hair of *Equisetum hyemale* with net-axial microtubules. *b* Immunofluorescence micropgraph of root hair of *Equisetum hyemale* treated with 10 mM colchicine. The part of the hair that had been growing in the colchicine solution is wider and has abnormal microtubules or none. The width of the root hairs on the images is about 20μm

could be the result of activation of microtubule organizing centers at or near the nuclear envelope.

In the legume *Vicia hirsuta*, Lloyd et al. (1987) found bundles of endoplasmic microtubules, which link the nuclear region to the apical dome, where they fountain out upon the cortex. The microtubules were visualized by chemical fixation and immunolocalization. Such a localization has never been reported for other species. Nuclei seem to follow the advancing tip of growing root hairs (Ridge 1992). Endoplasmic microtubules around the nucleus may function in nuclear movement. In a number of studies in different organisms, varying from yeast to higher plant cells, a relationship has been found between nuclear movement and the microtubule cytoskeleton. In budding yeast, interphase nuclear movement can be inhibited by nocodazole treatment (Shaw et al. 1997; review: Reinsch and Gönczy 1998). In sea urchin and *Xenopus* egg extracts, nuclei have been found to move over microtubules by dynein motor activity (reviewed in: Reinsch and Gönczy 1998). During branching of tip growing moss protonemata, endoplasmic microtubules are found to connect the nucleus to the prospective division site (Doonan et al. 1986), whereas in filamentous fungi cytoplasmic dynein has been found to be involved in nuclear migration (Xiang et al. 1994, 1995; Plamann et al. 1994; Bruno et al. 1996; Inoue et al. 1998). Additionally, microtubule nucleation sites have been found on nuclei of higher plants (maize cultured cells and tobacco BY-2 cells: Stoppin et al. (1994, 1996); tobacco cultured cells: Mizuno (1993); maize root in vivo: Baluska et al. (1992)). Lastly, in tip growing pollen tubes of *Nicotiana sylvestris*, the vegetative nucleus and generative cell dislocate when microtubules are depolymerized by colchicine treatment, indicating that microtubules are involved in their positioning (Joos et al. 1994). These observations suggest that endoplasmic microtubules may play a role in nuclear positioning in root hairs, but additional studies need to be done to obtain more clarity.

ACKNOWLEDGMENT

We wish to thank Dr. Daphne Wittich-Rainey for correcting the English text.

References

Bakhuizen R (1988) The plant cytoskeleton in the *Rhizobium*-Legume symbiosis. PhD thesis Rijksuniversiteit Leiden, The Netherlands

Baluska F, Parker JS and Barlow PW (1992) Specific patterns of cortical and endoplasmic microtubules associated with cell growth and tissue differentiation in roots of maize (*Zea Mays* L.). J Cell Sci 103:191-200

Baskin TI, Wilson JE, Cork A and Williamson RE (1994) Morphology and microtubule organization in *Arabidopsis* roots exposed to oryzalin or taxol. Plant Cell Physiol 35:935-942

Bibikova TN, Blancaflor EB and Gilroy S (1999) Microtubules regulate tip growth and orientation in root hairs of *Arabidopsis thaliana*. Plant J 17:657-665

Bokros CL, Hugdahl JD, Kim HH, Hanesworth VR, van Heerden A, Browning KS and Morejohn LC (1995) Function of the p86 subunit of eukaryotic initiation factor (iso)4F as a microtubule-associated protein in plant cells. Proc Nat Acad Sci USA 92:7120-7124

Bonnett HT Jr and Newcomb EH (1965) Polyribosomes and cisternal accumulations in root cells of radish. J Cell Biol 27: 423-32

Bruno KS, Tinsley JH, Minkle PF and Plamann M (1996) Genetic interactions among cytoplasmic dynein, dynactin, and nuclear distribution mutants of *Neurospora crassa*. Proc Nat Acad Sci USA 93: 4775-4780

Chang Jie J, and Sonobe S (1993) Identification and preliminary characterization of a 65 kDa higher-plant microtubule-associated protein. J Cell Sci 105: 891-901

Cyr RJ (1994) Microtubules in plant morphogenesis: role of the cortical array. Annu Rev Cell Biol 10: 153-80

Doonan JH, Jenkins GI, Cove DJ and Lloyd CW (1986) Microtubules connect the migrating nucleus to the prospective division site during branch formation in the moss *Physcomitrella patens*. Eur J Cell Biol 41:157-164

Durso NA, Leslie, JD and Cyr RJ (1996) In situ immunocytochemical evidence that a homolog of protein translation elongation factor EF-1 alpha is associated with microtubules in carrot cells. Protoplasma 190: 141-150

Emons AMC (1982) Microtubules do not control microfibril orientation in a helicoidal cell wall. Protoplasma 133: 85-87

Emons AMC (1987) The cytoskeleton and secretory vesicles in root hairs of *Equisetum* and *Limnobium* and cytoplasmic streaming in root hairs of *Equisetum*. Annals of botany 60: 625-632

Emons AMC (1988) A comparison of methods for visualization of the cell wall microfibrils. Acta Bot Neerl 37: 31-38

Emons AMC (1989) Helicoidal microfibril deposition in a tip-growing cell and microtubule alignment during rip morphogenesis: a dry-cleaving and freeze-substitution study. Can. J Bot 67: 2401-2408

Emons AMC and Derksen JHM (1986) Microfibrils, microtubules and microfilaments of the trichoblast of *Equisetum hyemale*. Acta Bot Neerl 35: 311-320

Emons AMC, Derksen J and Sassen MMA (1992) Do microtubules orient plant cell wall microfibrils? Phys Plantarum 84: 486-493

Emons AMC and Wolters-Arts AMC (1983) Microtubules and microfibrils in the cell wall of root hairs of *Equisetum hyemale*. Protoplasma 117:68-81

Emons AMC, Wolters-Arts AMC, Traas JA and Derksen J (1990) The effect of colchicine on microtubules and microfibrils in root hairs. Acta Bot Neerl 39:10-27

Emons AMC and van Maaren N (1987) Helicoidal cell wall texture in root hairs. Planta 170:145-151

Fisher DD and Cyr RJ (1998) Extending the microtubule/microfibril paradigm. Cellulose synthesis is required for normal cortical microtubule alignment in elongating cells. Plant Physiol 116:1043-1051

Galway ME, Heckman JW Jr, Schiefelbein JW (1997) Growth and ultrastructure of *Arabidopsis* root hairs: the rhd3 mutation alters vacuole enlargement and tip growth. Planta 201:209-218

Gertel ET and Green PB (1977) Cell growth pattern and wall microfibrillar arrangement. Plant Physiol 60:247-254

Giddings TH Jr and Staehelin LA (1991) Microtubule-mediated control of microfibril deposition: a re-examination of the hypothesis. In: Lloyd CW (Ed) The cytoskeletal basis of plant growth and form. Academic Press, San Diego, CA. pp 303-322

Green PB (1962) Mechanism for plant cellular morphogenesis. Science 138: 1404-1405

Green PB, Erickson RO and Richmond PA (1970) On the physical basis of cell morphogenesis. Ann. NY Acad Sci 175: 712-731

Heath IB (1974) A unified hypothesis for the role of membrane bound enzyme complexes and microtubules in plant cell wall synthesis. J Theor Biol 48: 445-449

Hepler PK and Fosket DE (1971) The role of microtubules in vessel member differentiation in *Coleus*. Protoplasma 72: 213-236

Inoue S, Turgeon BG, Yoder OC and Aist JR (1998) Role of fungal dynein in hyphal growth, microtubule organization, spindle pole body motility and nuclear migration. J Cell Sci 111: 1555-1566

Joos U, van Aken, J and Kristen U (1994) Microtubules are involved in maintaining the cellular polarity in pollen tubes of *Nicotiana sylvestris*. Protoplasma 179: 5-15

Ledbetter MC and Porter KR (1963) A 'microtubule' in plant fine structure. J Cell Biol 19: 239-250

Lloyd CW (1983) Helical microtubular arrays in onion root hairs. Nature 305: 311-313

Lloyd CW, Pearce KJ, Rawlins DJ, Ridge RW and Shaw PJ (1987) Endoplasmic microtubules connect the advancing nucleus to the tip of legume root hairs, but F-actin is involved in basipetal migration. Cell Motility and Cytoskeleton 8:27-36.

Lloyd CW and Wells B (1985) Microtubules are at the tips of root hairs and form helical patterns corresponding to inner wall fibrils. J Cell Sci 75: 225-238

Meekes HTHM (1985) Ultrastructure, differentiation and cell wall texture of trichoblasts and root hairs of *Ceratopteris thalictroedes* (L.) Brongn. (Parkeriaceae). Aquatic Botany 21: 347-362

Miller DD, de Ruijter NCA, Bisseling T and Emons AMC (1999) The role of actin in root hair morphogenesis: studies with lipochito-oligosaccharide as a growth stimulator and cytochalasin as an actin perturbing drug. Plant J 17: 141-154

Mizuno K (1993) Microtubule-nucleation sites on nuclei of higher plant cells. Protoplasma 173: 77-85

Newcomb EH and Bonnett HT (1965) Cytoplasmic microtubule and cell wall microfibril orientation in root hairs of radish. J Cell Biol 27: 575-589

Pluymaekers, HJ (1982) A helicoidal cell wall texture in root hairs of *Limnobium stoloniferum*. Protoplasma 112: 107-116

Plamann M, Minke PF, Tinsley, JH and Bruno KS (1994) Cytoplasmic dynein and actin-related protein Arp1 are required for normal nuclear distribution in filamentous fungi. J Cell Biol 127: 139-149

Reinsch S and Gönczy P (1998) Mechanisms of nuclear positioning. J Cell Sci 111: 2283-2295

Ridge RW (1992) A model of legume root hair growth and *Rhizobium* infection. Symbiosis 14: 359-373

Robinson DG and Quader H (1982) The microtubule-microfibril syndrome. In: Lloyd CW (Ed) The cytoskeletal basis of plant growth and form. Academic Press, San Diego, CA. pp 109-126

Rutten T, Chan J, Lloyd CW (1997) A 60-kDa plant microtubule-associated protein promotes the growth and stabilization of neurotubules in vitro. Proc Natl Acad Sci USA 94: 4469-4474

Schmid VHR and Meindl U (1992) Microtubules do not control orientation of secondary cell wall microfibril deposition in *Micrasterias*. Protoplasma 169: 148-154

Schnepf E, Röderer G and Herth W (1975) The formation of fibrils in the lorica of *Poteriochromonas stipitata*: tip growth, kinetics, site, orientation. Planta 125: 45-62

Seagull RW and Heath IB (1980) The organization of cortical microtubule arrays in the radish root hair. Protoplasma 103: 205-229

Shaw SL, Yeh E, Maddox P, Salmon ED and Bloom K (1997) Astral microtubule dynamics in yeast: a microtubule-base searching mechanism for spindle orientation and nuclear migration into the bud. J Cell Biol 139: 985-994

Stoppin V, Lambert A-M and Vantard M (1996) Plant microtubule-associated proteins (MAPs) affect microtubule nucleation and growth at plant nuclei and mammalian centrosomes. Eur J Cell Biol 69: 11-23

Stoppin V, Vantard M, Schmit A-C, Lambert A-M (1994) Isolated plant nuclei nucleate microtubule assembly: the nuclear surface in higher plants has centrosome-like activity. Plant Cell 6: 1099-1106

Tominaga M, Morita K, Sonobe S, Yokota E and Shimmen T (1997) Microtubules regulate the organization of actin filaments at the cortical region in root hair cells of *Hydrocharis*. Protoplasma 199: 83-92

Traas JA, Braat P, Emons AMC, Meekes H and Derksen J (1985) Microtubules in root hairs. J Cell Sci 76: 303-320

Upadhyaya MK and Noodén LD (1977) Mode of dinitroaniline herbicide action I. Analysis of the colchicine-like effects of dinitroaniline herbicides. Plant Cell Physiol 18: 1319-1330

van Amstel ANM and Derksen J (1993) The complex helical texture of the secondary cell wall of *Urtica dioica* root hairs is not controlled by microtubules: a quantitative analysis. Acta Bot. Neerl. 42: 141-151

van Spronsen PC, van Brussel AA and Kijne JW (1995) Nod factors produced by *Rhizobium leguminosarum* biovar viciae induce ethylene-related changes in root cortical cells of *Vicia sativa* ssp. nigra. Eur J Cell Biol 68: 463-469

Xiang X, Beckwith SM and Morris NR (1994) Cytoplasmic dynein is involved in nuclear migration in *Aspergillus nidulans*. Proc Nat Acad Sci USA 91: 2100-2104

Xiang X, Roghi C and Morris NR (1995) Characterization and localization of the cytoplasmic dynein heavy chain in *Aspergillus nidulans*. Proc Nat Acad Sci USA 92: 9890-9894

Chapter 3
The Root Hair Actin Cytoskeleton as Backbone, Highway, Morphogenetic Instrument and Target for Signalling

John Esseling, Norbert de Ruijter, and Anne Mie C. Emons

Department of Plant Sciences, Laboratory of Experimental Plant Morphology and Cell Biology, Wageningen University, Arboretumlaan 4, 6703 BD, Wageningen, The Netherlands

1 Introduction

The plant actin cytoskeleton is a structural cell element that connects cytoplasmic components with each other, with the cell surface and with neighbouring cells through plasmodesmata. It is the dynamic backbone of cytoplasmic strands. Along this backbone, cell components move and are targeted to specific sites in the cell. The specific targeting of cell organelles is instrumental in cell morphogenesis. Moreover, the actin cytoskeleton may connect to receptors, transduce signals and mediate cell responses. Actin monomers and filaments are bound to actin binding proteins that determine actin's function in place and time. Since root hairs are accessible for the application of signal molecules and drugs, and can be visualized and manipulated in the living state, these tip-growing cells are ideal for the study of the functions of the actin cytoskeleton.

In this review we will discuss what is known, mainly from fixed cells, about the actin cytoskeleton of root hairs. The first topic is a summary of the methods that have been used to visualize actin in root hairs. As a second topic, the actin cytoskeleton's function as intracellular backbone of cytoplasmic strands will be discussed. Third, the typical types of cytoplasmic streaming in the different developmental stages of root hairs and the way in which these patterns change after treatment with actin perturbing drugs will be covered. The streaming patterns of organelles will be related to their transport system, that is, the configuration of actin filaments found during these stages. The fourth topic is the role of actin in morphogenesis. Alone, the actin filaments can not exert all these functions. Apart from polymerisation kinetics, the function of the actin cytoskeleton is determined by actin binding proteins (ABPs). In the fifth section we will discuss the ABPs

Key words. actin cytoskeleton, cytoplasmic streaming, lipochito-oligosaccharide, Nod-factor, *Rhizobium*, root hair known to occur in root hairs. The last section of this review will address the way nodulation signals, well-characterized lipochito-oligosaccharides (LCOs) excreted by *Rhizobium* bacteria, affect the actin cytoskeleton of root hairs.

2 Actin Visualization

To study the cytoskeleton, a good visualization method is a prerequisite. Over time, different techniques have been developed for the visualization of actin filaments. Most techniques include fixation and several cell wall degradation steps. Recent achievements however, allow the study of the actin cytoskeleton and its dynamics in living cells.

At the ultrastructural level, actin filaments have been visualized with heavy meromyosin fragments (radish root hairs: Seagull and Heath 1979) and immunogold labelling (*Nicotiana* pollen tubes: Lancelle and Hepler 1991). Heavy meromyosin fragments are very useful in determining the polarity of the actin filaments. In fact, the terminology barbed and pointed end for the growing and non-growing ends respectively, is based on the fishbone-like pattern of heavy meromyosins on individual actin filaments in electron micrographs. Microfilaments, which probably are actin filaments, can also be observed in the electron microscope after glutaraldehyde osmium tetroxide fixation and uranyl acetate lead citrate staining (*Equisetum hyemale* root hairs: Emons 1987). Such contrast staining is a good method to show the orientation of microfilaments in the electron microscope, only if the nature of the filaments is known to be actin. The disadvantage of these methods is that they all need fixation steps, which may more or less disturb the cytoarchitecture and probably the arrangement of the cytoskeleton. Cryofixation followed by freeze substitution is preferred over chemical fixation with aldehydes, because it preserves the cytoarchitecture much better (root hairs: Emons and Derksen 1986; Emons 1987; Emons 1989; Ridge 1988).

In the light microscope, actin filaments can be visualized with fluorescently labelled phalloidin (Wieland and Faulstich 1978) and with anti-actin antibodies (for plant actin: Andersland et al. 1994). A common disadvantage of anti-actin antibodies is that they also label actin monomers. This may mask labelling of filamentous actin. For root hairs, anti-actin labelling has been performed in vetch (Miller et al. 1999).

Phalloidin is a toxic bicyclic heptapeptide from the poisonous mushroom *Amanita phalloides*. It is a relatively small molecule, with an approximate diameter of 12-15 Å and a molecular weight <1500 Dalton. From in vitro experiments it is known that phalloidin only binds to the actin monomers within an actin filament, and, if present in a high concentration, shifts actin's monomer/polymer equilibrium towards the polymer, lowering the critical concentration for polymerization up to 30-fold (Cooper 1987; Miki et al. 1987). Also with these methods, the fixation step is the most critical; the cytoarchitecture of the fixed cell has to resemble that of a living cell. Phalloidin may not bind to all filaments present in a cell, since it shares its actin-binding site with the actin binding protein cofilin / actin depolymerizing factor (ADF) (Nishida et al. 1987; Ressad et al. 1998).

Recently, freeze substitution was adapted for a light microscope study of the actin cytoskeleton of vetch root hairs, and a quicker aldehyde fixation method followed by a mild plasma membrane permeabilization was developed, which gave similar results (Miller et al. 1999). In the last method, stabilization of the actin cytoskeleton with m-maleimidobenzoyl N-hydroxysuccinimide ester (MBS ester; Sonobe and Shibaboka 1989) was followed by fixation with a mixture of paraformaldehyde and glutaraldehyde. For plasma membrane permeabilization, the fixed root hairs were treated in increasing concentrations of L-α-lysophosphatidylcholine, after which the actin cytoskeleton was labelled with fluorescein phalloidin. The advantages of this method are the facts that en bloc staining of all hairs along the root is possible, and that the method is fast. However, it has to be optimized for every cell type and compared to actin configurations obtained after freeze substitution. Figure 1a shows the actin cytoskeleton in a root hair of *Vicia sativa*, chemically fixed without a pre-treatment with MBS ester and with enzymatic wall degradation. Figure 1b shows the actin cytoskeleton in a root hair prepared with the optimized method in which the roots were pre-treated with MBS ester before chemical fixation and no enzymes for wall digestion were used. The optimized method preserves the cytoarchitecture better (Fig. 1d). This becomes clearly seen when one compares root hairs prepared with the two types of fixation (Fig. 1c and 1d respectively) with living root hairs (Fig. 1e), using the differential interference contrast (DIC) microscope.

For the study of cytoskeletal dynamics in response to internal and external stimuli, living cells are the most interesting. One approach to visualize actin dynamics in plant cells is microinjection of fluorescent phalloidin. For *Tradescantia* stamen hair cells two methods were developed. When injected in high concentrations (20 μM needle concentration), phalloidin immediately stabilizes the actin cytoskeleton (Valster and Hepler, 1997). A disadvantage of this "snap shot method" is that the cell dies. Actin remains functional when injection is done at lower concentrations of phalloidin (8-9 μM needle concentration), as shown by Staiger et al. (1994) and Valster et al. (1997). Also this method has to be optimized for each cell type. Cárdenas et al. (1998) injected 20 μM needle concentration FITC phalloidin in living *Phaseolus* root hairs (the "snap shot method") and reported stabilization of actin filament bundles followed by root hair death, generally 10 min after the injection. Microinjection of 0.66 μM needle concentration Alexa-488 phalloidin in *Medicago truncatula* and *Arabidopsis thaliana* root hairs is not lethal to the root hairs and shows an actin staining pattern in the first 10 min after injection. However, within 30 min after injection, this pattern is broken down and fluorescence accumulates in aggregates, which move along with the cytoplasmic streaming (J. Esseling, unpublished results). Figure 1f shows the actin cytoskeleton 3 min after Alexa-488 phalloidin injection in a growing *Arabidopsis thaliana* root hair. Figure 1g shows a growing *Arabidopsis thaliana* root hair, 30 min after Alexa-488 phalloidin injection.

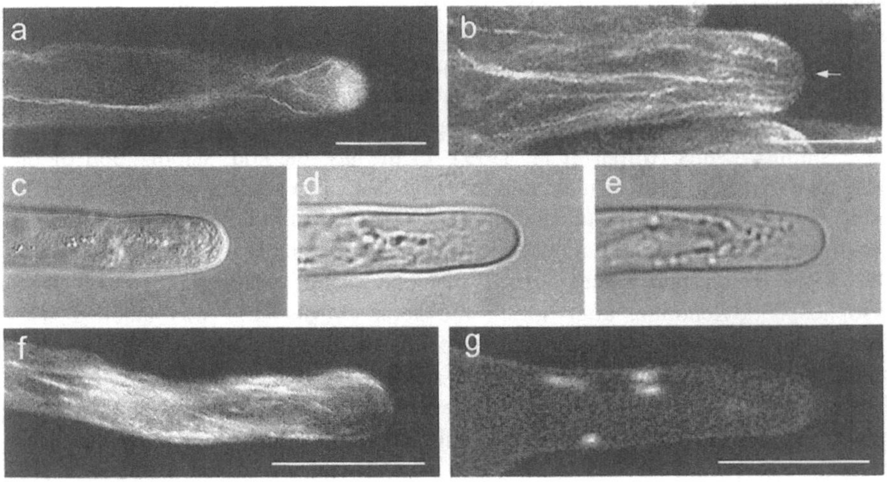

Fig. 1. *a-e*. Effects of fixation and staining procedure on the actin filament distribution, and cytoarchitecture of growing root hairs of *Vicia sativa*. *a* The actin cytoskeleton visualized with rhodamine phalloidin after conventional fixation with 3% paraformaldehyde and 0.05% glutaraldehyde without stabilization with MBS ester, followed by enzymatic wall digestion. Note the accumulation of actin filaments in the tip. *b* Actin cytoskeleton visualized with fluorescein phalloidin after an optimized fixing-staining procedure. F-Actin was stabilized with 300 μM m-maleimido benzoyl N-hydroxysuccinimide ester (MBS ester) in growth medium, and fixed by stepwise increasing the aldehyde concentration to 4% paraformaldehyde and 0.1% glutaraldehyde in Actin Stabilizing Buffer (ASB: 100 mM Pipes pH 6.8, 1 mM $MgCl_2$, 1 mM $CaCl_2$, 75 mM KCl) with 1 mM 4-2 aminoethylbenzene sulfonyl fluoride (AEBSF). Root hairs were permeabilized with 100 μg/ml L-α-lyso-phosphatidylcholine in ASB and actin filaments were stained with 0.33 μM fluorescein phalloidin in ASB supplemented with 0.05% acetylated BSA (ester-aldehyde-choline procedure). Actin filament bundles are net-axially aligned, are denser in the subapical area and the apex is devoid of actin filament bundles (*arrow*). *c* Differential interference contrast (DIC) image of (*a*). Note the poor preservation of the cytoarchitecture in the tip. *d* DIC image after the ester-aldehyde-choline procedure (as in *b*). Note the presence of a smooth area in the tip. *e* DIC image of living, growing root hair showing the "reference" cytoarchitecture with subapical dense cytoplasm and a smooth area at the tip. *f Arabidopsis thaliana:* Actin filament bundles in living, growing root hair at 3 min after microinjection with Alexa-488 phalloidin *g A. thaliana:* Fluorescence in living root hair at 30 min after microinjection with Alexa-488 phalloidin, showing accumulation of signal in aggregates in the cytoplasm. Magnification is the same for *a, c, d, e*. Bar in (*a, b, f, g*) = 20 μm

The most promising method is the use of GFP-fusion proteins. GFP (Green Fluorescent Protein) is a protein isolated from the jellyfish *Aequorea victoria* (Chalfie et al. 1994). Its gene can be coupled to any gene of interest. In a number of studies, GFP-actin fusion proteins have been used to visualize actin localization in mammalian cells (Ballestrem et al. 1998), yeast (Doyle and Botstein 1996) and

Dictyostelium discoideum (Aizawa et al. 1997b; Neujahr et al. 1997; Westphal et al. 1997). However, GFP-actin fusion proteins also have their problems. Expression of GFP-actin fusion proteins will show both actin filaments and monomers, which could mask actin filaments. Also, labelling of only a subpopulation of GFP-actin filaments has been reported (Doyle and Botstein 1996). Moreover, expression of GFP-actin can severely affect morphology and function of cells (Aizawa et al. 1997b).

Another approach is the use of a fusion between GFP and full length actin binding proteins or actin binding domains, which was used in *Drosophila melanogaster* (Edwards et al. 1997), *Dictyostelium discoideum* (Aizawa et al. 1997a; Pang et al. 1998), yeast (Doyle and Botstein 1996) and, most recently, in plants (Kost et al. 1998). Kost and colleagues used the actin binding domain of a mouse talin gene coupled to GFP, which they introduced in tobacco pollen tubes and tobacco BY-2 cells via particle bombardment and in *Arabidopsis thaliana* via *Agrobacterium* transformation (Kost et al. 1998). However, they report that pollen tubes with a high level of expression, terminate growth and die.

3 Actin: The Backbone

Plant cells contain vacuoles. Mature cells have one large vacuole, which fills most of the cell. So-called cytoplasmic strands traverse this vacuole. The cytoplasmic strands are continuous with the cortical cytoplasm, which covers the whole plasma membrane. Together, cortical cytoplasm and cytoplasmic strands are one connected system. Immunocytochemistry shows that cytoplasmic strands contain actin filament bundles. Moreover, when the actin filaments are depolymerized by, for instance, microinjection of the actin monomer binding protein profilin, the cytoplasmic strands become thinner and snap back into the cortical cytoplasm. This has been shown in mature (Staiger et al. 1994) and growing (Valster et al. 1997) *Tradescantia* stamen hair cells. Likewise, in root hairs, partial depolymerization of the actin filaments, in this case performed by use of cytochalasin D at relatively high concentrations (1-5 μM), causes the formation of one large vacuole in growing root hairs of *Vicia sativa*, which had cytoplasmic strands before application of the drug. However, bundles of actin filaments remain present in the cortical cytoplasm, after profilin injection in *Tradescantia* stamen hair cells (Valster et al. 1997), as well as after treatment of root hairs with high concentrations of cytochalasin D (Fig. 2a,b). Actin filament bundles appear to be the backbones of cytoplasmic strands.

4 Actin: The Highway

When living plant cells are studied with the light microscope, an obvious feature is the movement of the organelles inside cytoplasmic strands, also known as cytoplasmic streaming. In rotational streaming (Kamiya 1981), or cyclosis, all organelles follow one direction. In the more common circulation streaming, the pattern is less organized; tracks in which organelles move in opposite directions

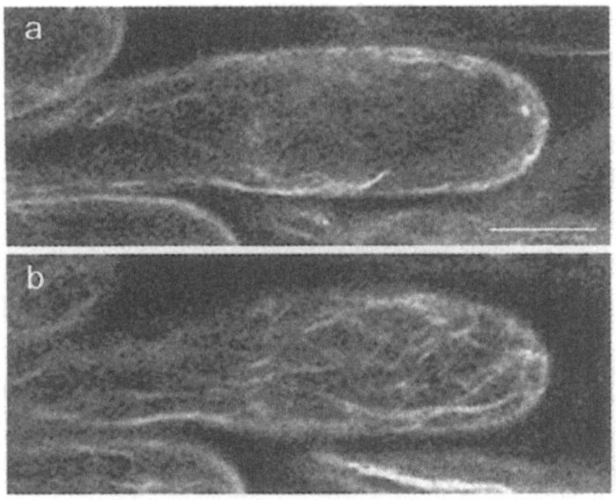

Fig. 2. Effect of cytochalasin D on the cytoarchitecture and actin cytoskeleton of *Vicia sativa* root hairs. *a,b* Confocal optical planes through a growing root hair treated with 2.0 μM cytochalasin D for 30 min, stained after the ester-aldehyde-choline procedure with fluorescein phalloidin and imaged at the mid plane (*a*) and cortical plane (*b*). Filamentous actin is not present in the mid plane of the cell, but actin filament bundles have remained in the cortical cytoplasm. The magnification is the same for *a,b*. Bar in (*a*) = 15 μm

may be located next to each other in the same cytoplasmic strand (Lichtscheidl and Url, 1987; Lichtscheidl and Weiss, 1988). In fountain streaming, a central track of organelles is running upward, splitting up and then reversing its orientation. Tip growing cells have reverse-fountain streaming. Iwanami (1956) introduced this term for pollen tubes. In wide pollen tubes (*Lilium*) or root hairs (*Hydrocharis*: Tominaga et al. 1997; Tominaga et al. 1998) an upward flow of cytoplasm to the tip at the cell borders and downward flow in a transvacuolar cytoplasmic strand in the cell centre characterize this type of streaming. The upward flow never reaches the plasma membrane at the tip of a growing root hair or pollen tube, but reverses just below the so-called vesicle-rich region. In thinner root hairs (such as from *Arabidopsis*, *Medicago* and *Vicia*) and pollen tubes (*Nicotiana tabacum*: de Win et al. 1999; *Papaver rhoeas*: A. Geitmann, personal communication, 1999), the downward streaming may be in the centre or at the side of the hair or tube. However, the most outstanding feature of reverse fountain streaming remains: the reversal of the upward streaming before the hair tip is reached.

The speed of cytoplasmic streaming has been measured in different regions of growing and full- grown root hairs of *Equisetum hyemale* (Emons 1987). In the tip of a 2 day old *Equisetum hyemale* root hair the speed of cytoplasmic streaming is 0 μm/s. Below the tip, the speed of streaming increases to a more steady level of about 3 μm/s. The cytoplasmic streaming pattern and speed in full-grown hairs is completely different compared to growing hairs. The flow does not reverse before reaching the tip of the hair as it does in growing hairs, but streams through the very tip in thin strands of cytoplasm between the large central vacuole and the plasma membrane. This type of cytoplasmic streaming is of the circulation type (Tominaga et al. 1998). In full-grown root hairs of *Equisetum hyemale*, the speed of cytoplasmic streaming is different when compared to the speed in growing hairs. Throughout the whole full-grown root hair, the speed of cytoplasmic streaming is between 4 and 6 μm/s (Emons 1987). So, an organelle of 1 μm moves about 5 times its own length per second. For comparison, if a 2 m tall sprinter runs 5 times his/her own length per second, his/her speed would be 36 km/h. Another type of organelle movement has been described as saltatory movement. It is observed when single organelles move in a non co-ordinated individual way over only short distances (review: Emons et al. 1991). This also occurs in root hairs (unpublished results).

In plant cells, cytoplasmic movement is actin based. After injection of the G-actin ABP profilin into *Tradescantia* stamen hair cells, cytoplasmic streaming ceases (full-grown cells: Staiger et al. 1994; growing cells: Valster et al. 1997). In these experiments, a second injection with fluorescent phalloidin confirmed that the stop of cytoplasmic streaming was caused by depolymerization of actin filaments. Moreover, actin perturbing drugs, such as the actin filament capping fungal toxin cytochalasin D, stops cytoplasmic streaming in *Vicia sativa* root hairs at concentrations above 5 μM (de Ruijter et al. personal communication 1999).

5 Actin, the Morphogenetic Instrument

In higher plants, cell morphogenesis is attained by local cell growth. One can discriminate between three types of growth: isodiametric-, intercalary- and tip-growth. Insertion of new plasma membrane and exocytosis of cell wall material through Golgi vesicles (Roberts 1994) determine plant cell growth. Thus, in the case of growing cells, the exocytotic vesicle is the unit of plant cell growth. The location of exocytosis determines where and when growth will take place. The most common type of plant cell growth is intercalary growth of elongating cells such as cells of the root elongation zone. In tip-growing cells, exocytosis occurs at one side of the cell. Well-known examples of tip growing cells in higher plants are pollen tubes (Schnepf 1986) and root hairs (Derksen and Emons 1990).

Root hairs develop as tubular extensions of specialized root epidermal cells, the trichoblasts (Dolan et al. 1994). The onset of root hair growth from a trichoblast can be seen as a local expansion of the trichoblast, the bulge. The bulge contains a peripheral layer of cytoplasm with cytoplasmic strands around a large vacuole (Fig. 3a) (Miller et al. 1999). The actin cytoskeleton in bulges is characterized by filament bundles running in the cytoplasm along the plasma membrane of the bulge, longitudinal to the root axis (Fig. 3b) in the same orientation as in the trichoblasts that have not formed a bulge yet (Miller et al. 1999).

Growing root hairs have a polar cytoarchitecture with a central vacuole in the basal part of the cell and a cytoplasmic dense region at the cell apex (Fig. 3c). Electron microscopic studies have shown that the first 1-3 μm from the tip contains almost exclusively secretory vesicles (chemical fixation: Bonnett and Newcomb 1966; freeze substitution: Emons 1987; Ridge 1988; Sherrier and Vandenbosch 1994; Galway et al. 1997). We call this the vesicle rich region. In the differential interference microscope it appears as a very smooth thin layer of

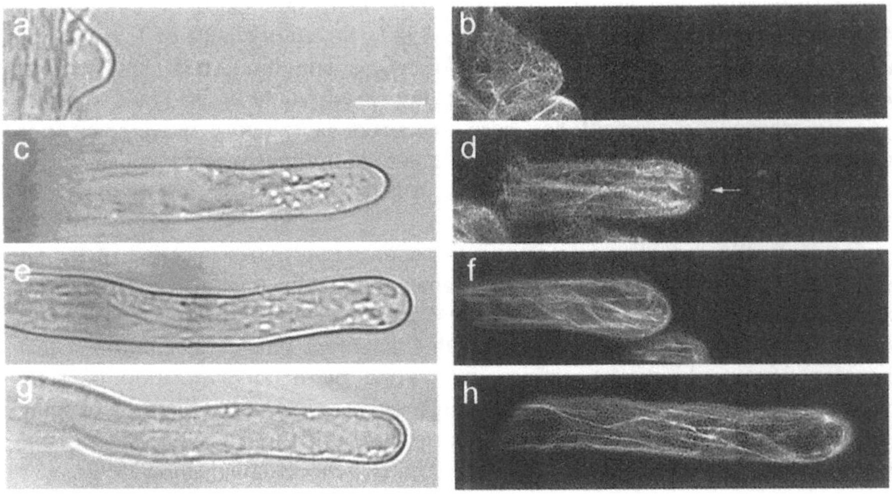

Fig. 3. The cytoarchitecture and actin filament configuration of subsequent developmental stages of *Vicia sativa* root hairs. *a,c,e,g* DIC images of living root hairs, representative for each developmental stage. *b,d,f,h* Representative fluorescence images showing the actin filament distribution in each developmental stage; actin was visualized with fluorescein phalloidin after the ester-aldehyde-choline procedure. The images are projections of confocal z-series. *a,b* Bulge containing a major central vacuole. Actin filament bundles are randomly oriented and pass through the peripheral cytoplasm at the tip. *c,d* Growing root hair with a smooth area at the very tip, and a subapical cytoplasmic dense region. The actin filament bundles are longitudinally oriented in the hair. In the subapical region, dense fine actin filament bundles are present (FB-actin), the tip is devoid of actin filament bundles (*arrow*). *e,f* Root hair in which growth is terminating. Small vacuoles are present close to the tip with small cytoplasmic strands between them. The area devoid of actin is absent, and fine actin filament bundles are present in the tip. *g,h* Full-grown root hair with large vacuole and peripheral cytoplasm. Thick actin filament bundles are present in strands looping through the extreme tip. Magnification is the same in all images. Bar = 20 μm

cytoplasm at the tip (Fig. 3c). During growth, the cytoplasmic dense region remains at the tip, while the vacuole increases in size. In these growing hairs, bundles of actin filaments are present in cytoplasmic strands around and traversing the central vacuole (Miller et al. 1999). These bundles lie longitudinally and helically in the root hair, perpendicular to their orientation in the epidermal part of the trichoblast (Fig. 3a). In the cytoplasmic dense region, the filament bundles flare out into thinner and thinner bundles and maybe even into single filaments (Fig. 3d). We call these subapical net-axially aligned fine bundles of actin filaments, FB-actin. The very apex of the growing hair is devoid of actin filament bundles (*arrow* Fig. 3d). In vetch, this region is approximately 2-6 μm deep and 5-8 μm wide (Miller et al. 1999) and coincides with the vesicle rich region.

The cytoarchitecture of root hairs that terminate growth is characterized by a shortening of the cytoplasmic dense region with an enlargement of the main vacuole further into the tip-region (Fig. 3e). The area with FB-actin is very short and a region devoid of actin in the extreme tip is not present any more (Fig. 3f) (Miller et al. 1999, de Ruijter et al. 1999).

In full- grown hairs, cytoplasmic polarity has clearly been lost; the central vacuole extends into the tip, and is surrounded with only a thin layer of peripheral cytoplasm (Fig. 3g). Actin filament bundles are found in this thin peripheral layer of cytoplasm just as elsewhere in the cell cortex, with some bundles looping through the tip and other bundles turning in the cortical cytoplasm before they reach the tip (Fig. 3h). FB-actin is not present in the subapex and there is no apical area devoid of filamentous actin.

From the actin cytoskeleton configuration during root hair development, we hypothesized that the role of FB-actin is to retain the Golgi vesicles in the subapical area and to deliver them to the base of the Golgi vesicle rich region (Miller et al. 1999). We had two ways to test this. First, the inhibition of actin filament elongation by the F-actin capping drug cytochalasin-D should change the cytoarchitecture of a growing root hair into that of a hair that is terminating growth, and, after longer treatment, into that of a full-grown hair. On the other hand, a signal molecule that reinitiates growth in hairs that are terminating growth, should induce FB-actin.

When cytochalasin-D is used at low concentrations (<1 μM), the bundles of actin filaments in the hair tube remain present and are still functional, as shown by the fact that cytoplasmic streaming continues. However, indeed, as predicted by our hypothesis, when FB-actin disappears (compare Fig. 4a before and 4b after cytochalasin-D application to growing hairs), hairs obtain the cytoarchitecture of hairs that are terminating growth (Fig. 5).

To reinitiate growth in legume root hairs, host specific lipochito-oligosaccharides (LCOs) can be used. LCOs, also known as Nod-factors, are well-characterized signal molecules excreted by *Rhizobium* bacteria that induce reinitiation of tip growth in legume root hairs that are terminating growth (Fig. 6). Indeed, the application of LCOs induces the formation of FB-actin, prior to new tip growth (Fig. 3f before and 4c after LCO application; Miller et al. 1999).

Since a growing root hair elongates in one direction, and actin filaments are continuous in the whole hair, actin polymerization in the direction of growth has to take place during root hair elongation. During root hair elongation, the area with FB-actin has a more or less constant length. The most logical explanation for this phenomenon is that filaments elongate at the tip of the FB-actin zone and bundle at the base of this zone. Then, actin monomers have to be incorporated into filaments at the tip of the zone with FB-actin. The situation is clearly different from that at the leading edge of motile animal cells. There, it is widely accepted that polymerization of new filaments and membrane protrusion are tightly coupled (Welch et al. 1997). Here, in the root hair, polymerization of new filaments would be coupled to Golgi vesicle retention and/or Golgi vesicle delivery to the vesicle rich area at the extreme tip of the root hair.

In plant cells, growth itself is the incorporation of Golgi vesicles. The study of actin in root hairs shows that FB-actin is one of the factors that determine where and when Golgi vesicles are incorporated into the plasma membrane. Therefore,

FB-actin can be seen as one of the morphogenetic instruments of plant cells. It could well be that the same morphogenetic instrument is involved in other types of growth, including bulge formation prior to tip-growth. Results from studies by Foissner et al. (1996), in which the actin cytoskeleton after cell wounding was investigated, demonstrated the involvement of fine filaments of actin in recruiting vesicles for cell wall repair. Because vesicle incorporation in isodiametrically and intercalary growing cells is far less localized, and therefore less abundant, than in tip growing cells, the area with fine bundles of actin filaments should be a thin layer along the plasma membrane. Such a thin peripheral network of actin filaments is difficult to image, but has been reported for, among other cells, potato suspension cells (Collings and Emons 1999).

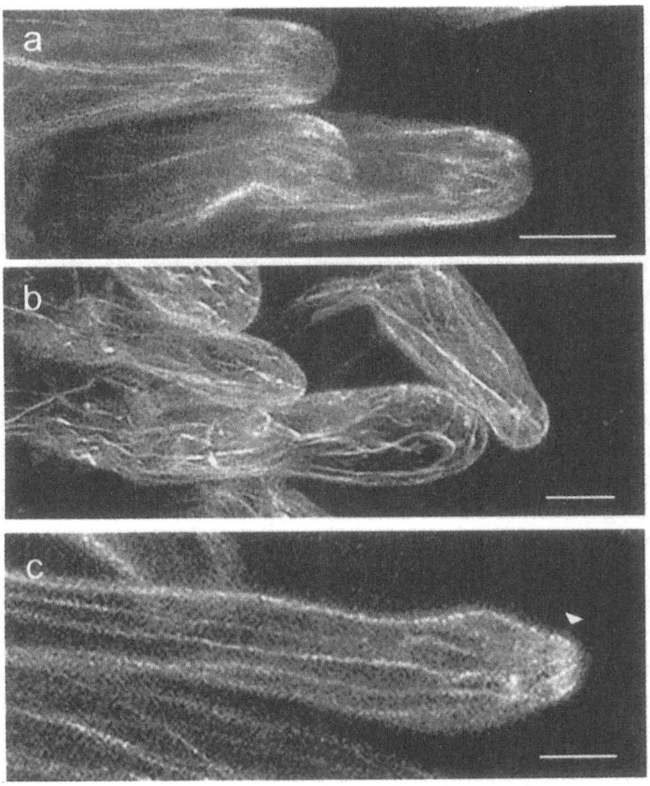

Fig. 4. The presence of dense subapical fine actin filament bundles (FB-actin) correlates with tip growth in root hairs of *Vicia sativa*. *a* Growing root hairs with FB-actin in the subapical region, and the very tip devoid of actin filament bundles. *b* Growing root hairs treated with 1.0 μM cytochalasin D for 30 min with an actin cytoskeleton typical for hairs terminating growth. Note the disappearance of FB-actin and the presence of actin filament bundles in the very tip. *c* Root hair that was terminating growth treated with 10^{-10} M LCO for 75 min, with a new actin cytoskeleton typical for growing hairs, FB-actin (*arrowhead*), and a small area devoid of actin at the extreme tip. Bar = 15 μm

6 The Regulation of the Actin Cytoskeleton

The changing configurations of the actin cytoskeleton during root hair development and correlated patterns of cytoplasmic streaming are indications for a highly dynamic actin cytoskeleton. The outstanding question in understanding actin dynamics is the role of actin polymerization and the regulation of polymerization. The fuel for actin polymerization has to be a large cellular pool of unpolymerized actin, which should be maintained at concentrations well above the critical concentration for actin polymerization. To maintain this pool, cells employ actin binding proteins (ABPs; review of plant ABPs: Ayscough 1998; plant cells: de Ruijter and Emons 1999). Much of our understanding about the function of ABPs comes from in vitro research (Ruhlandt et al 1994; Carlier et al. 1997) and work on ABPs from animal cells (Puius et al. 1998).

In plants, profilin is the best-characterized ABP (reviewed by Staiger et al. 1997). Originally, it was identified in plants as a pollen allergen (Valenta et al. 1991) and has now been isolated from a wide variety of plant species (Staiger et al. 1993; Mittermann et al. 1995; Darnowski et al. 1996; Huang et al. 1996; Clarke et al. 1998). In *Arabidopsis thaliana*, thus far 10 gene members have been isolated (Huang et al. 1996) and maize is predicted to have at least six profilin genes (Staiger et al. 1993). Profilin was originally identified as an actin monomer sequestering protein that can inhibit actin filament growth in vitro (Carlsson et al. 1977) and can be placed in the group of ABPs with assembly/disassembly function. Indeed, when microinjected into plant cells, i.e., *Tradescantia* stamen hair cells, profilin causes depolymerization of actin filaments in a dose- dependent manner (Staiger et al. 1994; Valster et al. 1997), consistent with an actin monomer binding function of profilin. Another actin cytoskeleton regulating function of profilin is the shift in equilibrium from ADP-actin to ATP-actin at low profilin concentrations. Since ATP-actin binds more efficiently to the barbed end of actin filaments, lowering of the profilin concentration promotes actin polymerization (Hartwig and Kwiatkowski 1991; Aderem 1992; Goldschmidt-Clermont et al. 1992). Like many other ABPs, profilin is a multifunctional protein. It can reversibly bind to poly-L-proline and to the phospholipid phosphatidyl-inositol 4,5-biphosphate (PIP_2) (Goldschmidt-Clermont et al. 1990; Goldschmidt-Clermont et al. 1991; Sohn and Goldschmidt-Clermont 1994). The binding to PIP_2 provides a link between the phosphatidyl-inositol signal transduction pathway and the actin cytoskeleton (Aderem 1992). In vitro, human platelet-derived profilin binds to PIP_2 and inhibits its hydrolysis by unphosphorylated PLC (Goldschmidt-Clermont et al. 1991). When phosphorylated by a tyrosine kinase, PLC is activated and is able to hydrolyze the PIP_2-profilin complex, resulting in the release of profilin (Goldschmidt-Clermont et al. 1991). In plants, experiments in *Papaver rhoeas* pollen tubes indicate a possible role for profilin in signal transduction. In these cells, profilin regulates protein kinase and protein phosphatase activity and actin based cytoskeletal protein assembly (Clarke et al. 1998). Also a controlling role of profilin in phospho-inositide turnover in plant cells has been demonstrated (Drøbak et al. 1994), and recently it has been shown

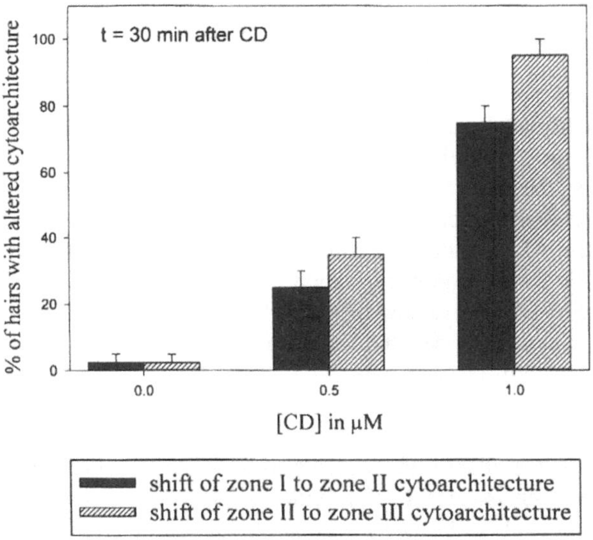

Fig. 5. Cytochalasin D (CD) effects on the cytoarchitecture of *Vicia sativa* root hairs. Histogram showing the percentage of root hairs in zone I and II that had obtained the cytoarchitecture of respectively zone II and III after 30 min (y-axis), related to the concentration of CD in μM (x-axis). The shift to a cytoarchitecture of the next developmental stage increases with the concentration of CD; zone II hairs are more sensitive than zone I hairs. *Zone I:* Growing root hairs; *Zone II:* Root hairs terminating growth; *Zone III:* full-grown root hairs

that in microspores and mature pollen, profilin is associated with the plasma membrane (Von Witsch et al. 1998). All these data indicate that profilin is an interesting ABP in plant cells, since it links signal transduction directly to the actin cytoskeleton. Unfortunately, there are no publications yet on the role of profilin in root hairs. In growing pollen tubes (Vidali and Hepler 1997) and root hairs (Braun et al. 1999) profilin is localized throughout the cytoplasmic dense region.

To maintain continuous polymerization at one end, actin must be depolymerized at the other end, and the subunits recycled. ADF / cofilin (Maciver et al. 1998) could be an important protein for the regulation of actin depolymerization. Plant-ADF was first identified in *Lilium longiflorum* pollen tubes (Kim et al. 1993), after which Rozycka et al. (1995) and Lopez et al. (1996) found ADF in *Zea mays* pollen tubes. Plant ADF appears to be modulated by pH (Lopez et al. 1996), interacts with phosphoinositides (Gungabissoon et al. 1998) and promotes actin dynamics in vitro (Jiang et al. 1997b). When pH increases (to about 8.0), the ability of ADF to sever and depolymerize actin bundles increases (Staiger et al. 1997; Maciver et al. 1998). In vitro experiments with maize ADF indicate that it binds to actin monomers in a 1:1 complex (Staiger et al. 1997). It also promotes treadmilling by accelerating the association rate at barbed ends and the dissociation rate at pointed ends of filaments (Carlier et al. 1997). In *Dictyostelium*, data suggest that ADF (here named cofilin) has a role in remodelling the cortical actin meshwork into actin filament bundles (Aizawa et al.

1997a). ADF localization in maize root hairs, in the apex and subapex (Jiang et al. 1997a), where FB-actin is present (Miller et al. 1999) suggests that ADF has a role in remodelling the actin cytoskeleton in growing root hairs. In bulging trichoblasts, which do not have visible FB-actin (Miller et al. 1999), ADF is localized in the cytoplasm without specific distribution (Jiang et al. 1997a).

A third possible actin binding protein known to occur in root hairs, is a protein that labels with anti-spectrins from human and chicken origin (de Ruijter et al. 1998). Although the gene has not been isolated yet, the anti-spectrin antibodies recognize an interesting protein product, since it marks tip growth very efficiently. The epitope is not accumulated in the trichoblast, is present in the apex of a growing root hair, disappears from the tip when growth stops, re-appears at the plasma membrane of the swelling after application of LCOs, and is again present

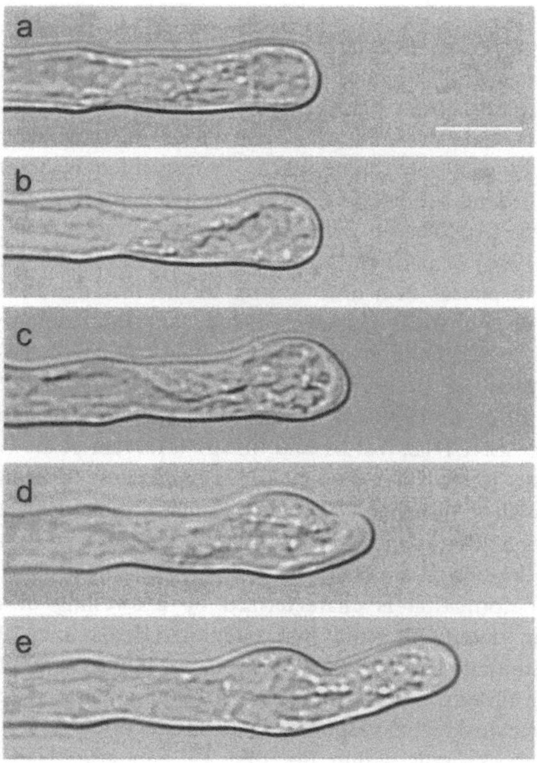

Fig. 6. Root hair deformation after LCO application to roots of *Vicia sativa*. *a-e* A time-series of DIC images from one living root hair that was terminating growth, showing a swelling and reinitiation of a new outgrowth from that swelling after application of 10^{-10} M LCO. *a* Typical cytoarchitecture of a hair terminating growth. *b* An initial swelling is visible at the tip at 1 h after LCO application. *c* At one side of the swelling cytoplasm accumulates at 1 h 15 min after LCO application, a smooth area becomes visible, and tip-polarized growth is initiated. *d* Cytoplasmic strands orient towards the new tip at 1 h 30 min after LCO application, and renewed tip growth is apparent. *e* The cytoarchitecture of the outgrowth (at t = 2 h 30 min after LCO) has similar characteristics as a normal growing hair (*see Fig. 1e, 3c*). Magnification is the same for *a-e*. Bar (in *a*) = 20 μm

at the tip of the outgrowth (de Ruijter et al. 1998). It has a punctuate pattern in the vesicle rich region at the root hair tip (Miller et al. 1997) and, therefore, may be a membrane protein that is inserted into the plasma membrane by Golgi vesicles. Spectrin-like proteins may indeed be involved in membrane sorting and Golgi dynamics (Lippincott-Schwartz 1998). The same antibodies label tips of pollen tubes (Derksen et al. 1995) and hyphae (Kaminsky and Heath 1995) and localize to the plasma membrane of growing cells (de Ruijter and Emons 1993). Since a plant spectrin gene has not been reported yet, we can only speculate about the function of this spectrin-like plant protein. Recently, a gene of the spectrin family has been found in *Arabidopsis* (McCurdy and Kim 1998).

7 Actin as Target for Nod-Factor Signalling

Nitrogen fixing *Rhizobium* bacteria can live in symbiosis with legume plants. In the legume root, *Rhizobia* induce the formation of so-called root nodules, where the nitrogen fixation takes place (Mylona et al. 1995). For this root nodule formation, the bacteria grow into the root via an infection thread, which is first formed inside the root hair (Kijne 1992). During infection thread formation, the root hair curls around the bacteria. This root hair curling means modification of root hair growth. Application of very low concentrations (as low as 10^{-12} M, Heidstra et al. 1994) of purified host-specific LCOs to liquid grown legume roots induce several responses in a time course. In the first minutes after application, LCOs trigger physiological responses, such as alkalinization of the root hair (Ehrhardt et al. 1992; Felle et al. 1996), depolarization of the root hair plasma membrane (Ehrhardt et al. 1992; Felle et al. 1996; Kurkdjian 1995) and calcium spiking (Ehrhardt et al. 1996). It induces the expression of early nodulin genes, such as ENOD5, ENOD12, VsLb1 and Mtrip1 (Horvath et al. 1993; Journet et al. 1994; Cook et al. 1995; Heidstra et al. 1997a), and causes root hair deformation in 80-90% of growth terminating root hairs (Heidstra et al. 1997b; de Ruijter et al. 1998) about 1-2 h after application and, finally, induces mitotic activity in cells of the root cortex (Mylona et al. 1995).

Root hair deformation is characterized by a swelling of the root hair tip, followed by an outgrowth from that swelling (Heidstra et al. 1994). In fact, outgrowth is reinitiation of tip growth from the swelling, with all characteristics of normal root hair tip growth (de Ruijter et al. 1998; Fig. 6a-e).

The earliest cytoskeleton responses to LCOs reported so far are fragmentation of the actin cytoskeleton within the first 15 min after Nod-factor application (*Medicago sativa*: Allen and Bennet 1996; *Phaseolus vulgaris*: Cárdenas et al. 1998). This fragmentation was reported as discontinuities in the longitudinal actin bundles and formation of discrete foci of phalloidin staining at the root hair tip. FITC phalloidin microinjected *Phaseolus* root hairs show a fluorescent glow at the tip of the hairs within 5 min after Nod-factor treatment (Cárdenas et al. 1998). This glow has been interpreted as actin filament fragmentation. As discussed above, actin filament breakdown causes the disappearance of cytoplasmic strands (full-grown cells: Staiger et al. 1994; growing cells: Valster et al. 1997; vetch root

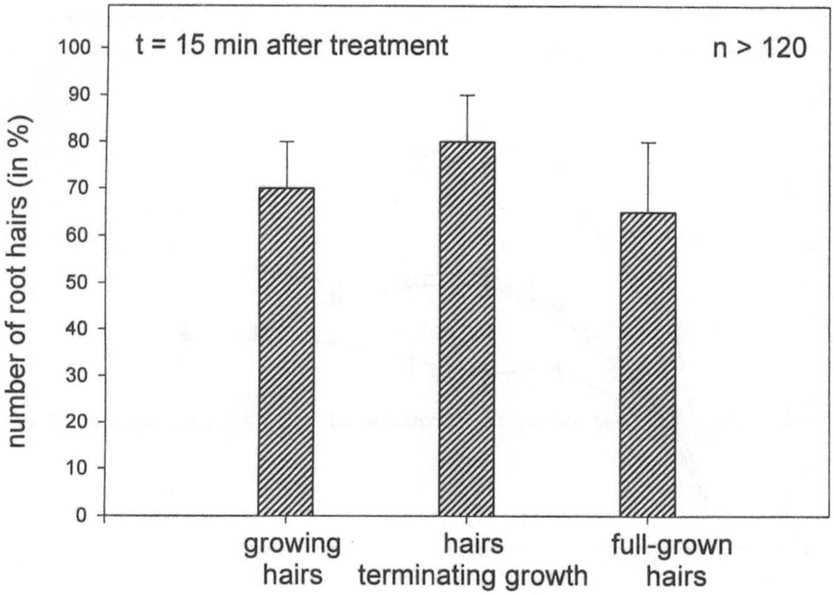

Fig. 7. Percentage of root hairs of *Vicia sativa* with a LCO-induced increase in the density of sub-apical actin filament bundles. The percentage of *Vicia sativa* root hairs with an increase in the number of sub-apical actin filament bundles at 15 min after application of 10^{-10} M *Rhizobium leguminosarum* bv *viciae* Nod RIv V [Ac, C18:4] (LCOs) (y-axis) is displayed for each developmental stage (x-axis). Root hairs in each developmental stage respond to LCOs. The density of actin filament bundles is compared to the density of actin filament bundles in control root hairs, to which fresh growth medium without LCOs was applied. A random population of root hairs from at least 12 different roots was examined in at least 6 different experiments, and the same amount was used for the controls. Error bars indicate standard error (SE)

hairs Fig. 2a,b). Thus, the results of Cárdenas et al. (1998) would suggest that actin filament fragmentation by Nod-factor should cause the breakdown of cytoplasmic strands in root hairs.

For *Vicia sativa* root hairs, it has been shown that, consistent with the idea that cytoplasmic strands can not exist without filamentous actin, actin filaments remain intact after Nod-factor application (Miller et al. 1997; de Ruijter et al. 1999). However, the number of fine bundles of actin filaments increased in the subapical area of the root hair within 3-15 min after Nod-factor application, in 80% of the all root hairs (Fig. 7). This increase in the number of fine bundles of actin filaments occurred in all stages of root hair development. Thus, root hairs that do and do not respond to Nod-factor by root hair deformation, show a uniform response in the actin cytoskeleton, which is an increase in the number of subapical fine bundles of actin filaments (Fig. 8). Figure 9a is an example of hairs in the developmental

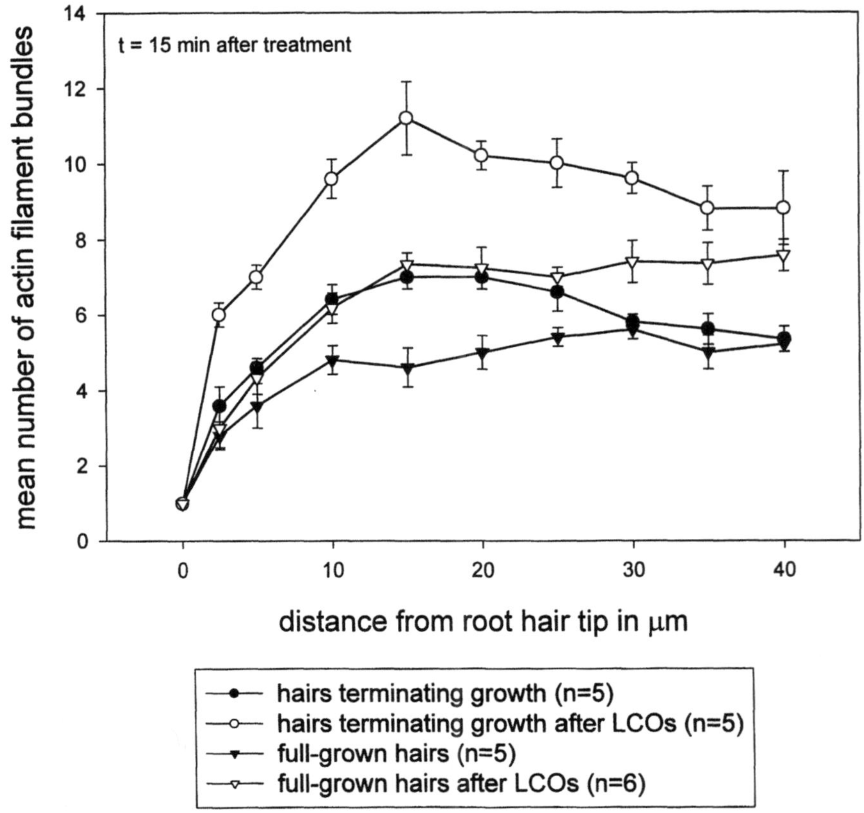

Fig. 8. Mean densities of actin filament bundles in the subapical area of control and LCO treated *Vicia sativa* root hairs. Mean densities of actin filament bundles (y-axis) are displayed versus the distance from the tip (x-axis). In all root hairs, the density profile shows an increase in actin filament bundles after application of LCOs (Shown for root hairs terminating growth and root hairs terminating growth). Note that the increase in full-grown hairs does not exceed the critical FB-actin density that correlates with growth. Random sampling was done within each zone at 3-15 min after treatment, z-series were processed, and means of numbers of actin filaments displayed. Error bars indicate SE

stage of growth termination taken before, and Fig. 9b taken 3 min after Nod-factor application. The density of subapical net-axial fine bundles of actin filaments has clearly increased. After quantification (for procedure see legend of Fig. 9c) of this increase in fine bundles of actin filaments, an interesting difference appears (de Ruijter et al. 1999). After Nod-factor treatment, the number of fine bundles of actin filaments in full-grown hairs did not exceed the threshold level of that seen in control growth terminating hairs (Fig. 8), which are the cells that still have some capacity for growth and which do respond with root hair deformation.

Among other factors, like an already stiff secondary cell wall at tips of full- grown vetch root hairs (still to be proven), the limited increase of FB actin in full- grown hairs may be the reason why those hairs do not respond with root hair deformation. In Fig. 10 we have mapped the occurrence of the reaction of the actin cytoskeleton to Nod-factor in relation to the changes in calcium in the root hair (Ehrhardt et al. 1996; de Ruijter et al. 1998; Felle et al. 1998; see also Chapter 9 this volume).

The increase in the number of fine bundles of actin filaments in the subapical area of the root hair suggest a signal transduction pathway involving the activation of several ABPs. Since the in vitro binding of profilin and ADF / cofilin to PIP_2 provides a link between the phosphatidyl-inositol signal transduction pathway and the actin cytoskeleton (Yonezawa et al. 1990; Aderem 1992; Staiger et al. 1997), it

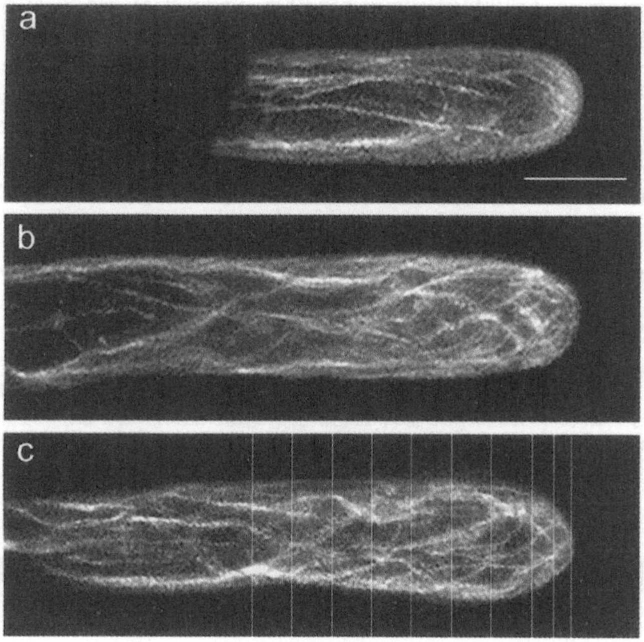

Fig. 9. Increase in the density of bundles of actin filaments in *Vicia sativa* root hairs that were terminating growth at the moment of LCO application. *a* Typical distribution and density of actin filament bundles in a control hair after replacing growth medium (t = 6 min), with actin filament bundles up to the tip, and only few sub-apical fine bundles of actin filaments. *b,c* Typical distribution and density of actin filament bundles after application of LCOs (t = 3 min, t = 6 min respectively). Note that dense sub-apical FB-actin has appeared. These fine bundles of actin filaments often had a wider range of orientations in relation to the long axis of the hair, than growing hairs after LCO application. *c* To compare data, all settings during imaging and processing were similar. Confocal z-series were made at 1 μm steps. To quantify the number of actin filament bundles, a grid with equally spaced lines was superimposed onto the projected image of a root hair. The grid was perpendicular to the long axis of the hair and the zero-line was always positioned at the very tip. The number of all fine and thick actin filament bundles hit by these cross-lines was counted on a high-resolution screen. Bar = 15 μm

can be hypothesized that this signal transduction pathway is involved in Nod-factor signalling. Recent experiments support this hypothesis (Pingret et al. 1998). Inhibitors of phospholipase C (PLC) applied simultaneously with Nod-factor to *Medicago truncatula* roots inhibit the expression of a MtENOD12-GUS construct in the root epidermis, indicating that the activity of PLC increases after LCO treatment (Pingret et al. 1998). In vitro, PLC cleaves PIP_2 into inositol (1,4,5)-trisphosphate (IP_3) and diacylglycerol (DAG) (Dicheva and Irvine 1995), suggesting a rapid increase in IP_3 after Nod-factor application.

The increase of cytoplasmic pH after LCO application to *Medicago sativa* root hairs (Ehrhardt et al. 1992; Felle et al. 1995; Felle et al. 1996) suggests another correlation between ADF regulated changes in the actin cytoskeleton and Nod-factor treatment. In LCO treated *Medicago sativa* root hairs, the pH increases to levels (Ehrhardt et al. 1992; Felle et al. 1995; Felle et al. 1996) at which plant-ADF is able to sever and depolymerize actin filaments (Staiger et al. 1997; Maciver et al. 1998). Moreover, in growing maize root hairs, the localization of maize ADF3 coincides more or less (Jiang et al. 1997a) with the region of *Vicia sativa* root hairs in which an increase in fine bundles of actin filaments is seen shortly after Nod-factor treatment (de Ruijter et al. 1999). Until now, the distribution of ADF in legume root hairs is unknown, but it can be hypothesized

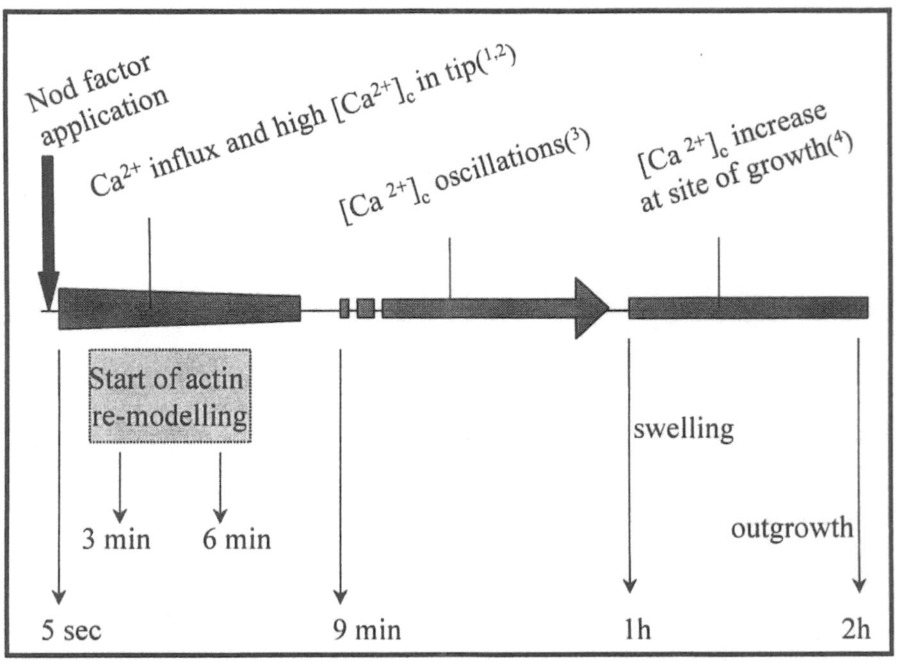

Fig. 10. Time scale correlating the increase in FB-actin with the changes in cytoplasmic free calcium ion concentration in root hairs responding to LCOs. [1]Felle et al. (1998); [2]Gehring et al. (1997); [3]Ehrhardt et al. (1996); [4]de Ruijter et al. (1998)

that Nod-factor induced cytoplasmic pH changes influence the severing/depolymerizing activity of ADF, which may result in a contribution to the increase of subapical fine bundles of actin filaments.

Profilin and ADF are, based on their activities, distribution and involvement in signal transduction pathways, good candidates for ABPs involved in the short term actin rearrangements after Nod-factor treatment. In analogy to knowledge about ABP function in animal cells and yeast, we expect that more ABPs are involved in the rearrangement of the actin cytoskeleton by Nod-factor.

8 Future Prospects

To investigate the complex signal transduction pathways leading from a signal molecule like the Nod-factor to changes in cell morphogenesis, a bioassay is a useful tool. We have used the occurrence of root hair deformation by Nod-factor as a bioassay to study the influence of external factors on this process (Heidstra et al. 1997b, Miller et al. 1999). Full root hair deformation is swelling of the root hair tip followed by root hair outgrowth from that swelling. In vetch, this whole process takes 2-3 h (Fig. 6). The discovery of the rapid, and quantifiable increase of FB-actin in root hairs after Nod-factor application (Fig. 8, 9), provides an additional bioassay, like the measurement of calcium dynamics does. Instead of waiting for 2 h to score root hair deformation, one can score the rapid effect Nod-factor has on the actin cytoskeleton and on calcium dynamics (See Fig. 10). By using drugs that affect the cytoskeleton and signal transduction cascades in combination with Nod-factor, one will be able to discriminate between the early factors in the signal transduction cascade that change the cytoskeleton and later factors that for instance need calcium spiking and induction of new gene transcription. Ultimately, this has to be done in living cells. Therefore, GFP-technology and microinjection techniques will be necessary, and should be applied in combination with molecular genetic and biochemical approaches.

ACKNOWLEDGMENT

We would like to thank Dr. Daphne Wittich-Rainey for correcting the English text.

References

Aderem A (1992) Signal transduction and the actin cytoskeleton: the roles of MARCKS and profilin. Trends Biochem Sci 17: 438-443

Aizawa H, Fukui Y, Yahara I (1997a) Live dynamics of *Dictyostelium* cofilin suggests a role in remodelling actin latticework into bundles. J Cell Sci 110: 2333-2344

Aizawa H, Sameshima M, Yahara I (1997b) A green fluorescent protein-actin fusion protein dominantly inhibits cytokinesis, cell spreading, and locomotion in *Dictyostelium.* Cell Struct Funct 22: 335-345

Allen NS, Bennett MN (1996) Electro-optical imaging of F-actin and endoplasmic reticulum in living and fixed plant cells. Scanning Microsc Suppl 10:177-187

Andersland JM, Fisher DD, Wymer CL, Cyr RJ, Parthasarathy MV (1994) Characterization of a monoclonal antibody prepared against plant actin. Cell Motil Cytoskeleton 29: 339-344

Ayscough KR (1998) In vivo functions of actin-binding proteins. Curr Opin Cell Biol 10: 102-111

Ballestrem C, Wehrle-Haller B, Imhof BA (1998) Actin dynamics in living mammalian cells. J Cell Sci 111: 1649-1658

Bonnett HT, Newcomb EH (1966) Coated vesicles and other cytoplasmic components of growing root hairs of radish. Protoplasma 62: 59-75

Braun M, Baluska F, Von Witsch M, Menzel D (1999) Redistribution of actin, profilin and phosphatidylinositol-4,5-bisphosphate in growing and maturing root hairs. Planta 209: 435-443

Cárdenas L, Vidali L, Domínguez J, Pérez H, Sánchez F, Hepler PK, Quinto C (1998) Rearrangement of actin microfilaments in plant root hairs responding to *Rhizobium etli* nodulation signals. Plant Physiol 116: 871-877

Carlier MF, Laurent V, Santolini J, Melki R, Didry D, Xia GX, Hong Y, Chua NH, Pantaloni D (1997) Actin depolymerizing factor (ADF / cofilin) enhances the rate of filament turnover: implication in actin-based motility. J Cell Biol 136: 1307-1322

Carlsson L, Nystrom LE, Sundkvist I, Markey F, Lindberg U (1977) Actin polymerizability is influenced by profilin, a low molecular weight protein in non-muscle cells. J Mol Biol 115: 465-483

Chalfie M, Tu Y, Euskirchen G, Ward WW, Prasher DC (1994) Green fluorescent protein as a marker for gene expression. Science 263: 802-805

Clarke SR, Staiger CJ, Gibbon BC, Franklin-Tong VE (1998) A potential signalling role for profilin in pollen of *Papaver rhoeas*. Plant Cell 10: 967-979

Collings D, Emons AMC (1999) Microtubule and actin filament organization during acentral divisions in potato suspension culture cells. Protoplasma 207: 156-168

Cook D, Dreyer D, Bonnet D, Howell M, Nony E, VandenBosch K (1995) Transient induction of a peroxidase gene in *Medicago truncatula* precedes infection by *Rhizobium meliloti*. Plant Cell 7: 43-55

Cooper JA (1987) Effects of cytochalasin and phalloidin on actin. J Cell Biol 105: 1473-1478

de Ruijter NCA, Emons AMC (1993) Immunodetection of spectrin antigens in plant cells. Cell Biol Int 17: 169-182

de Ruijter NCA, Emons AMC (1999) Actin-binding proteins in plant cells. Plant Biology 1: 26-35

de Ruijter NCA, Bisseling T, Emons AMC (1999) *Rhizobium* Nod-factors induce an increase in subapical fine bundles of actin filaments in *Vicia sativa* root hairs within minutes. Mol Plant Microbe Interact 12: 829-832

de Ruijter NCA, Rook MB, Bisseling T, Emons AMC (1998) Lipochito-oligosaccharides re-initiate root hair tip growth in *Vicia sativa* with high calcium and spectrin-like antigen at the tip. Plant J 13: 341-350

de Win AH, Pierson ES, Derksen J (1999) Rational analyses of organelle trajectories in tobacco pollen tubes reveal characteristics of the actomyosin cytoskeleton. Biophys J 76: 1648-1658

Darnowski DW, Valenta R, Parthasarathy MV (1996) Identification and distribution of profilin in tomato (*Lycopersicon esculentum* Mill.). Planta 198: 158-161

Derksen J, Emons AMC (1990) Microtubules in tip growth systems. In: Heath IB (Ed) Tip growth in plant and fungal cells. Academic Press, pp 147-181

Derksen J, Rutten T, Van Amstel T, de Win A, Doris F, Steer M (1995) Regulation of pollen tube growth. Acta Bot Neerl 44: 93-119

Dicheva N, Irvine RF (1995) Phospholipid signalling. Cell 80: 269-278

Dolan L, Duckett CM, Grierson C, Linstead P, Schneider K, Lawson E, Dean C, Poethig S, Roberts, K (1994) Clonal relationships and cell patterning in the root epidermis of *Arabidopsis*. Development 120: 2465-2474

Doyle T, Botstein D (1996) Movement of yeast cortical actin cytoskeleton visualized in vivo. Proc Natl Acad Sci USA 93: 3886-3891

Drøbak BK, Watkins PAC, Valenta R, Dove SK, Lloyd CW, Staiger CJ (1994) Inhibition of plant plasma membrane phosphoinositide phospholipase C by the actin-binding protein, profilin. Plant J 6: 389-400

Edwards KA, Demsky M, Montague RA, Weymouth N, Kiehart DP (1997) GFP-moesin illuminates actin cytoskeleton dynamics in living tissue and demonstrates cell shape changes during morphogenesis in *Drosophila*. Dev Biol 191: 103-117

Ehrhardt DW, Wais R, Long SR (1996) Calcium spiking in plant root hairs responding to *Rhizobium* nodulation signals. Cell 85: 1-20

Ehrhardt DW, Atkinson EM, Long SR (1992) Depolarization of alfalfa root hair membrane potential by *Rhizobium meliloti* Nod-factors. Science 256: 998-1000

Emons AMC (1987) The cytoskeleton and secretory vesicles in root hairs of *Equisetum* and *Limnobium* and cytoplasmic streaming in root hairs of *Equisetum*. Ann Bot 60: 625-632

Emons AMC (1989) Helicoidal microfibril deposition in a tip-growing cell and microtubule alignment during tip morphogenesis: a dry-cleaving and freeze-substitution study. Can J of Botany 67: 2401-2408

Emons AMC, Derksen JWM (1986) Microfibrils, microtubules and microfilaments of the trichoblast of *Equisetum hyemale*. Acta Bot Neerl 35: 311-320

Emons AMC, Pierson ES, Derksen JWM (1991) Cytoskeleton and intracellular movement in plant cells. In: PN Cheremisinoff and LM Ferrante (Eds) Biotechnology Current Progress 1, Technomic Publishing Company Inc., Lancaster UK. pp 311-335

Felle HH, Kondorosi É, Kondorosi Á, Schultze M (1995) Nod signal-induced plasma membrane potential changes in alfalfa root hairs are differently sensitive to structural modifications of the lipochito-oligosaccharide. Plant J 7: 939-947

Felle HH, Kondorosi É, Kondorosi Á, Schultze M (1996) Rapid alkalinization in alfalfa root hairs in response to rhizobial lipochito-oligosaccharide signals. Plant J 10: 295-301

Felle HH, Kondorosi É, Kondorosi Á, Schultze M (1998) The role of ion fluxes in Nod-factor signalling in *Medicago sativa*. Plant J 13: 445-463

Foissner I, Lichtscheidl IK, Wasteneys GO (1996) Actin-based vesicle dynamics and exocytosis during wound wall formation in characean internodal cells. Cell Motil Cytoskeleton 35: 35-48

Galway ME, Heckman JW Jr, Schiefelbein JW (1997) Growth and ultrastructure of *Arabidopsis* root hairs: the rhd3 mutation alters vacuole enlargement and tip growth. Planta 201: 209-218

Gehring CA, Irving HR, Kabbara AA, Parish RW, Boukli NM, Broughton WJ (1997) Rapid, plateau-like increases in intracellular free calcium are associated with Nod-factor-induced root-hair deformation. Mol Plant Microbe Interact 10: 791-802

Goldschmidt-Clermont PJ, Furman MI, Wachsstock D, Safer D, Nachmias VT, Pollard TD (1992) The control of actin nucleotide exchange by thymosin beta 4 and profilin. A potential regulatory mechanism for actin polymerization in cells. Mol Biol Cell 3: 1015-1024

Goldschmidt-Clermont PJ, Kim JW, Machesky LM, Rhee SG, Pollard TD (1991) Regulation of phospholipase C-gamma 1 by profilin and tyrosine phosphorylation. Science 251: 1231-1233

Goldschmidt-Clermont PJ, Machesky LM, Baldassare JJ, Pollard TD (1990) The actin-binding protein profilin binds to PIP2 and inhibits its hydrolysis by phospholipase C. Science 247: 1575-1578

Gungabissoon RA, Jiang C-J, Drøbak BK, Maciver SK, Hussey PJ (1998) Interaction of maize actin-depolymerizing factor with actin and phosphoinositides and its inhibition of plant phospholipase C. Plant J 16: 689-696

Hartwig JH, Kwiatkowski DJ (1991) Actin-binding proteins. Curr Opin Cell Biol 3: 87-97

Heidstra R, Geurts R, Franssen H, Spaink HP, van Kammen A, Bisseling T (1994) Root hair deformation activity of nodulation factors and their fate on *Vicia sativa*. Plant Physiol 105: 787-797

Heidstra R, Nilsen G, Martinez-Abarca F, van Kammen A, Bisseling T (1997a) Nod-factor-induced expression of leghemoglobin to study the mechanism of NH4NO3 inhibition on root hair deformation. Mol Plant Microbe Interact 10: 215-220

Heidstra R, Yang WC, Yalcin Y, Peck S, Emons AM, van Kammen A, Bisseling T (1997b) Ethylene provides positional information on cortical cell division but is not involved in Nod-factor-induced root hair tip growth in *Rhizobium*-legume interaction. Development 124: 1781-1787

Horvath B, Heidstra R, Lados M, Moerman M, Spaink HP, Prome JC, van Kammen A, Bisseling T (1993) Lipo-oligosaccharides of *Rhizobium* induce infection-related early nodulin gene expression in pea root hairs. Plant J 4: 727-733

Huang S, McDowell JM, Weise MJ, Meagher RB (1996) The *Arabidopsis* profilin gene family. Evidence for an ancient split between constitutive and pollen-specific profilin genes. Plant Physiol 111: 115-126

Iwanami Y (1956) Protoplasmic movement in pollen grains and pollen tubes. Phytomorphology 6: 288-295

Jiang C-J, Weeds AG, Hussey PJ (1997a) The maize actin-depolymerizing factor, ZmADF3, redistributes to the growing tip of elongating root hairs and can be induced to translocate in the nucleus with actin. Plant J 12: 1035-1043

Jiang C-J, Weeds AG, Khan S, Hussey PJ (1997b) F-actin and G-actin binding are uncoupled by mutation of conserved tyrosine residues in maize actin depolymerizing factor (ZmADF). Proc Natl Acad Sci USA 94: 9973-9978

Journet EP, Pichon M, Dedieu A, de Billy F, Truchet G, Barker DG (1994) *Rhizobium meliloti* Nod-factors elicit cell-specific transcription of the ENOD12 gene in transgenic alfalfa. Plant J 6: 241-249

Kaminsky SGW, Heath IB (1995) Integrin and spectrin homologues and cytoplasm-wall adhesion in tip growth. J Cell Sci 108: 849-856

Kamiya N (1981) Physical and chemical basis of cytoplasmic streaming. Ann Rev Plant Physiol 32: 205-236

Kijne J (1992) The *Rhizobium* infection process. In: G Stacey, RH Burris, HJ Evans (Eds) Biological Nitrogen Fixation. Chapman and Hall, New York. pp 349-398

Kim SR, Kim Y, An G (1993) Molecular cloning and characterization of anther-preferential cDNA encoding a putative actin-depolymerizing factor. Plant Mol Biol 21: 39-45

Kost B, Spielhofer P, Chua NH (1998) A GFP-mouse talin fusion protein labels plant actin filaments in vivo and visualizes the actin cytoskeleton in growing pollen tubes. Plant J 16: 393-401

Kurkdjian AC (1995) Role of differentiation of root epidermal cells in Nod-factor (from *Rhizobium meliloti*)-induced root hair depolarization of *Medicago sativa*. Plant Physiol 107: 783-790

Lancelle SA and Hepler PK (1991) Association of actin with cortical microtubules revealed by immunogold localization in *Nicotiana* pollen tubes. Protoplasma 165: 167-172

Lichtscheidl IK, Url WG (1987) Investigation of the protoplasm of *Allium cepa* inner epidermal cells using ultraviolet microscopy. Eur J Cell Biol 43: 93-97

Lichtscheidl IK, Weiss DG (1988) Visualization of submicroscopic structures in the cytoplasm of *Allium cepa* inner epidermal cells by video-enhanced contrast light microscopy. Eur J Cell Biol 46: 376-382

Lippincott-Schwartz (1998) Cytoskeletal proteins and Golgi dynamics. Curr Opin Cell Biol 10:52-59

Lopez I, Anthony RG, Maciver SK, Jiang CJ, Khan S, Weeds AG, Hussey PJ (1996) Pollen specific expression of maize genes encoding actin depolymerizing factor-like proteins. Proc Natl Acad Sci USA 93: 7415-7420

Maciver SK, Pope BJ, Whytock S, Weeds AG (1998) The effect of two actin depolymerizing factors (ADF / cofilins) on actin filament turnover: pH sensitivity of F-actin binding by human ADF, but not of *Acanthamoeba actophorin.* Eur J Biochem 256: 388-397

McCurdy DW, Kim M (1998) Molecular cloning of a novel fimbrin-like cDNA from *Arabidopsis thaliana.* Plant Mol Biol 36: 23-31

Miki M, Barden JA, dos Remedios CG, Phillips L, Hambly BD (1987) Interaction of phalloidin with chemically modified actin. Eur J Biochem 165: 125-130

Miller DD, de Ruijter NCA, Bisseling T, Emons AMC (1999) The role of actin in root hair morphogenesis: studies with lipochito-oligosaccharide as a growth stimulator and cytochalasin as an actin perturbing drug. Plant J 17: 141-154

Miller DD, de Ruijter NCA, Emons AMC (1997) From signal to form: aspects of the cytoskeleton-plasma membrane-cell wall continuum in root hair tips. J Exp Bot 48: 1881-1896

Mittermann I, Swoboda I, Pierson E, Eller N, Kraft D, Valenta R, Heberle-Bors E (1995) Molecular cloning and characterization of profilin from tobacco (*Nicotiana tabacum*): increased profilin expression during pollen maturation. Plant Mol Biol 27: 137-146

Mylona P, Pawlowski K, Bisseling T (1995) Symbiotic nitrogen fixation. Plant Cell 7: 869-885

Neujahr R, Heizer C, Albrecht R, Ecke M, Schwartz JM, Weber I, Gerisch G (1997) Three-dimensional patterns and redistribution of myosin II and actin in mitotic *Dictyostelium* cells. J Cell Biol 139: 1793-1804

Nishida E, Iida K, Yonezawa N, Koyasu S, Yahara I, Sakai H (1987) Cofilin is a component of intranuclear and cytoplasmic actin rods induced in cultured cells. Proc Natl Acad Sci USA 84: 5262-5266

Pang KM, Lee E, Knecht DA (1998) Use of a fusion protein between GFP and an actin-binding domain to visualize transient filamentous-actin structures. Curr Biol 8:405-408

Pingret JL, Journet EP, Barker DG (1998) *Rhizobium* nod-factor signalling. Evidence for a G protein-mediated transduction mechanism. Plant Cell 10: 659-672

Puius YA, Mahoney NM, Almo SC (1998) The modular structure of actin-regulatory proteins. Curr Opin Cell Biol 10: 23-34

Ressad F, Didry D, Xia GX, Hong Y, Chua NH, Pantaloni D, Carlier MF (1998) Kinetic analysis of the interaction of actin-depolymerizing factor (ADF)/cofilin with G- and F-actins. Comparison of plant and human ADFs and effect of phosphorylation. J Biol Chem 273: 20894-20902

Ridge RW (1988) Freeze-substitution improves the ultrastructural preservation of legume root hairs. Botanical Magazine Tokyo 101: 427-441

Roberts K (1994) The plant extracellular matrix: in a new expansive mood. Current Opinion in Cell Biology 6: 688-694.

Rozycka MD, Khan S, Lopez I, Greenland AJ, Hussey PJ (1995) A *Zea mays* pollen cDNA encoding a putative actin-depolymerizing factor. Plant Physiol 107: 1011-1012

Ruhlandt G, Lange U, Grolig F (1994) Profilins purified from higher plants bind to actin from cardiac muscle and to actin from a green alga. Plant Cell Physiol 35: 849-854

Schnepf E (1986) Cellular polarity. Annu Rev Plant Physiol 37: 23-47

Seagull RW, Heath IB (1979) The effects of tannic acid on the in vivo preservation of microfilaments. Eur J Cell Biol 20: 184-188

Sherrier DJ, Vandenbosch KA (1994) Secretion of cell wall polysaccharides in *Vicia* root hairs. Plant J 5: 185-195

Sohn RH, Goldschmidt-Clermont PJ (1994) Profilin: at the crossroads of signal transduction and the actin cytoskeleton. Bioessays 16: 465-472

Sonobe S, Shibaoka H (1989) Cortical fine actin filaments in higher plant cells visualized by rhodamine phalloidin after pre-treatment with m-maleimidobenzoyl N-hydroxysuccinimide ester. Protoplasma 148: 80-86

Staiger CJ, Gibbon BC, Kovar DR, Zonia LE (1997) Profilin and actin-depolymerizing factor: modulators of actin organization in plants. TIPS 2: 275-281

Staiger CJ, Goodbody KC, Hussey PJ, Valenta R, Drobak BK, Lloyd CW (1993) The profilin multigene family of maize: differential expression of three isoforms. Plant J 4: 631-641

Staiger CJ, Yuan M, Valenta R, Shaw PJ, Warn RM, Lloyd CW (1994) Microinjected profilin affects cytoplasmic streaming in plant cells by rapidly depolymerizing actin microfilaments. Curr Biol 4: 215-219

Tominaga M, Morita K, Sonobe S, Yokota E, Shimmen T (1997) Microtubules regulate the organization of actin filaments at the cortical region in root hair cells of *Hydrocharis*. Protoplasma 199: 83-92

Tominaga M, Sonobe S, Shimmen T (1998) Mechanism of inhibition of cytoplasmic streaming by auxin in root hair cells of *Hydrocharis*. Plant Cell Physiol 39: 1342-1349

Valenta R, Duchene M, Pettenburger K, Sillaber C, Valent P, Bettelheim P, Breitenbach M, Rumpold H, Kraft D, Scheiner O (1991) Identification of profilin as a novel pollen allergen; IgE autoreactivity in sensitized individuals. Science 253: 557-560

Valster AH, Hepler PK (1997) Caffeine inhibition of cytokinesis: Effect on the phragmoplast cytoskeleton in living *Tradescantia* stamen hair cells. Protoplasma 196: 155-166

Valster AH, Pierson ES, Valenta R, Hepler PK, Emons AMC (1997) Probing the plant actin cytoskeleton during cytokinesis and interphase by profilin microinjection. Plant Cell 9: 1815-1824

Vidali L, Hepler PK (1997) Characterization and localization of profilin in pollen grains and tubes of *Lilium longiflorum*. Cell Motil Cytoskeleton 36: 323-338

Von Witsch M, Baluska F, Staiger CJ, Volkmann D (1998) Profilin is associated with the plasma membrane in microspores and pollen. Eur J Cell Biol 77: 303-312

Welch MD, Mallavarapu A, Rosenblatt J, Mitchison TJ (1997) Actin dynamics in vivo. Curr Opin Cell Biol 9: 54-61

Westphal M, Jungbluth A, Heidecker M, Muhlbauer B, Heizer C, Schwartz JM, Marriott G, Gerisch G (1997) Microfilament dynamics during cell movement and chemotaxis monitored using a GFP-actin fusion protein. Curr Biol 7: 176-183

Wieland T, Faulstich H (1978) Amatoxins, phallotoxins, phallolysin, and antamanide: the biologically active components of poisonous *Amanita* mushrooms. CRC Crit Rev Biochem 5: 185-260

Yonezawa N, Nishida E, Iida K, Yahara I, Sakai H (1990) Inhibition of the interactions of cofilin, destrin and deoxyribonuclease-I with actin by phosphoinositides. J Biol Chem 265: 8382-8386

Chapter 4
Cell Cleaving: A Method for Studying the Extracellular Matrix–Cytoskeleton–Plasma Membrane Continuum in Root Hairs

Anne Mie C. Emons[1] and Jan A. Traas[2]

[1]Laboratory of Experimental Plant Morphology and Cell Biology, Wageningen University, Arboretumlaan 4, NL-6703 BD Wageningen, The Netherlands
[2]Laboratoire de Biologie Cellulaire, Institut National de la Recherche Agronomique, Route de Saint Cyr, 78026 Versaille Cedex, France

1 Introduction

The cell surface is an important domain due to its involvement in the interaction between cell and environment, cell morphology and cell signalling. The fundamental importance of the cell surface underlies the need for its careful study. It functions in processes such as signal transduction, endocytosis, exocytosis, cell division, cell polarity and cell expansion. Cleaving techniques can be especially informative in studies using easily accessible root hairs. In this chapter, we describe cell cleaving techniques and their use in the study of the cell surface of root hairs.

The cell surface includes all major structural elements of the cortical membrane-bound cytoplasm, such as cytoskeletal elements, membrane components endoplasmic reticulum, coated pits, vesicles and the proteins associated with these structures (Boyles and Bainton 1979; Heuser 1980; Mesland et al. 1981; Aggeler et al. 1983; for review see Nermut 1982; Traas and Ramaekers 1986). On the other hand, cell cleaving is the only technique which reliably allows visualization of the nascent microfibrils of the cell wall (Sassen et al. 1985). In other preparations for cell walls, it is not clear whether or not the last deposited microfibrils are exposed. Generally, with cleaving techniques the sample is so thin that no replica has to be made and the sample can be examined directly in the transmission electron microscope after labelling, staining or shadowing with platinum. Since no embedding is used, epitopes are accessible for immunocytochemistry (Traas and Kengen 1986; review Traas 1989). Dry cleaving techniques have been used successfully for scanning force microscopy of

Key words. cell cleaving method, cell surface, cell wall, cellulose microfibrils, critical point drying, coated pits, cortical microtubules, dry-cleaving, electron microscopy, embedding-free technique, extracelluar matrix, freeze fixation, immunolabeling, shadow casting, wet-cleaving cytoskeletal elements (Pietrasanta et al. 1994) and for atomic force microscopy of cellulose microfibrils (van de Wel et al. 1996). Especially in the case of easily accessible root hairs, cleaving techniques can be useful. We will discuss two major variations of cleaving techniques: (1) dry cleaving with shadowing or without shadowing and making replicas and (2) wet cleaving, with or without immunolabelling.

2 Cleaving in Comparison to Other Techniques

Cleaving techniques are useful complements to existing methods. Thin sectioning can produce only small stretches of tangential sections through the cortical cytoplasm, and freeze fracturing primarily splits the two halves of the unit membrane and exposes neither the exoplasmic nor the cytoplasmic side of the plasma membrane. The quick freeze-deep etch technique is another technique that exposes the membrane bound structures (Hirokawa and Heuser 1981) and the extracellular matrix (McCann et al. 1990). For instance, these techniques have been instrumental in the study of coated pits and coated vesicles (Heuser and Kirschner 1980). Cleaving techniques for electron microscopy require a critical point drying or freeze-drying step, while with deep etching a replica is made right after etching when the material is still frozen. Therefore, deep etching (McCann et al. 1990) is preferred over dry cleaving if exact dimensions between extracellular matrix constituents are required. For exact dimensions of the structures themselves, at the molecular level, the Pt/C coating should be taken into account. An advantage of the dry cleaving technique over deep etching, when cleaving in order to study cell wall deposition, is that it can be determined whether the cleavage plane is adjacent to the plasma membrane. This is important for the study of nascent cellulose microfibrils. In addition, the method allows the exposure of large stretches of cortical cytoplasm, in contrast to other techniques. Since the preparations are large, the dimensions and orientations of cell structures, such as cortical microtubules and cellulose microfibrils, in relation to the long axis of the cell, can be determined. This technique allowed statistical analyses of cellulose microfibril reorientation during cell elongation in *Petunia* styles (Wolters-Arts and Sassen 1991).

3 Cleaving Procedures

In cleaving procedures, the moment of cleaving varies from just before fixation (wet cleaving before fixation: Nicol et al. 1987), just after fixation (Aggeler et al. 1983; Traas and Ramaekers 1986) or just after drying (dry cleaving: Mesland et al. 1981; Mesland and Spiele 1983; Traas 1984; Emons 1986; Emons and Derksen 1986; Emons and Traas 1986; Emons and Van Maaren 1987; Emons et al. 1990). Cleaving may be followed by shadowing (Sassen et al. 1985; Emons 1989; Traas and Derksen 1989), or may be combined with immunogold labelling after dry cleaving (Roos et al. 1985; Wiegant et al. 1986), or after wet cleaving (Traas and Kengen 1986; Traas 1989). We will discuss two essential variations of the cleaving technique: dry cleaving and wet cleaving. We have taken most examples

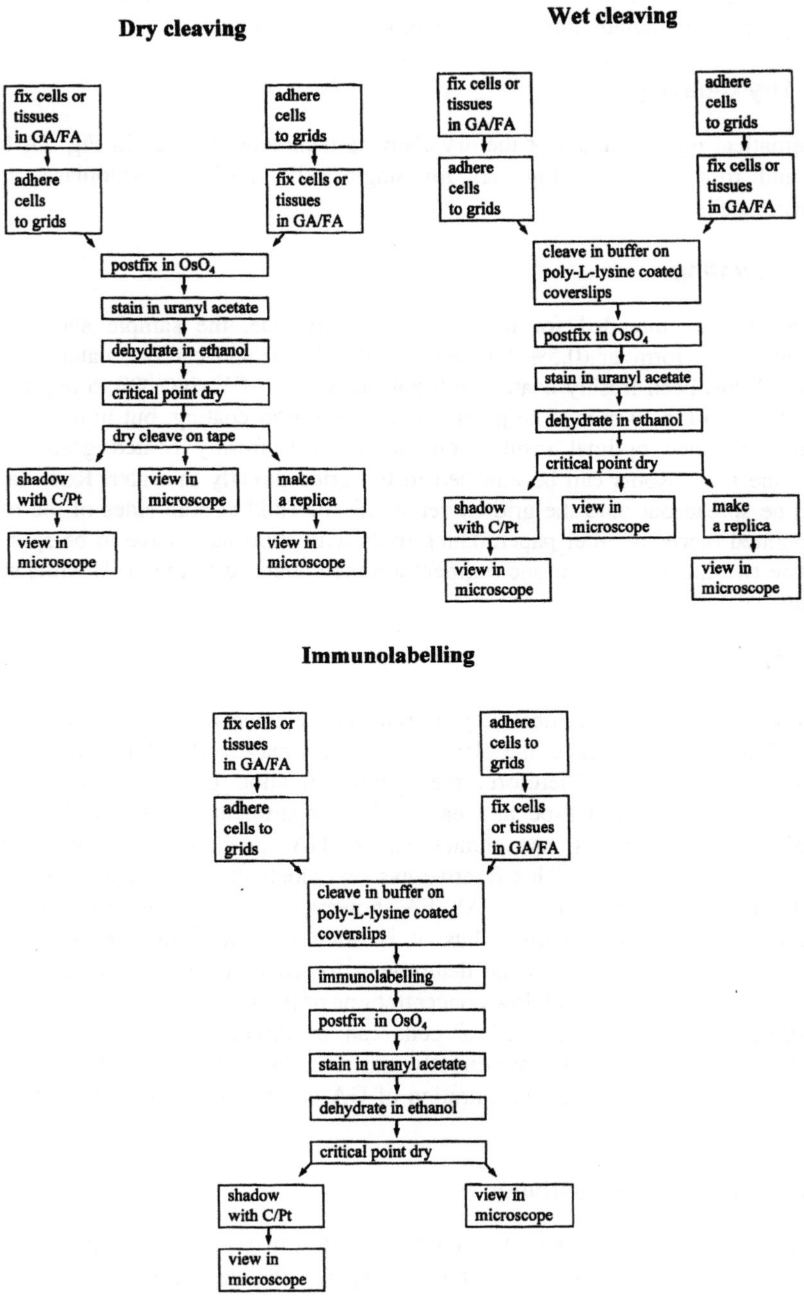

Fig. 1. Schematic flow sheet demonstrating the procedures for dry cleaving, wet cleaving, and cleaving in combination with immunolabelling

from root hairs. Figure 1 shows schematic flow sheets for dry cleaving, wet cleaving, and immunolabelling in combination with cleaving.

3.1 Dry Cleaving

A schematical representation of the dry cleaving technique is given in Fig. 2A for roots and root hairs and Fig. 2B for single cells, such as protoplasts and suspension cells.

3.1.1 Mounting

For preparations intended for the electron microscope, the sample should be mounted onto a formvar (0.5% formvar in chloroform) and carbon coated nickel grid which has been freshly coated with poly-L-lysine (MW>300.000, 5 mg/ml in distilled water) for 20 min. The grids can be dried after coating, but in our hands we have obtained optimal results consistently with freshly coated grids used within one hour. Roots can be attached to the grids directly in water. Root hairs should be spread out onto the grid by retrieving the fluid at both sides of the root with pointed pieces of filter paper. Thick roots with short hairs have to be split in two. The two halves of the tissue segment are then attached to the grids while still in water.

3.1.2 Fixation

Fixation is an extremely critical step. Certain cell elements are very sensitive to certain fixatives; for instance, the plant actin system may be depolymerized in the presence of aldehydes. Therefore, the optimal fixation protocols have to be determined for each cell type and each cell compartment to be visualized. In general, it is preferable to use glutaraldehyde (GA) in combination with short fixation times (up to 1 hr). This fixative has two aldehyde groups and is a better cross-linker than formaldehyde (FA). FA, however, is a smaller molecule and penetrates tissues more rapidly. Thus, it is often used in a mixture with GA. Optimal concentrations have to be determined for each cell type, but in general better results are obtained with low concentrations of both fixatives.

Different buffers can be used or cells can be directly fixed in the growth medium. Roots with root hairs are directly immersed in the fixative and fixed for 1 h or more if necessary. The concentration of GA can be increased if preservation is poor.

3.1.3 Improving Accessibility

After fixation, cells are washed thoroughly in buffer. In general, five washes of 10 min are sufficient; tissue blocks can be left longer in the wash buffer. Plant tissues have to be treated with cellulase (1-2% of cellulase for 20-25 min) in order to facilitate cleaving. Longer treatments can sometimes be necessary, but they tend to make the tissue fragile. Some cellulase samples can contain other enzymatic

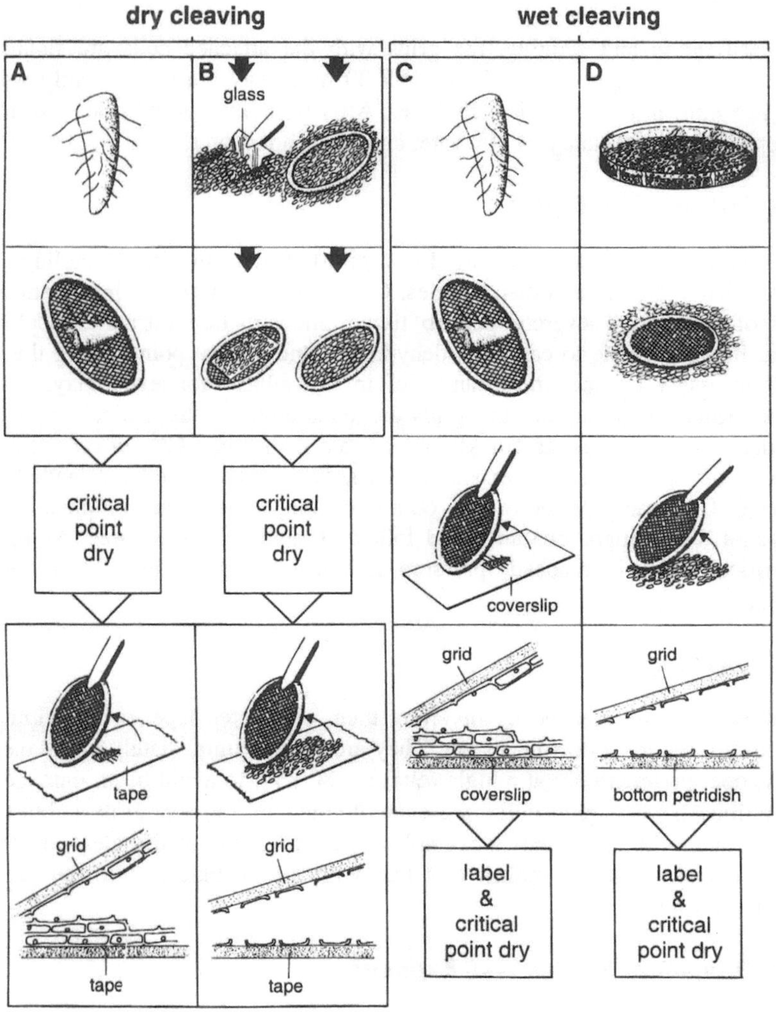

Fig. 2. Drawing showing the procedures used for roots with root hairs and single cells: *A* dry cleaving of roots with root hairs; *B* dry cleaving of single cells; *C* wet cleaving of roots with root hairs, *D* wet-cleaving of single cells

activities (proteases); therefore, prolonged treatments should be avoided. We use driselase or cellulysin. If necessary other cellulases can be used, in combination with pectinase (0.5%). After enzyme treatments the plants are washed thoroughly in buffer. These treatments are usually done in Petri dishes under the dissecting microscope.

3.1.4 Staining

For post-fixation and staining, the grids with the attached cells are floated on drops of 0.5% OsO_4 in distilled water for 30 min They are subsequently washed well (2x5 min) in a Petri dish with water. After post-fixation the grids are floated on drops of uranyl acetate (1%) in water and washed in water.

3.1.5 Critical Point Drying

For dehydration, grids are transferred to a grid holder to facilitate handling. They are then dehydrated in an ethanol series. Complete dehydration is important, since traces of water can severely perturb tissue fine structure during critical point drying. It is preferable to carry out dehydration and critical point drying the same day. If necessary, the specimens can be left in 70% ethanol for several days.

For critical point drying, the grids are transferred to the critical point dryer container, which is filled as quickly as possible with liquid CO_2 in order to avoid drying of the specimens. It is essential to remove all traces of ethanol with 10 - 15 washes of CO_2 over a period of 45 - 60 min or longer. After the complete removal of ethanol, the preparations are dried following standard procedures. As soon as the critical point is reached, pressure is released very slowly at a constant temperature.

3.1.6 Cleaving

The dried cells are cleaved by inverting them on Scotch tape, or by wiping the surface of the grids with filter paper. They are viewed immediately in the electron microscope, preferentially at a high voltage (>80 kV), to avoid beam damage. The preparations can be stored in the presence of silica gel for several days, but this is better avoided. Figure 3 shows examples of micrographs of dry-cleaved cells: a: cortical cytoskeleton of a radish root hair with microtubules, b: coated pits in a growing radish root hair.

3.1.7 Procedure for Cell Wall Exposure

In order to expose the last deposited cellulose microfibrils of the plant cell wall, the material is plasmolysed before fixation and no enzymatic cell wall degradation is done. Further steps are the same as described above. The dry or wet cleaved cells can be shadowed with carbon/platinum under an angle of 45°, which is useful to visualise the cellulose microfibrils. Figure 3c shows an example of a cell wall preparation obtained with dry cleaving and shadowing of the cell wall of an *Equisetum hyemale* root hair. In Fig. 3d, the procedure has been done such that microtubules in the cortical cytoskeleton and microfibrils in the cell wall of an *Equisetum hyemale* root hair can be seen in the same preparation.

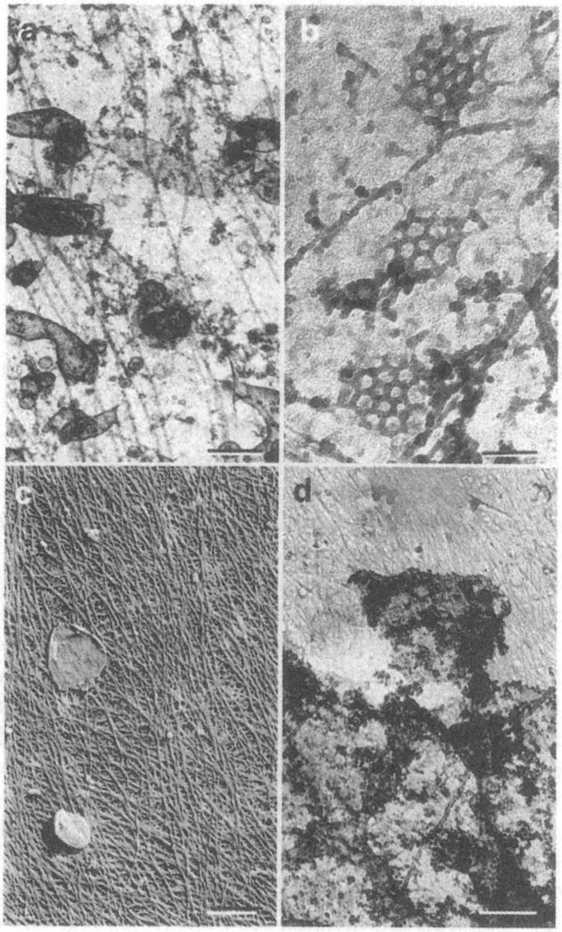

Fig. 3. Micrographs of cleaved cells: *a* cortical cytoskeleton of radish root hair, showing microtubules (*bar* 500 nm); *b* coated pits in growing radish root hair (*bar* 200 nm); *c* cell wall of *Equisetum hyemale* root hair, shadowed with platinum and carbon; note the two patches of plasma membrane, showing that the underlying microfibrils are the last deposited ones (*bar* 500 nm); *d Equisetum hyemale* root hair showing microtubules in the cortical cytoskeleton and microfibrils in the cell wall (*bar* 500 nm)

3.1.8 Replication

Replicas can be made, either of dry- or of wet-cleaved cells. For this purpose, the cells are dried on small coverslips and rotary shadowed using carbon and platinum. In principle, no post fixation or staining is required. The replicas are isolated by slowly immersing the slides in HF following standard replica procedures. When wet cleaving is used, much of the soluble material is removed from the cytoplasm during the washes and mainly the cytoskeletal and membrane associated elements remain.

3.2 Wet Cleaving and Immunolabelling After Cleaving

Cells can also be broken open while still in buffer. This can be useful if immunolabelling is required, if it is not possible to detach cells from their substrate, or, if it is necessary to keep part of the cells for other applications, i.e., light microscopy. Figures 2C (roots with root hairs) and 2D (single cells) show the procedure for wet cleaving. Most of the steps of the procedure are the same as for dry cleaving. However, after fixation, the tissue blocks are cleaved by gently pressing them onto the grid surface. The tissues and grids are blotted dry to enhance contact; the grids are removed and transferred to a Petri dish with water. Dehydration, critical point drying and observation are done as described above. If necessary, a control grid can be made before dehydration. In that case the grid is negatively stained with uranyl acetate and viewed in the microscope directly after drying. This will show if enough material has adhered to the grids. The preservation of the cytoplasm is not always satisfactory, although membranes, vesicles, and some cytoskeletal elements can be readily recognized.

The exposed part of the cytoplasm can be labelled with antibodies. For this purpose, the cells are wet-cleaved as described above. Non-specific binding is blocked by incubation of the cells in 5% bovine serum albumin (BSA) in buffer for 10 min This has given satisfactory blocking for most of the antibodies we used. If BSA is not sufficient, the free aldehyde groups have to be reduced with $NaBH_4$ (1mg/ml in buffer). Grids with cleaved cells are transferred to this solution for 10 min and washed in buffer. Although this technique has given good results, it can give distortion of fine cellular structure, which is to be avoided.

For the immunolabelling procedure, the primary antibody is diluted in buffer, after the optimal dilution has been determined with immunofluorescence. We currently use Tris Buffered Saline (TBS) containing 1 % BSA, but other buffers can be used also. The grids are inverted on drops of antibody on multiwell slides. The slides with the grids are incubated at room temperature or if necessary at 37°C in plastic boxes at 100 % relative humidity for 1-3 hours, under stationary or shaking conditions. Grids can be incubated overnight, in which case the antibody must be diluted 2x more and the cells have to be labelled at 4°C. After labelling the cells are washed thoroughly by immersing them in Petri dishes with TBS. Then, the cells are incubated (again on multiwell slides) with gold-labelled secondary antibody in TBS/BSA. We use antibodies with 5- or 10 nm gold particles at the lowest possible concentration (between 1/10 to 1/100). Incubations are at 20°C-37°C for 2-4 h or overnight at 4°C. After labelling, the cells are again washed thoroughly in TBS.

If the cells were fixed in FA, they must be postfixed in 0.5% GA in buffer (TBS) for 30 min After fixation in GA the grids are washed in buffer or in water and finally transferred to a Petri dish with water. Immunolabelled, FA/GA fixed cells are postfixed in OsO_4 and stained in uranyl acetate as described earlier. The labelled cells are dehydrated and critical point dried as described above and viewed directly in the microscope or left under vacuum in the presence of silica gel. Figure 4 is a micrograph of a wet-cleaved cortex cell of *Equisetum hyemale*, in which the cortical microtubules were immunogold labelled with anti-tubulin.

4 Conclusions and Evaluation of the Technique

The cleaving methods described here are valuable techniques for studying the cortical cytoplasm. The procedures expose large areas of the membrane and associated cytoskeletal elements and the innermost layer of the cell wall, while leaving them free of embedding materials. Areas of nascent cell wall of individual root hairs of more than 1 mm long have been exposed and analyzed (Emons 1989). At a magnification of 30,000 times, platinum shadowed cellulose microfibrils are easily visible (Fig. 3c,d). Cleaving techniques have been used very successfully for animal cells (Boyles and Bainton 1979; Mesland et al. 1981; Aggeler et al. 1983; for review see Nermut 1982) and for plant cells. For the latter they have proven to be a useful tool for studying the cortical cytoskeleton (Fig. 3a; Traas et al. 1985), coated pits and vesicles (Fig.3b; Emons and Traas 1986) and the cell wall (Fig. 3c; Sassen et al. 1985; Emons 1989; Wolters-Arts and Sassen 1991).

A great advantage of the technique is that large surface areas of many cells are usually left on the grid (see also Mesland and Spiele 1984; Traas 1984), making quantitative analyses possible. Dry cleaving, followed by shadow casting of the root hair cell walls of *Equisetum hyemale* allowed measurement of enough parameters to formulate a model for cellulose microfibril deposition (Emons 1994; Emons and Kieft 1994). These parameters are: (i) distance between cellulose microfibrils within lamellae, (ii) net orientation of microfibrils within a lamella in relation to cell axis, and (iii) net orientation of microfibrils within any lamella in relation to both adjacent lamellae (Emons 1994). The model has general applicability (Emons and Mulder 1997; 1998; 2000).

The combination of dry cleaving techniques makes simultaneous visualization of cell wall and cortical cytoskeleton possible (Fig. 3d). This type of visualization is important in the study of the putative role of the cortical microtubules in cellulose microfibril orientation. If combined with labelling of cell surface proteins the technique allows the study of plasma membrane-cortical cytoskeleton interactions (Roos et al. 1985). Figure 4 shows how the wet cleaving technique was successfully applied in combination with immunolabelling to show the presence of microtubules in cortex cells of *Equisetum hyemale*. This technique has not been sufficiently utilized and holds great potential for elucidating structural aspects of signal transduction events that take place at the plasma membrane.

Every type of cleaving technique has it own advantages and limitations. The direct examination of cleaved cells has several advantages over the use of replicas; because the preparation is rapid, all material on the grid can be studied, and the cells remain coherent. A disadvantage, however, is that these preparations are vulnerable to moisture and electron beam damage in the electron microscope (Bridgeman and Reese 1984). The technique was mainly designed to study

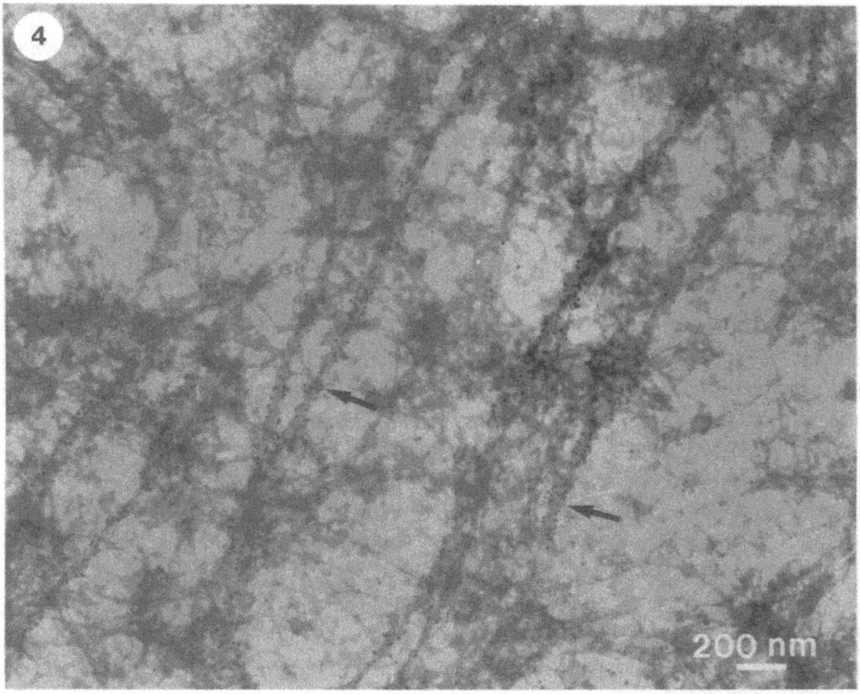

Fig. 4. Micrograph of a wet cleaved cortex cell of *Equisetum hyemale*, in which the cortical microtubules were immunogold labelled with antitubulin (*bar* 200 nm)

membrane-associated and cytoskeleton associated structures. It is less well suited for the analysis of elements that are only loosely attached, because these components could be washed away during the procedure. Another problem, when immunocytochemical techniques are used, can be the accessibility of the antigen to the antibody. For such specimens, extraction methods have to be included in the procedure, which will disturb the cell structure.

Signal transduction and the cytoskeleton are rapidly evolving topics in cell biology studies, and root hairs are important cells for this type of research. To understand the role of molecules implicated in the complicated processes involving signalling cascades and the cortical cytoskeleton, the elucidation of the structural aspects of these phenomena is one of the prerequisites. Cleaving techniques are the methods of choice herein. For the study of many other cell processes, the technique is a valuable tool. We have mentioned cell expansion, cell division, endocytosis, exocytosis and intracellular transport, but the elucidation of every process that involves structural aspects of the cell surface may profit from these techniques.

For electron microscopy, freeze fixation has been proven superior to chemical fixation. For a critical review of artefacts in fixed critical point dried cells see Heuser and Kirschner (1980) and Ris (1985). However, there is no reason why cleaving techniques should not be used in combination with rapid freezing, in a way comparable to the use of rapid freezing in combination with thin sectioning in the freeze substitution technique. The advantage of the dry cleaving technique is that cell surfaces can be studied. The cleaving techniques allow the study of the intact membrane-cytoskeleton complex, including associations of cytoskeleton-associated proteins with the lipid bilayer.

ACKNOWLEDGMENTS

We thank Dr Daphne Wittich-Rainey for critical reading of the manuscript and A. Haasdijk, S. Massalt and P. van Snippenburg for artwork and photography.

References

Aggeler J, Takemura R, Werb Z (1983) High-resolution three dimensional view of membrane associated clathrin and cytoskeleton in critical point dried macrophages. J Cell Biol 97:1452

Bridgeman PC and Reese TS (1984) the structure of cytoplasm in directly frozen cultures cells. 1. Filamentous networks and the cytoplasmic ground substance. J Cell Biol 99:1655-1668

Boyles J, Bainton D (1979) Changing patterns of plasma membrane-associated filaments during the initial phases of polymorphonuclear adherence. J Cell Biol 82:347

Emons AMC (1986) Cell wall texture in root hairs of the genus *Equisetum.* Can J Bot 64:2201-2206

Emons (1989) Helicoidal microfibril deposition in a tip-growing cell an microtubule alignment during tip morphogenesis: a dry cleaving and freeze substitution study. Can J Bot 67:2401-2408

Emons AMC (1994) Winding threads around plant cells: a geometrical model for microfibril deposition. Plant Cell and Env 17:3-14

Emons AMC, Derksen J (1986) Microfibrils, microtubules and microfilaments of the trichoblast of Equisetum hyemale. Acta Bot Neerl 35: 331-320

Emons AMC, Kieft H (1994) Winding threads around plant cells: applications of the geometrical model for microfibril deposition. Protoplasma 180:59-69

Emons AMC, Mulder BM (1997) The architecture of the plant cell wall. Comments on Theor Biol 4:115-131

Emons AMC, Mulder BM (1998) The making of the architecture of the plant cell wall: How cells exploit geometry. Proc Natl Acad Sci USA 95:7215-7219

Emons AMC, Mulder BM (2000) How the deposition of cellulose microfibrils builds cell wall architecture. Trends in Plant Sci 5:35-40

Emons AMC, Traas JA (1986) Coated pits and coated vesicles on the plasma membrane of plant cells. Eur J Cell Biol 41:57-64

Emons AMC, Van Maaren N (1987) Helicoidal cell wall texture in root hairs. Planta 170:145-151

Emons AMC, Wolters-Arts AMS, Traas J, Derksen J (1990) The effect of colchicine on microtubules and microfibrils in root hairs. Acta Bot Neerl 39:19-27

Heuser J (1980) Three dimensional visualization of coated vesicle formation in fibroblasts. J Cell Biol 84:560-583

Heuser J, Kirschner M (1980) Filament organization revealed in platinum replicas of freeze dried cytoskeletons. J Cell Biol 86:212-234

Hirokawa N, Heuser JE (1981) Quick-freeze, deep-etch visualization of the cytoskeleton beneath surface differentiations of intestinal epithelial cells. J Cell Biol 91:399-409

McCann MC, Wells B, Roberts K (1990) Direct visualization of cross-links in the primary plant cell wall. J Cell Sci 96:323-334

Mesland D, Spiele H (1983) Plasma membrane associated filament systems in cultured cells visualized by dry cleaving. J Cell Sci 64:351-364

Mesland D, Spiele H (1984) Brief extraction with detergent induces the appearance of many plasma membrane associated microtubules in hepatotic cells. J Cell Sci 68:113

Mesland D, Spiele H, Roos E (1981) Membrane associated cytoskeleton and coated vesicles in cultured hepatocytes visualized by dry cleaving. Exp Cell Res 132:169

Nermut MV (1982) The cell monolayer technique in membrane research. Eur J Cell Biol 28:160

Nicol A, Nermut M, Doeinck A, Robenek H, Wiegant C, Jokusch BM (1987) Labeling of structural elements at the ventral plasma membrane of fibroblasts with the immunogold technique. J Histochem Cytochem 35:499-506

Pietrasante LI, Schaper A, Jovin TM (1994) Imaging subcellular structures of rat mammary carcinoma cells by scanning force microscopy. J Cell Sci 107:2427-2437

Ris H (1985) The cytoplasmic filament system in critical point dried whole mounts and plastic embedded sections. J Cell Biol 100:1474-1487

Roos E, Spiele H, Feltkamtp CA, Huisman H, Wiegant FA, Traas JA, Mesland DAM (1985) Localization of cell surface glycoproteins in membrane domains associated with the underlying filament network. J Cell Biol 101:1817

Sassen MMA, Traas JA, Wolters-Arts AMC (1985) Deposition of cellulose microfibrils in cell walls of root hairs. Eur J Cell Biol 37:21-26

Traas JA (1984) Visualization of the membrane bound cytoskeleton and coated pits of plant cells by means of dry cleaving. Protoplasma 119:212

Traas JA (1989) Colloidal gold labeling of microtubules in cleaved whole mounts of cells. In: Hayat (Ed) Colloidal gold. Vol.2. Academic Press, pp 227-241

Traas JA, Braat P, Emons AMC. Meekes H, Derksen J (1985) Microtubules in root hairs. J Cell Sci 76:303-320

Traas J, Derksen J (1989) Microtubules and cellulose microfibrils in plant cells; simultaneous demonstration on dry cleave preparations. Eur J Cell Biol 48:159-164

Traas JA, Kengen H (1986) Gold labeling of microtubules in cleaved whole mounts of cortical cells. J Histochem Cytochem 34:1501

Traas J, Ramaekers F (1986) The membrane associated cytoskeleton in cultured lens cells. Electron microscopical visualization in cleaved whole mount preparations and platinum replicas. Exp Eye Res 43:519-528

van de Wel NN, Putman CAJ, van Noort SJT, de Grooth BG, Emons AMC (1996) Atomic force microscopy of pollen grains, cellulose microfibrils and protoplasts. Protoplasma 194:29-39

Wiegant F, Blok F, Defize L, Linnemans W, Verkley A, Boonstra J (1986) Epidermal growth factor associated with cytoskeletal elements of epidermoid carcinoma (A431) cells. J Cell Biol 103:87-94

Wolters-Arts AMC, Sassen MMA (1991) Deposition and reorientation of cellulose microfibrils in elongating cells of *Petunia* stylar tissue. Planta 185:179-189

Chapter 5
Advanced Fluorescence Microspectroscopic Methods for the Study of Single Living Root Hairs

Joachim Goedhart and Theodorus W.J. Gadella Jr.

Microspectroscopy Center Wageningen, Department Biomolecular Sciences, Agricultural University Wageningen, Dreijenlaan 3, 6703 HA Wageningen, The Netherlands

1 Introduction

Root hairs are very suitable for fluorescence microscopy, as they consist of a single cell, have low autofluorescence (because of the lack of chloroplasts) and are not covered by other cell layers. For many plants, transferring seedlings to liquid medium contained between a microscopy slide and a coverslip and growth for one or two days at room temperature is sufficient for microscopic observation of root hairs. Cytoplasmic streaming provides an easy check for viability.

Fluorescence microscopy has the advantage that it is sensitive, non-invasive, can be used for the study of living cells and it provides the possibility to monitor specifically fluorescently tagged molecules with very high spatial and temporal resolution. Fluorescence microscopy not only enables the study of localization and dynamics of fluorescent (bio)molecules but also enables probing of the local microenvironment of the fluorescent molecule. This can be achieved by carefully monitoring the spectral properties of the fluorophores in the microscopic object. The integration of spectroscopic modalities into fluorescence microscopes is referred to as fluorescence microspectroscopy. This chapter will highlight a number of such techniques and their use for studying biomolecular behavior in single living root hairs.

For each microspectroscopic technique, the principle, the implementation onto a fluorescence microscopy system, and an application to the study of root hairs will be described. We will show that in addition to localization in root hairs, we also can probe the local environment of fluorescent (bio)molecules, reporting on absolute concentration, diffusional mobility, local pH, hydrophobicity, and

Key words. BODIPY, CCD-camera, cSNAFL-2, diffusion, fluorescence correlation microscopy (FCM), fluorescence lifetime imaging microscopy (FLIM), fluorescence ratio imaging microscopy (FRIM), fluorescence resonance energy transfer (FRET), green fluorescent protein (GFP), lipochitooligosaccharides, Nod-factor, root hair, spectral imaging microscopy (SPIM), *Vicia sativa* molecular proximity of (bio)molecules, all of which influence the spectroscopic properties of the probe fluorescence.

Some of the microspectroscopic techniques will be illustrated by showing the results obtained after application of fluorescent analogs of nodulation (Nod) factors. Nod-factors are lipochito-oligosaccharides that are secreted by *Rhizobium* bacteria and are essential for accomplishing symbiosis between the bacteria and the root system of legumes. Nod-factors consist of a tetra- or pentamer of *N*-acetylglucosamine residues and an acyl chain attached to the non-reducing sugar residue (Lerouge et al. 1990; Spaink et al. 1991). Purified Nod-factors are active at subnanomolar concentrations, induce morphological changes on root hairs of legumes (Heidstra et al. 1994) (see also Chapter 15 this volume), and activate several signaling events in root hairs including membrane depolarization (Ehrhardt et al. 1992; Felle et al. 1995), calcium spiking (Ehrhardt et al. 1996), alkalinization (Felle et al. 1996) and activation of genes possibly through the activation of G protein linked phospholipase C (Pingret et al. 1998). Fluorescent Nod-factors (Gadella et al. 1997b) allow us to study the binding sites for these molecules on living root hairs. However, to obtain results at physiologically relevant concentrations of these factors, advanced microspectroscopic techniques are of vital importance, e.g., for discriminating root hair autofluorescence from Nod-factor fluorescence. In this chapter we will discuss the Nod-factor studies only with respect to the microspectroscopic technology used for monitoring their in situ behavior on living root hairs. Although these studies have significant biological relevance in relation to the *Rhizobium*-legume interaction, these aspects are described elsewhere (Goedhart et al. 1999; Goedhart et al. 2000). In this way we intend to stress the more general technological applicability of the microspectroscopic methodology for the study of root hairs. Of course, the microspectroscopic techniques have a wider applicability then the study of root hairs. So if appropriate, references will be made to study of other (plant) systems.

2 Fluorescence Ratio Imaging Microscopy (FRIM)

2.1 Principle

Fluorescence ratio imaging microscopy (FRIM) employs the property of fluorescent probes that change their absorbance or fluorescence spectral properties in the presence of ions (Bright et al. 1989), voltage changes (Gross and Loew 1989; Bullen and Saggau 1999) or upon complexation to other molecules. By far the most popular is the application of ratio imaging to the measurement of intracellular ion concentrations. Ratio probes can be divided into two groups, the so-called excitation ratio dyes and the emission ratio dyes. An example of a calcium sensitive excitation ratio dye is fura-2 which has an absorption maximum at 362 nm in the absence of calcium, whereas the calcium bound state has a maximum at 335 nm (Grynkiewicz et al. 1985). Both forms of the molecule emit at green wavelengths (around 510 nm). Consequently, the ratio of the intensity obtained with 335 nm excitation to the intensity obtained with 362 nm is directly related to the calcium concentration.

An example of a calcium sensitive emission ratio dye is indo-1, which absorbs maximally around 340 nm. The fluorescence emission is maximal at 482 nm in the absence of calcium, whereas the emission maximum is at 398 nm for the calcium bound form (Grynkiewicz et al. 1985). The use of ratiometric dyes can easily be combined with imaging microscopy, allowing studies of single cells. There are also non-ratiometric fluorescent indicators of which only the magnitude of fluorescence (or quantum yield) can be influenced by external factors such as ions or hydrophobicity. These probes can be used for fluorescence lifetime imaging applications (see section 4).

The great advantage of ratiometric over non-ratiometric dyes is that by the ratioing procedure (pixel-by-pixel division of the fluorescence intensities at the respective wavelengths), the local dye concentration, optical pathlength or spatial excitation light distribution divide out completely. In other words, the ratio-images only reflect the relative spectral state of the fluorophore and as a result can be directly related to the indicator function of the dye.

By performing a calibration procedure, it is possible to quantitatively relate the experimental ratio values to an ion concentration (or other parameter). The ratio, R, is measured at different (known) ion-concentrations, yielding a sigmoid curve of ratio versus ion concentration. The calibration curve shows the spectral response of the dye to the ion concentration, [ion], and can be described by:

$$[\text{ion}] = K' \cdot [(R - R_{min})/(R_{max} - R)] \qquad (1)$$

or logarithmic:

$$p[\text{ion}] = pK' - \log [(R - R_{min})/(R_{max} - R)] \qquad (2)$$

The three parameters are the maximal ratio R_{max}, the minimal ratio R_{min} and the K' (the ion concentration at which the ratio is exactly halfway R_{min} and R_{max}). This last value is usually close to the actual K of the indicator (being the ion concentration at which half of the indicator is free and the other half is ion-bound) but depends on instrumental factors. The sensitivity of the probe is highest around the (p)K'.

For illustrating the basic principles of ratio measurements and the calibration procedure, we show the in vitro spectral properties of the pH sensitive dye, cSNAFL-2 (Whitaker et al. 1991; Haugland 1996) in Fig. 1. As can be inferred from this Figure, the excitation spectra depend on the pH: at low pH the probe is most effectively excited at 515 nm, whereas the excitation maximum is shifted to 550 nm at alkaline pH. A calibration curve is constructed by calculating the ratio (emission observed at 620 nm after excitation at 550 nm divided by the emission observed at 620 nm after excitation at 515 nm) for every pH value. After fitting the calibration curve (inset of Fig. 1) according to equation 2, a pK' value of 8.0 was obtained. If, however, the detection wavelength is changed from 620 to 600 or to 640 nm, apparent pKs of 8.5 and 7.8 were obtained, respectively. On the one hand, this illustrates the dependence of the pK' on the instrumental settings, and the necessity to acquire new calibration curves when changing emission and

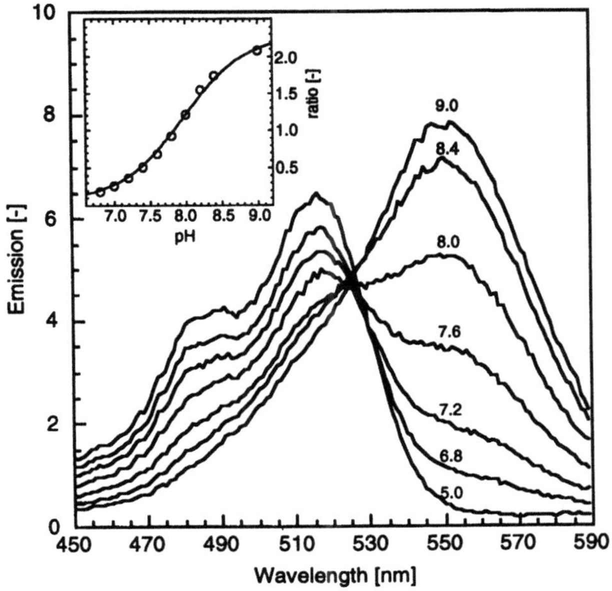

Fig. 1. Excitation spectra of 0.1 μM of cSNAFL-2 in 0.1 M phosphate buffer at different pH. The excitation spectra were acquired with a SPF 500 spectrofluorimeter (SLM Instruments, Urbana IL) by detecting fluorescence emission at 610 nm (slit 5 nm) and scanning the excitation from 450 to 590 nm (*slit* 1 nm). The *inset* shows the calibration curve obtained by calculating the ratios of the fluorescence intensities acquired upon excitation at 550 nm and 515 nm. The *smooth* curve represents the fit according to equation 2, yielding the parameters R_{max}=2.29, R_{min}=0.064, pK'=7.97, with regression R=0.999

excitation wavelengths (filters), or other optical components in the FRIM system (such as dichroic mirrors). On the other hand, it illustrates that by only changing some optical components in the (microscopy) system, the apparant pK can be adjusted to one's convenience. This can be very useful for optimizing the responsiveness of an indicator to a certain ion concentration.

2.2 Implementation of FRIM onto a Fluorescence Microscope

Excitation ratio-imaging can be easily implemented on wide-field fluorescence digital imaging microscopy (FDIM) systems by incorporating a filter wheel (with several different excitation bandpass filters) between the excitation source (Hg or Xe Arc lamps, or multiline laser) and the microscope. After implementation of an emission filter changer, emission ratio-imaging on an FDIM system can also be done. However, this can be problematic due to registration aberrations (pixel shift) on the imaging detector (Bright 1993). Emission ratio-imaging is more easily done with confocal laser scanning microscopy (CLSM), using dual channel detection of the fluorescence emission (Pawley 1995). Alternatively, spectral imaging (discussed in section 3) can be used to monitor emission ratio dyes. The different implementation modes of FRIM and their advantages and disadvantages are reviewed by Bright (1993) and Fricker et al. (1999).

We have implemented excitation-FRIM by the incorporation of a Lep filter wheel (Ludl electronic products, München, Germany with 6 different bandpass filters) between a Leica DMR epifluorescence microscope (Leitz, Wetzlar, Germany) and its 100 W Hg excitation source. For FRIM of cSNAFL-2, we used an Omega 520DF40 and an Omega 577DF10 bandpass filter (Omega, Brattleboro, VT, USA) in the filter wheel, and an Omega 595DRLP dichroic mirror with a longpass RG610 filter (Schott, Mainz, Germany) filter (both mounted into a filter cube of the microscope) for selecting the fluorescence emission. The images were captured using a Quantix CCD camera (Photometrics, with an Kodak KAF1400 grade 1 CCD sensor, Tucson, AZ, USA) which was mounted on the phototube exit (f-mount) port of the microscope. The CCD and the Lep filter wheel were controlled simultaneously using the IPLab spectrum software package (Signal Analytics, Vienna VA, USA) on a Macintosh PowerPC 8500 (Apple computer Cupertino, CA, USA).

A complete ratio image acquisition (including exposures at the two excitation wavelengths and changing the excitation filter position) on the instrument described above can be achieved within 0.6 s, but usually takes about 2-3 s depending on the integration/exposure time needed by the CCD to obtain a satisfactory signal to noise values. The resulting two images, only differing in excitation wavelength, are used for the data processing/ratioing procedure. Data processing includes background subtraction (camera bias and background), calculation of the ratio image through pixel-by-pixel division of the fluorescence intensities of the two images, and the calculation of the in situ ion-concentration using equation 2. The procedures are implemented within the IPlab package using home-written macros, and within the SCILimage package (TPD, Delft, The Netherlands) on a Silicon Graphics Indy workstation (Silicon Graphics, Mountain View, CA, USA).

2.3 Applications

FRIM has found a wide application to measuring ion activities in single cells. The major application consists of measuring Ca^{2+} and pH (Read et al. 1992; Gilroy 1997). Here we will focus on the measurement of pH in single living root hairs. Furthermore we will discuss a novel generation of indicators, which are based on fusion proteins containing the green fluorescent protein from *Aequoria victoria.*

2.3.1 Imaging Root Hair pH with cSNAFL-2

In Fig. 2, we show the use of the ratiometric pH indicator cSNAFL-2 for imaging cytosolic pH in *Vicia sativa* (vetch) root hairs. The indicator (10 μM) is applied to the roots in an esterified, non-fluorescent form (cSNAFL-2 diacetate from Molecular Probes, Eugene, OR, USA) for 30 min. Intracellular esterases cleave off the acetate groups and produce the fluorescent, negatively charged and membrane impermeable active form of the indicator. In this way, the indicator becomes highly concentrated in the root hair cytosol and the nucleus. Not all fluorescent indicators can be loaded as conveniently and efficiently as cSNAFL-2. Some end up in the vacuole, e.g., BCECF (Brauer et al. 1995; Brauer et al. 1996), while others are incompletely or not hydrolyzed (e.g., Calcium Green AM, Fluo-3 AM, J.Goedhart unpublished observations 1999). Acid loading, microinjection or electroporation with or without dye-coupling to dextrans can be used under these circumstances. For a recent review on fluorescent dye loading strategies and protocols for living plant cells, we refer to Fricker et al. (1999).

As shown in Fig. 2, the root hairs loaded with cSNAFL-2 show an even distribution of the dye in the top left fluorescence image, which is acquired with an excitation bandpass filter of 520DF40 nm (Fig. 2a). More fluorescence is visible in the cytosol in the top right image which is excited with the 557DF10 nm bandpass filter (Fig. 2b), which preferentially excites the deprotonated form of cSNAFL-2. Therefore in this image the more alkaline compartments will show higher fluorescence intensities as compared to the other image. The excitation ratio at every pixel of the image is determined (Fig. 2c), and can be converted into a pH image (Fig. 2e) by using the parameters obtained from the calibration curve shown in Fig. 2d and equation 2. It is of note that in situ calibrations are generally more reliable than the in vitro calibration curves presented here (see Fricker et al. 1999 for extensive description of in situ calibration protocols).

The pH distribution in a root hair shows a cytosolic pH of around 7.3, and is comparable to values measured in root hairs by others with either pH sensitive microelectrodes (Felle et al. 1996) or FRIM (Bibikova et al. 1998). The cytosolic pH value is close to the pK' of cSNAFL-2 with these settings, which is the preferable situation. The pH in the vacuole region was much lower (around 6.8). It is of note that this value represents a mixed measurement of fluorescence arising from the acidic vacuole with the more alkaline cytosol in which the SNAFL-2 is more concentrated. With BCECF, which specifically sequesters into the vacuole of maize root hairs, Brauer et al. (1995) found a vacuolar pH of 5.8.

2.3.2 GFP-Based Ratio Indicators

Recently, a new type of ion-sensitive probes has been developed, based on green fluorescent proteins (GFP) (Tsien 1998). A fluorescent calcium sensitive chimeric protein, cameleon, based on calmodulin and GFP has been constructed by Miyawaki et al. (Miyawaki et al. 1997; Miyawaki et al. 1999). Two mutants of GFP, a cyan (CFP) and yellow fluorescing mutant (YFP) are attached to either side of a calmodulin and a calmodulin binding protein. The binding of calcium changes the relative orientation of the two fluorescent proteins, thereby causing an increased degree of intramolecular fluorescence resonance energy transfer (FRET, see also 4.3.3). Because of the increased FRET efficiency, the YFP/CFP

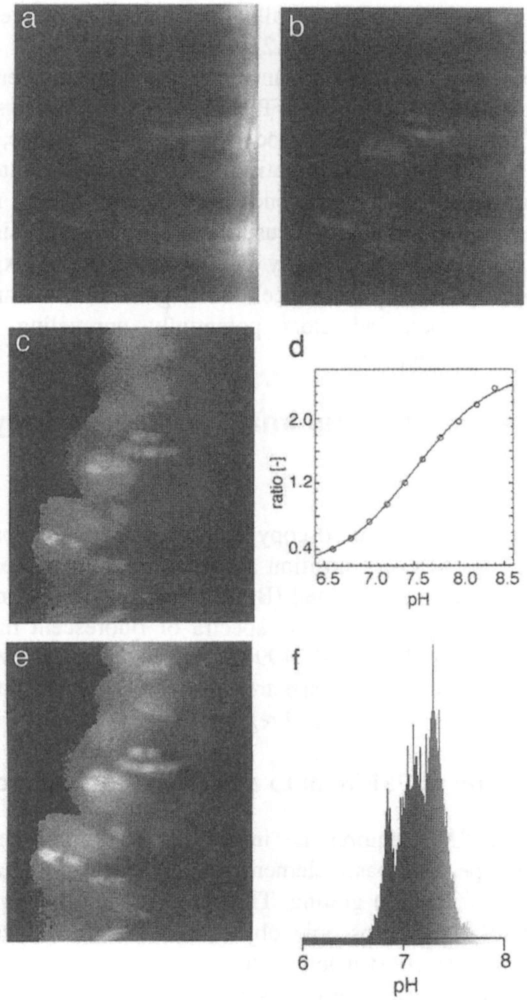

Fig. 2. Root hair pH measurement by fluorescence ratio imaging microscopy (FRIM). The root system of *Vicia sativa* (vetch) grown in Fåhreus slides (Heidstra et al. 1994) was incubated with plant growth medium (Goedhart et al. 1999) containing 10 μM cSNAFL-2 diacetate (esterified cSNAFL-2, Molecular Probes, Eugene, OR) for half an hour. The individual images taken at the two excitation wavelengths: $\lambda1$=520DF40 nm (*a*) and $\lambda2$=577DF10 nm (*b*) are shown. After subtraction of background the ratio (*c*) was determined for every pixel by dividing the images ($\lambda2/\lambda1$). To translate the ratio image into a pH image (*e*), a calibration was performed by ratio-imaging of microcuvettes containing 5 μM cSNAFL-2 in 0.1 M phosphate buffer set at various pH values, with identical microscope settings. The calibration curve (*d*) shows the response of the dye to pH. The smooth curve represents the fit (R=0.9995) according to equation 2 with pK'=7.5, Rmax=2.6 and Rmin=0.14. The pH in image (*e*) is represented by a *gray-scale* ranging from black (pH=6.0) to white (pH=8.0) as can also be inferred from the histogram (*f*) that is derived from the pH image

fluorescence emission ratio is increased, allowing the use of the yellow cameleon as an emission ratio indicator (see also 3.3.2).

Additionally, pH-sensitive GFP mutants can provide an alternative way to measure pH. A pH-sensitive mutant of GFP is available which has similar spectral properties and pK as BCECF (Miesenböck et al. 1998). Hence, this so-called pHluorin, can also be used for excitation ratio-imaging of intracellular pH.

An important advantage of the gene-encoded indicators is that no dye loading or microinjection is required as the cells use their own biosynthesis machinery to produce the indicator. Furthermore, they can be specifically expressed in the cytosol or targeted to a variety of subcellular organelles. We anticipate many future applications of such indicators potentially unraveling detailed ionic activities in all plant cell organelles.

3 Fluorescence Spectral Imaging Microscopy (FSPIM)

3.1 Principle

Fluorescence spectral imaging microscopy (FSPIM) combines spatial resolution with spectral resolution. At every position across a line in a microscopic object a complete emission spectrum is obtained (Balaban et al. 1986). Consequently, this technique can be used to yield emission spectra of fluorescent molecules inside living cells (Martínez-Zaguilán et al. 1994). Hereby, the FSPIM measurement provides much more detail about the spectroscopic origin of the fluorescence than FRIM, which only employs two averaged regions of the spectra of fluorophores.

3.2 Implementation of FSPIM onto a Fluorescence Microscope

Spectral imaging (FSPIM) requires an imaging spectrograph mounted onto a fluorescence microscope. The basic elements of the spectrograph are an entrance slit and an astigmatic diffraction grating. The entrance slit, effectively reduces the spatial information of the microscopic object into one dimension (see Fig. 3). Imaging spectrographs are constructed in such a way that the direction at which the photons are diffracted by the grating is perpendicular to the orientation of the entrance slit. As a result, an image is produced at the exit port of the spectrograph with combined spatial and spectral resolution which is captured by a CCD camera. The photosensor of the camera is aligned to the spectrograph in such a way that columns of pixels are parallel to the entrance slit and rows of pixels are parallel to the diffraction direction of the grating. Consequently, the image captured by the CCD camera contains spectral information in the horizontal direction and spatial information in the vertical direction.

Our FSPIM instrument is built around a Leica DMR epifluorescence microscope and uses a laser (see section 4.2) or a 100 W Hg lamp/ filter wheel system (see section 2.2) as excitation source. The detector unit consists of i) a Chromex 250is (Chromex Inc., Albuquerque, NM, USA) f/4 image spectrograph mounted at the phototube exit (c-mount) of the microscope with ii) a slow scan series 200 (CH250) CCD camera (Photometrics, Tucson, AZ, USA) mounted at

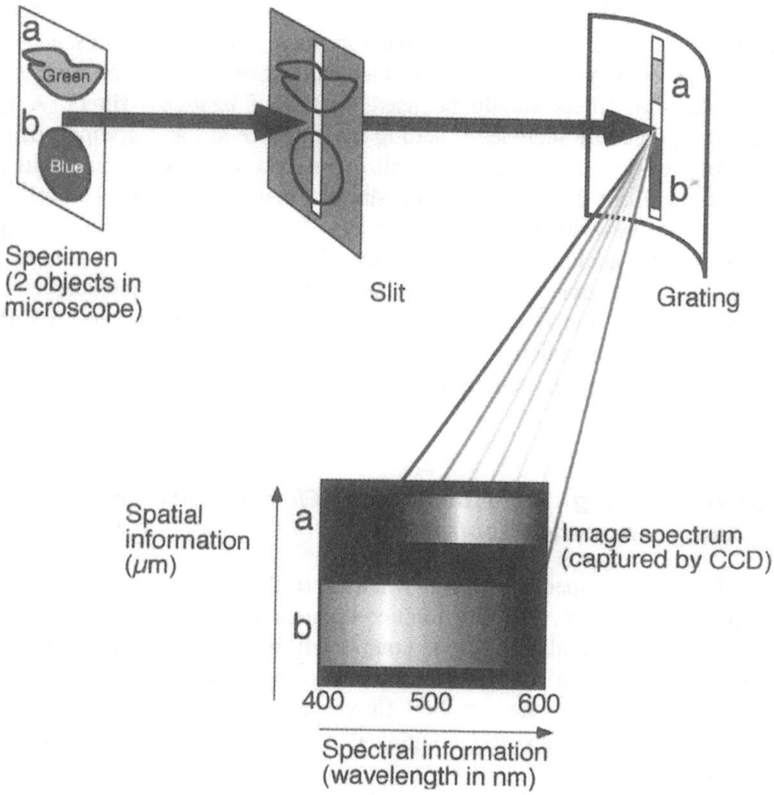

Fig. 3. The principle of fluorescence spectral imaging microscopy (FSPIM). For explanation *see section 3.2*. Note that both the position (*vertical axis*) and the emission maximum (*horizontal axis*) of green (*a*) and blue (*b*) fluorescing objects can be discerned from the image spectrum

the exit port using a home-built f-mount adapter. The spectrograph has a remote-controlled slit width (adjustable from 20-2000 μm), 3 user selectable gratings with 150, 300 or 600 grooves/mm, all with a central wavelength of 500 nm. The spectrograph settings can be conveniently adjusted using the Chromex manual control unit. The CCD-camera is interfaced onto the Nubus slot of a Macintosh PowerPC 7100 computer and controlled using the IPlab Spectrum image processing software (Signal Analytics, Vienna VA, USA). Alternatively, the CCD camera and the spectrograph can be interfaced to a PC using an ISA slot and RS232 serial link, respectively. The CCD and spectrograph control are then integrated using the MAPS 2.0 software package (Photometrics, Tucson, AZ, USA) running under Microsoft Windows 95. The spectral dimension is conveniently calibrated using monochromatic laser lines, or mercury arc lamp lines.

The actual FSPIM measurement is not more complicated than any conventional fluorescence imaging measurement. Normal fluorescence filter cubes can be employed, however for the acquisition of full emission spectra, longpass emission filters should be used instead of bandpass filters. A typical experimental procedure includes i) setting the entrance slit and grating position to the required position, ii) aligning the cell of interest with the entrance slit of the spectrograph, iii) opening the excitation shutter (manually or TTL-controlled), iv) taking an exposure with the CCD (typically 1s-10s exposures), v) saving and processing of the acquired image spectrum. Usually only steps ii-v need to be performed if the instrument settings are identical for different microscopic objects. In a time-series an image stack can be acquired by repeating steps iii) and iv) several times with an optional user controlled delay between the exposures. This procedure provides spatial, temporal and spectral information in one data set.

3.3 Applications

3.3.1 Decomposing Multicomponent Fluorescence Emission and Autofluorescence Subtraction

Spectral imaging is especially useful when more than one type of fluorophore is present in the specimen, yielding multicomponent fluorescence emission spectra (Martínez-Zaguilán et al. 1994). A component which is always present when studying living cells is autofluorescence. Especially at low fluorescent probe concentrations, where the probe fluorescence intensity approaches the autofluorescence levels, spectral imaging is a very useful tool to decompose the dual component spectra.

Growing plant root hairs display relatively low levels of autofluorescence. However, when fluorophores are studied at nanomolar or even lower concentrations (e.g., Nod-factors, see Chapter 15 this volume) the fluorescence intensity is close to the intensity of autofluorescence. A unique feature of root hairs is their elongated shape allowing optimal alignment of the total cell with the entrance slit of the imaging spectrograph. Therefore, root hairs are the most ideal cell type for spectrographic studies. Gadella et al. (1997b) described the use of FSPIM to be able to detect BODIPY FL labeled fluorescent Nod-factors on *Vicia sativa* roots. A procedure was developed to decompose the dual component spectra into autofluorescence and a BODIPY component spectrum. In this way, it was possible to locate the presence of Nod-factor on *Vicia sativa* root hairs when applied at a concentration of 1 μM and using an excitation wavelength of 475 nm.

The autofluorescence of *Vicia sativa* root hairs is significantly lower if the excitation is shifted from blue (475 nm, necessary for excitation of BODIPY FL) to green or yellow (510-580 nm) which is optimal for exciting recently developed red-shifted fluorescent Nod-factors (Goedhart et al. 1999). Upon application of plant growth medium containing 1 nM of the red-shifted BODIPY 581/591 tagged fluorescent Nod-factor to *Vicia sativa* plant roots grown in a Fåhreus slide (Heidstra et al. 1994), the fluorescence signal of the probe is comparable to autofluorescence. In Fig. 4a, an FSPIM analysis of both a labeled and unlabeled

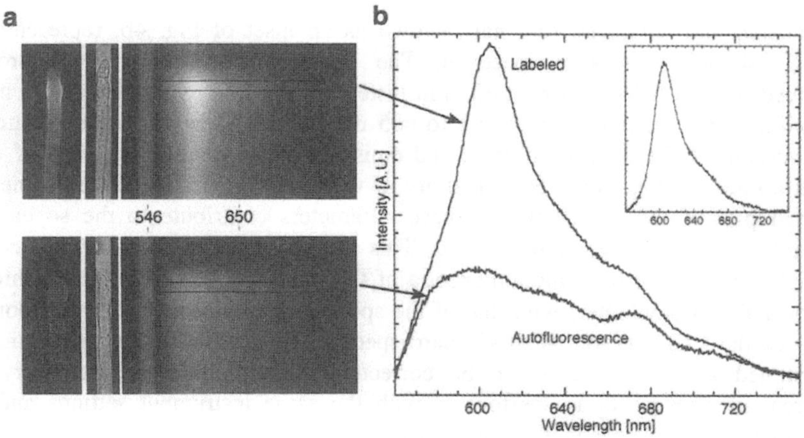

Fig. 4. Spectral images (*a*) of a *Vicia sativa* root hair labeled with 1 nM NodRlv-IV (BODIPY 581/591-C_{16}) (*upper*) and an unlabeled control (*lower*). A 10× PL fluotar NA 0.3 objective in combination with an Omega (Brattleboro, VT, USA) 546DF10 nm bandpass excitation filter, an Omega 555DRLP dichroic mirror and a Schott (Mainz, Germany) OG570 longpass emission filter were used. The slit width of the imaging spectrograph was 200 μm and the central wavelength was 650 nm. The acquisition time was 10 s. Averaged fluorescence emission spectra of the tip region (indicated by the *box*) of labeled and unlabeled root hairs are shown (*b*). The difference spectrum representing net BODIPY 581/591 fluorescence is shown in the inset

root hair is shown. Left, a fluorescence image and a phase contrast image of both hairs is shown, indicating the width of the entrance slit. Both images are captured using the spectrograph with the diffraction grating set at zero-order position. In this position, the grating functions as a mirror enabling conventional fluorescence imaging. The spectral images are shown on the right side of the phase contrast images. In the vertical (spatial) dimension of the spectral image of the labeled root hair, clearly the most intensely labeled region of the root hair can be identified. In the horizontal (spectral) direction of the image spectrum the maximal fluorescence intensity is observed between 546 and 650 nm. Also some excitation light is detected with the spectrograph as can be seen from the band with the width of the entrance slit centered around 546 nm (a Hg-line). The analysis of the unlabeled root hair shows a reduced fluorescence intensity, but also no clear maximum in the horizontal direction of the image spectrum can be identified. This "flatness" is typical for autofluorescence spectra. In Fig. 4b average emission spectra of the tips of both root hairs are shown. From the Figure it can be inferred that the BODIPY- and autofluorescence contribute equally to the total fluorescence. The spectrum acquired from the unlabeled root hair shows a broad spectrum without any sharp peaks. The bump around 670 nm was also observed by Gadella et al. as a large peak, and corresponds to porphyrin fluorescence emission, which is excited less effectively at the wavelength used here (546 nm). Contrarily, the emission spectrum of the labeled root hair does show a clear peak, but also the autofluorescence spectrum can be discerned. Indeed, the difference spectrum of the labeled and unlabeled root hair, shown as an inset of Fig. 4b, represents a typical BODIPY emission spectrum. The emission maximum of the probe observed on root hairs is around 605 nm instead of the true maximum at 591 nm. The shift of the maximum from 591 to 605 nm is mainly caused by the reduced transmission of the dichroic mirrors and emission filters of the blue edge of the spectrum and by the increased sensitivity of the CCD camera in the red to near-infrared part of the spectrum. All these parameters contribute to the so called spectral instrument response curve. This curve can be considered as a multiplication of the transmission spectra of the microscope optics (lens, dichroic mirror and emission filter) with that of the spectrograph and the spectral response curve of the CCD sensor. With standard spectra, the instrument response can be determined and the spectra can be corrected. Correction is not necessary if different measurements are performed with the same instrument settings and if only relative spectral changes are important.

3.3.2 Usage of FSPIM for Monitoring Fluorescence Emission Ratio Indicators

As mentioned before, FSPIM can also provide an alternative to emission ratio-imaging, as changes in emission spectra are easily detected by FSPIM. This even allows the simultaneous determination of Ca^{2+} and pH in single cells (Martínez-Zaguilán et al. 1994; Martínez-Zaguilán et al. 1996). Here, we will explain the principle of determining pH by FSPIM using cSNAFL-2. In Fig. 5, the pH dependent emission spectra of cSNAFL-2 as measured by FSPIM are shown. The deprotonated state (Fig. 5a) shows an emission maximum above 580 nm, whereas the emission maximum of the protonated dye (Fig. 5b) is below 580 nm. When a spectral image at intermediate pH is acquired (Fig. 5c), it can be inferred that this is a mixture of the spectrum at low pH representing the protonated probe with that at high pH representing the deprotonated probe spectrum. By fitting these intermediate spectra, using the spectra of the pure protonated and deprotonated probe, the contributions of both species can be determined very accurately. In this way, a calibration curve can be constructed which describes the contribution of the protonated and deprotonated state for different pH values. These calibration values can be used to exactly measure pH in living plant cells after the decomposition of the in situ spectra in the deprotonated and protonated component spectra. When a third fluorescent species (i.e., autofluorescence) also contributes to the total emission spectrum the fit procedure (Gadella et al. 1997b) should be extended to three or more components. Especially under those conditions, the FSPIM method is superior to FRIM which would produce aberrant ratios in the presence of a significant amount of autofluorescence.

Recently we successfully applied the FSPIM method for monitoring Ca^{2+} in living *Lotus* root hairs that were stably transformed with the yellow-cameleon calcium indicator (Miyawaki et al. 1997; Miyawaki et al. 1999), see also section 2.3.2. The CFP and YFP contributions from the yellow cameleon fluorescence emission spectrum could be fitted and indicated a clear positive cytosolic calcium gradient towards the root hair tip. In addition to the CFP and YFP components, it

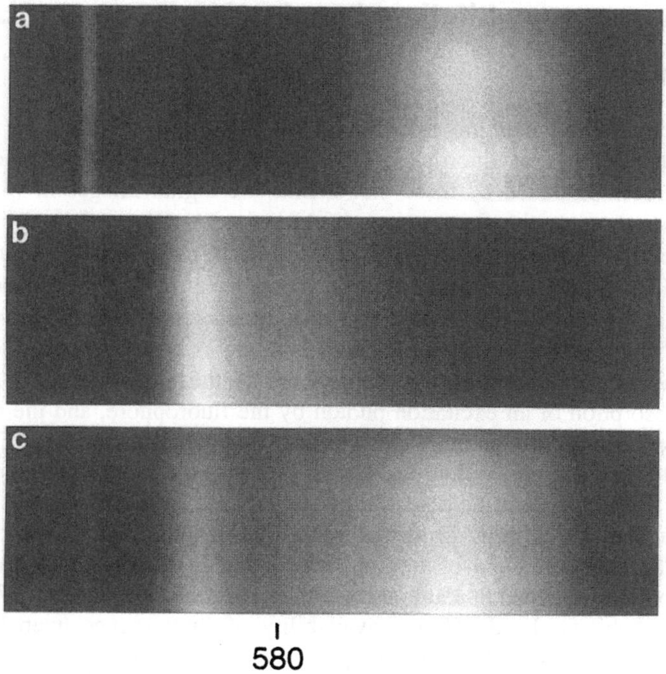

Fig. 5. FSPIM of microcuvettes filled with 5 μM cSNAFL-2 in an alkaline (pH=9.5) (*a*), acidic (pH=4.5) (*b*), or neutral (pH=7.0) (*c*) buffer. A 20× HC PL fluotar NA 0.5 objective in combination with an Omega 500RDF25 nm bandpass excitation filter, an Omega 525DRLP dichroic mirror and a Schott GG530 longpass emission filter were used. The slit width of the spectrograph was 200 μm and the central wavelength was 580 nm. Acquisition time was 10 s and 4× binning in the spatial direction was used on the CCD camera

is essential to fit an autofluorescence component and a background fluorescence component in order to obtain reliable data. The results and the four-component fit procedure are described in detail elsewhere (Gadella et al. 1999; Gadella, manuscript in preparation).

A clear advantage of FSPIM over FRIM is that all spectral components are acquired in one measurement. FRIM usually employs two consecutive images acquired with different filter combinations. Consequently, the FRIM method is dependent on the stability of the excitation source, and in case of emission ratio imaging also registration errors can occur. FSPIM does not suffer from registration problems. However, the superior accuracy obtained by FSPIM is obtained at the sacrifice of one spatial dimension. Given their elongated shape, this may be tolerable for root hairs. In this respect, it should be noted that by scanning the microscope stage it is possible to obtain two-dimensional spatial information with FSPIM.

4 Fluorescence Lifetime Imaging Microscopy (FLIM)

4.1 Principle

Fluorescence lifetime imaging microscopy (FLIM) is a technique for imaging excited state lifetimes of fluorophores by means of a fluorescence microscope. Lifetime images produced by these instruments are digital images in which each pixel value represents the fluorescence lifetime (τ). As opposed to conventional steady-state fluorescence microscopy, FLIM reports on a kinetic parameter (τ) which, unlike fluorescence intensity, is independent of probe concentration, excitation light intensity, moderate levels of photobleaching, direct absorption (or filtering) of fluorescence or on the optical path length in the microscope.

The fluorescence lifetime of a fluorophore is the average time that passes between absorption of an excitation photon by the fluorophore, and the release of the absorbed energy in the form of a red-shifted emission photon. Generally, excited state lifetimes are in the nanosecond (10^{-9} s) time range. The fluorescence lifetime is linearly proportional to the quantum yield Q of the fluorophore (the average number of photons produced by a fluorophore per absorbed photon, $0<Q<1$). The quantum yield of fluorophores is highly dependent on their direct chemical microenvironment. Parameters that change the fluorescence lifetime are local hydrophobicity, local pH, the availability of fluorescence quenchers, ions (e.g., Ca^{2+}) or the proximity of energy transfer acceptors. With steady-state fluorescence microscopy, quantum yields cannot be studied independent from probe concentration, because the local fluorescence intensity is a product of local probe concentration and the local fluorescence quantum yield, and several other parameters such as excitation light intensity, absorption cross-section, detector efficiency, and optical path length (Jovin and Arndt-Jovin 1989b; Jovin et al. 1990). Consequently, with conventional steady-state fluorescence microscopy, one cannot distinguish between a high fluorophore concentration with a low quantum yield and a low fluorophore concentration with a high quantum yield. FLIM, however, enables the generation of separate images of fluorescence intensity and fluorescence lifetimes and thereby provides the capability to image both the local fluorophore concentration and the local fluorophore microenvironment.

Fluorescence lifetime imaging microscopes can be implemented in many different ways, for a recent review see Gadella (1999). The two most widely used modes of implementation are the frequency domain (Lakowicz and Berndt 1991; Gadella et al. 1993; So et al. 1994; Gadella et al. 1997a) and time domain approaches (Wang et al. 1992; Oida et al. 1993; Periasamy et al. 1996; Sytsma et al. 1998). In the time domain, a very short pulse (fs-ps time duration) of light excites the sample, and time-correlated detection of the fluorescence emission provides the lifetime information. In the frequency domain (used in our laboratory), the specimen is continuously excited with intensity-modulated light. The intensity-modulation occurs at a radio frequency (typically in the tens of MHz region) in a sinusoid or other shaped way. As a result, also the induced fluorescence emission of the microscopic object will also be intensity-modulated.

However, due to the time difference between excitation and emission (i.e., lifetime of the probe), the emitted light will be demodulated and phase shifted with respect to the excitation light . The frequency domain principle of fluorescence lifetime estimation is explained in Fig. 6. The two observables (phase shift $\Delta\varphi$ and demodulation M) can be related to the fluorescence lifetime via two straightforward equations (Jameson et al. 1984; Gadella et al. 1994b):

$$\tau_{\varphi} = \frac{1}{2\pi f} tan(\Delta\varphi) \qquad (3)$$

$$\tau_M = \frac{1}{2\pi f}\sqrt{(1/M^2)-1} \qquad (4)$$

where f is the frequency of modulation. When there is more than one way of radiative deactivation of one (or multiple) excited state(s), the fluorescence emission process can be described by more than one fluorescence lifetime and can be considered as multicomponent emission. The formulas relating the observables $\Delta\varphi$ and M to these different lifetimes are more complicated in this case (see also Gadella et al. 1994b and Gadella 1999), but a multicomponent emission can be easily identified by the fact that the lifetimes determined from the phase (τ_φ) and from the modulation (τ_M) are different ($\tau_\varphi < \tau_M$).

4.2 Implementation of Fluorescence Lifetime Imaging Capabilities in a Microscopy System

The implementation of the frequency-domain method of fluorescence lifetime estimation in a microscope, requires the inclusion of specific components allowing fast modulation of the excitation light, an imaging detector capable of measuring fast changes in fluorescence intensities, and several electronics to control the modulation and phase settings. The set-up used in our laboratory has been described in detail by Gadella et al. (1997a), therefore only a basic description of the equipment will be given here, following the light path. The first component is a laser source (CW Ar/Kr laser) which puts out a single (user selectable) line which is passed through an acousto-optic modulator (AOM). The AOM modulates the laser light intensity up and down in MHz frequency range somewhere between 10-150 MHz. One of the diffraction spots produced by the AOM is selected by a diaphragm, and after expanding the beam it is incorporated into a Leica DMR/BE epifluorescence microscope. A dichroic mirror reflects the excitation light onto the sample. The fluorescence emission is transmitted through the dichroic mirror and an emission filter. The filtered fluorescence emission is focused onto the photocathode of the rf-frequency gain-modulated image intensifier (Hamamatsu, model C5825). The light emitted from the phosphor screen of the image intensifier is focused onto the chip of a CCD camera (Photometrics model Quantix with a

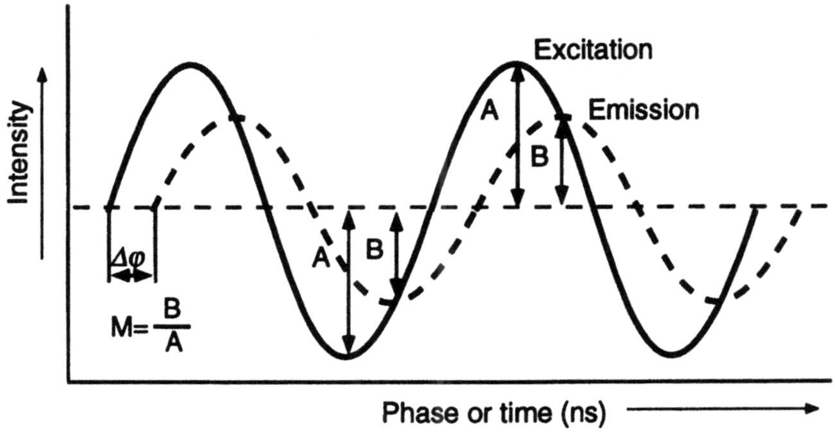

Fig. 6. Principle of fluorescence lifetime determination with the frequency-domain approach. The excitation light is sinusoidally intensity-modulated. The resulting fluorescence emission is also sinusoidally intensity-modulated but phase shifted (by $\Delta\varphi$ degrees, $0° < \Delta\varphi < 90°$) and demodulated (by a factor M, $0 < M < 1$) with respect to the excitation light

Kodak KAF1400 grade 1 CCD sensor, Tucson, AZ, USA) by means of a relay lens system. Two computer-controlled frequency synthesizers (Programmed Test Sources model 310, Littleton, MA, USA) drive the modulation on AOM and the image intensifier gain respectively. Phase image acquisition by the CCD, excitation shutter control and phase settings on the frequency synthesizers are computer controlled and integrated within the IPLab spectrum image acquisition/processing software package (Signal Analytics, Vienna VA, USA). Image analysis for constructing lifetime images from the phase images is done on a Silicon Graphics Indy workstation using self-written software (Gadella et al. 1994a).

We use a homodyne detection scheme in which the fluorescence signal and the detector (image intensifier) gain-modulation occurs at exactly the same frequency. When the gain-modulation of the image intensifier happens to be exactly in phase with the fluorescence intensity-modulation, a maximal signal will be integrated on the CCD. If the phase difference is 180° (exactly out of phase), a minimum signal will be integrated on the CCD. In a typical experiment, the phase between the image intensifier gain-modulation and the excitation light-modulation is varied from 0 to 360°, and at each phase difference an image is captured by the CCD. Typically, 4 to 10 of such phase images (0.1-2 s exposure time each) are acquired by the CCD camera. The spatially resolved lifetime information is extracted from the phase image stack by fitting a sinusoid for every single pixel and determining the phase shift $\Delta\varphi$ and demodulation M with an image analysis procedure (Gadella et al. 1993; Gadella et al. 1994b).

4.3 Applications

The fluorescence lifetime is a fundamental spectroscopic parameter, like wavelength or intensity, and hence, lifetime-resolved images can be used for many applications, for a review see Gadella (1999).

4.3.1 Imaging of Ion and Oxygen Concentrations

Several reports have appeared showing novel ways of detecting ion concentrations by employing non-ratiometric fluorescent indicators that change their lifetime rather than their fluorescence spectrum. For ratiometric determination of Ca^{2+} concentrations, using fura or indo (Haugland 1996), UV excitation has to be employed. UV, however, is generally hazardous to living cells. UV excitation can be avoided by using non-ratiometric lifetime-based fluorescent Ca^{2+} indicators absorbing at longer wavelengths such as Ca-Green (blue excitation) (Lakowicz et al. 1992; Sanders et al. 1994; So et al. 1995) or Ca-Crimson (orange excitation) (Periasamy et al. 1996).

Also the pH sensitive dye cSNAFL-2 shows a change in the fluorescence lifetime upon altering the pH. By using FLIM, we determined that the lifetime (τ_φ) of SNAFL-2 was 2.3 ns in alkaline solution (pH=9) and 4.0 ns in acidic solution (pH=5) upon excitation at 514 nm and detection with a 610DF20 nm emission bandpass filter. Similar experiments using 568 nm excitation, showed a relatively constant fluorescence lifetime around 1.0 ns both in alkaline and acidic solution, which is comparable to the value reported by others (Szmacinski and Lakowicz 1993). Most likely, only the protonated probe is excited at 568 nm, which has a lifetime of 1.0 ns.

Many other lifetime-based fluorescence indicators which can be ester-loaded into living cells are available, e.g., for detecting Mg^{2+} (Magnesium-Green), Na^+ (Sodium-Green) Cl^- (N-ethoxycarbonylmethyl-6-methoxyquinolinium bromide (MQAE) and heavy metals (bis-BTC), but still await application in lifetime resolved imaging (Szmacinski et al. 1994; Haugland 1996). By employing the ability of oxygen to decrease fluorescence lifetimes by collisional quenching, FLIM has also been successfully applied in imaging oxygen concentrations in living cells (Gerritsen et al. 1997).

4.3.2 Contrast Enhancement

In a situation where autofluorescence and probe fluorescence intensities and spectra are comparable, it is very difficult to distinguish between probe fluorescence and autofluorescence using steady-state FDIM. However, if the fluorescence lifetime of the probe is different from that of the autofluorescing environment, the lifetime image can specifically enhance the contrast for that specific probe. Such a situation occurs upon labeling *Vicia sativa* roots with fluorescent Nod-factors containing a BODIPY FL moiety (Gadella et al. 1997b). Upon excitation with the 488 nm laser line, the green fluorescence produced by the Nod-factor is very comparable with the root autofluorescence both in color and in intensity. Because the fluorescence lifetime of the autofluorescence is only about 1 ns, and the fluorescence of the BODIPY FL is about 5.5 ns, FLIM enables us to study the localization of this Nod-factor on living root hairs. In Fig. 7, the result of such an experiment is shown. The lifetime image shows that the fluorescence lifetimes vary between 1.5 and 3.0 ns, clearly representing a mixture of the two fluorescent components. Interestingly, the fluorescence lifetime increases when going towards the tips of the root hairs. This is due to the increased ratio of BODIPY FL fluorescence to autofluorescence. Hence this Nod-factor has a preference to bind at the root hair tips.

4.3.3 FRET and GFP: Probing Molecular Proximity

FLIM is also a very valuable tool for studying fluorescence resonance energy transfer (FRET). FRET is the process by which an excited donor fluorophore transmits its energy (radiationless) to a neighboring acceptor chromophore. FRET only occurs over a very short distance range: generally shorter than 8 nm. Hereby FRET provides an excellent way to monitor molecular interactions in complex environments (e.g., living cells) with a high spatial resolution. FRET is manifested by a decrease in donor fluorescence intensity and fluorescence lifetime and by an increased (sensitized) emission by the acceptor (if the acceptor is a fluorophore). For a detailed description of FRET, the reader is referred to (Förster 1948; Stryer 1978; Tsien et al. 1993; Wu and Brand 1994; Clegg 1995, 1996).

Because quantitative measurements of fluorescence intensities are difficult in a steady-state fluorescence microscope, FRET measurements using FDIM are complex and sensitive to experimental errors (Jovin and Arndt-Jovin 1989a; Gordon et al. 1998). FLIM, on the other hand, is a reliable and convenient method to quantify FRET in single living cells because it directly measures the decrease in donor-fluorescence lifetime upon the transfer of energy to a FRET acceptor. FRET-FLIM has been successfully applied to monitor proteolytic processing of protein kinase C (PKC) in intact cells (Bastiaens and Jovin 1996), for monitoring membrane fusion in cells (Oida et al. 1993) and to monitor oligomerization of cell surface receptors (Gadella and Jovin 1995). For each of these studies, alterations in the fluorescence lifetime reported on a change in proximity (on the nm distance scale) of two interacting molecules in single cells.

The green fluorescent protein (GFP) from the jellyfish *Aequoria victoria* is a very good probe for FLIM. The enhanced version (EGFP, S65T) has a lifetime of 2.67 ns and can be imaged in single living plant cells with a very high accuracy (coefficient of variation < 3.5%) (Gadella 1999). The availability of chromophore mutants of GFP with blue (BFP), cyan (CFP) and yellowish (YFP) colors enables us to perform colocalization and FRET studies in single living cells (for reviews see Tsien 1998; Ellenberg et al. 1999; Pollok and Heim 1999). A review on GFP-based FRET applications in plant cells is published elsewhere (Gadella et al. 1999).

Recently, we could show that the GFP/FRET based calcium indicator yellow-cameleon2 (Miyawaki et al. 1997; Miyawaki et al. 1999) can be expressed in the cytosol of root hairs of *Lotus japonicus*, using the RH2 root epidermis specific

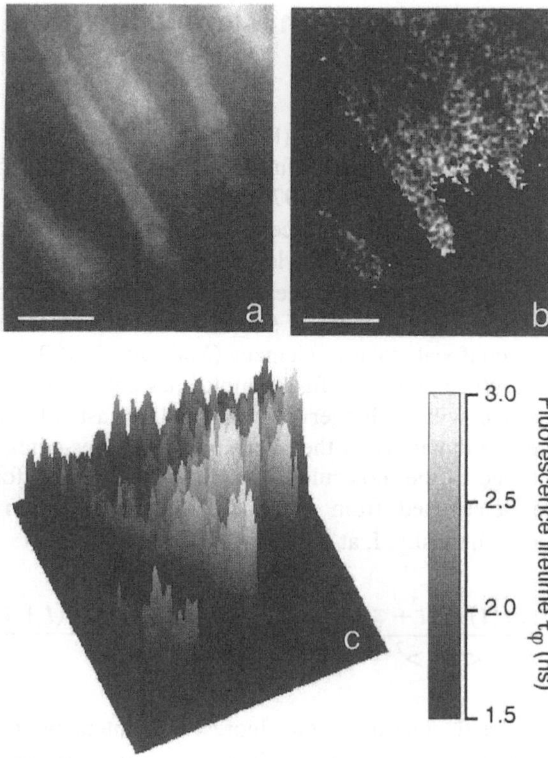

Fig. 7. FLIM analysis of *Vicia sativa* root hairs labeled with a Nod-factor containing a fatty acid tail with a BODIPY FL moiety. *a* (steady-state) fluorescence intensity image, *b* fluorescence lifetime image, *c* pseudo three-dimensional representation of the fluorescence lifetime image. The lifetime values are indicated in a *gray-scale* ranging from black (1.5 ns) to white (3 ns) as indicated next to *c*. The *white distance bar* represents 50 μm in the object plane. For labeling conditions and the Nod-factor structure *see ref.* (Gadella et al. 1997b). The specimen was excited at 488 nm modulated at 51.204 MHz; a 40× PL Fluotar oil immersion objective (NA 0.5-1.0, set at 1.0), an Omega 505DRLP dichroic mirror and an Omega 525DF20 nm bandpass emission filter were used. Other conditions are similar to those described in (Gadella et al. 1997a)

promoter (Gadella et al. 1999). We could detect an increased FRET efficiency in yellow-cameleon2 at the tips of growing root hairs which indicates a local positive calcium concentration gradient at the root hair tip (Gadella et al. 1999). Because of the relative ease of plant transformation, many GFP-based (FRET-) FLIM applications are possible for the near future. The plant root hair seems to be an almost ideal cell type for such studies. Hence, it is likely that there will be an important role for FLIM in the detection of molecular interactions in single living root hair cells using GFP fusion proteins, which undoubtedly will shed new light on subcellular signaling, trafficking and communication mechanisms.

5 Fluorescence Correlation Microscopy (FCM)

5.1 Principle

Fluorescence correlation spectroscopy (FCS) is a fluorescence microscopic technique that gives information on diffusion and concentration of fluorescent molecules (Rigler 1995; Schwille et al. 1997a).

FCS is based on the confocal fluorescence microscopic technique, but unlike confocal laser scanning microscopy (Pawley 1995), the position of the confocal excited volume is fixed throughout the measurement. FCS measures fluorescence intensity fluctuations due to movement of single fluorescent molecules in and out a small (1 fL) open confocal volume element (Maiti et al. 1997), schematically illustrated in Fig. 8a. When slowly diffusing molecules enter the confocal volume, the rise in fluorescence will be longer as compared to fast diffusing molecules (Fig. 8b). Therefore, the dynamics of the intensity fluctuations contain information on the diffusional speed of the molecules. The diffusion constant for translational movement can be determined from the autocorrelation function G(τ), which relates the fluorescence intensity, I, at a time t to that τ seconds later:

$$G(\tau) = \frac{\langle I(t)\cdot I(t+\tau)\rangle}{\langle I\rangle^2} = \frac{\langle I\rangle^2 + \langle \delta I(t)\cdot \delta I(t+\tau)\rangle}{\langle I\rangle^2} \tag{5}$$

Here δI denotes the fluctuation of the fluorescence intensity around the mean value <I>. In order to measure fluctuations in fluorescence intensity, it is essential that the number of fluorescent molecules inside the confocal volume element changes significantly over time. If there are more than 100 molecules on average in the volume element at any time, the change in fluorescence intensity due to the movement of a single molecule in or out of the detection volume will be very insignificant. Hence the number of molecules in the volume element must be low enough, determining the upper concentration limit at which FCS can be used (around 100 nM). The lower concentration limit is confined by background or the time that can be spent in order to obtain enough intensity fluctuations, and can be as low as 10^{-12} M (Rigler 1995). Assuming a Gaussian shaped laser focus in three dimensions, the autocorrelation function can be written as:

$$G(\tau) = 1 + \frac{\left(1 - F + F\cdot e^{-\lambda\tau}\right)}{N_m\left(1+\frac{4D_{tran}\cdot\tau}{\omega_{xy}^2}\right)\left(1+\frac{4D_{tran}\cdot\tau}{\omega_z^2}\right)^{1/2}} \tag{6}$$

where D_{tran} denotes the translational diffusion constant in $m^2\cdot s^{-1}$ and N_m indicates the average number of fluorescent particles in the detection volume. Figure 8c shows the shape of the autocorrelation function. Equation 6 also contains a term F,

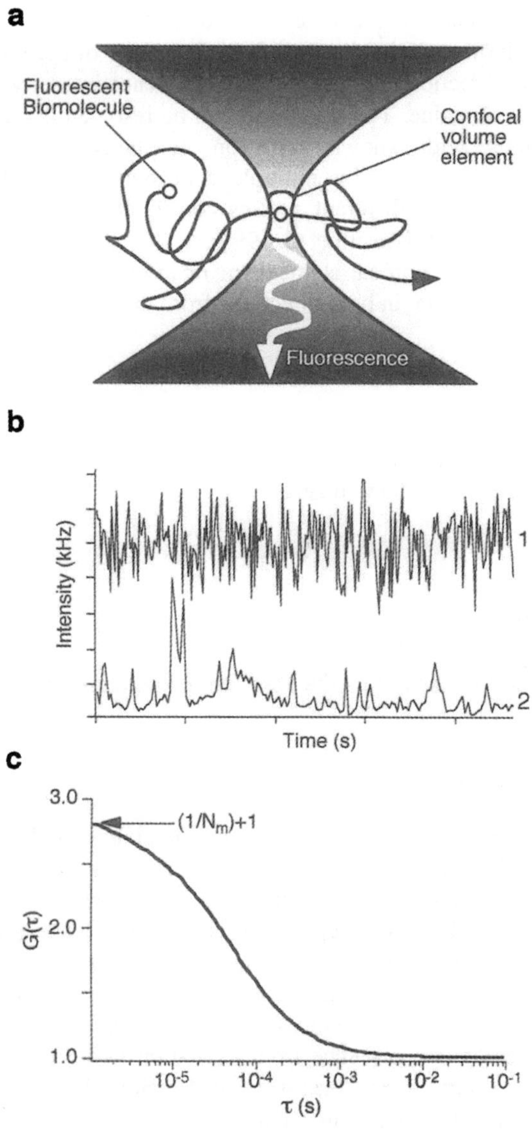

Fig. 8. Principle of fluorescence correlation spectroscopy (FCS). A fixed confocal volume is continuously illuminated by a focused laser beam (*a*). Fluorescent molecules entering the confocal volume will give rise to a burst of fluorescence emission (*b*), of which the duration is dependent on the time they spend in the confocal volume. This is illustrated by *curves 1 and 2*, representing intensity fluctuations caused by diffusion of small organic fluorophores (Rhodamine green in H_2O) or large multilamellar vesicles labeled with fluorescent lipids, respectively. Autocorrelation of the intensity-fluctuations gives the autocorrelation curve (*c*) which can be analyzed to yield the average number of particles N_m in the confocal volume, and the diffusion constant D_{tran} (*see equation 6*)

describing the fraction of molecules in the triplet state and the characteristic triplet decay rate λ. The constants ω_{xy} and ω_z describe the dimensions of the volume element. These are defined as the distance from the center of the laser focus in the radial, ω_{xy}, and axial direction, ω_z, where the fluorescence intensity has dropped to e^{-2} ($\approx$13.5%) of its peak value. The laser radii can be resolved by calibration using reference compounds with known diffusion constants. The autocorrelation function can easily be extended for multiple contributions (Brock et al. 1998). However, quantum yields of different species contributing to the autocorrelation should be taken into account (Meseth et al. 1999).

The volume, V, of the confocal element (m^3) can be approached by a cylinder with radius ω_{xy} and height $2\omega_z$ in both directions from the center and hence equals:

$$V = \pi\,\omega_{xy}^2 \cdot 2\,\omega_z \tag{7}$$

The hydrodynamic radius, r, of the fluorescent particles is related to the diffusion constant via the Stokes-Einstein equation (Edward 1970):

$$r = \frac{kT}{6\pi\eta D_{tran}} \tag{8}$$

In which η is the viscosity, T is the absolute temperature and k the Boltzmann constant.

5.2 Implementation of FCS

FCS measurements were performed on a commercially available system: the Zeiss-Evotec ConfoCor (Carl-Zeiss, Jena, Germany and Evotec Biosystems, Hamburg, Germany). The system consists of a Zeiss Axiovert inverted microscope with standard confocal epi-fluorescence microscope optics. An argon ion laser (488/514 nm) and a helium-neon laser (543 nm) are used as excitation sources. Zeiss dichroic mirrors (we used the 560 nm dichroic), and Zeiss emission filters (we used the 565-610 nm bandpass filter) mounted in a Zeiss filter slider are employed to separate fluorescence from excitation light. The light is focused in the sample using a high grade Zeiss water immersion objective (40×, 1.2 NA) with an adjustable ring for matching slight differences in refractive index (e.g., caused by variation in coverglass thickness). Confocal detection is achieved by a motor-controlled pinhole (diameter usually 40 μm). The detector behind the pinhole consists of an avalanche photodiode coupled to a fast digital correlator, which supplies a real-time display of the autocorrelation curve. The setup is described in detail elsewhere (Hink and Visser 1998; Hink et al. 1999).

Calibration, to obtain the dimensions of the confocal volume (ω_{xy} and ω_z in equation 7), is done by acquiring autocorrelation curves of fluorophores with a known diffusion constant such as rhodamine 6G or tetramethylrhodamine in H_2O (both have diffusion constants of $2.8 \cdot 10^{-10}$ $m^2 s^{-1}$) (Rigler et al. 1992).

5.3 Applications

FCS has been applied mainly to study molecular diffusion in solution. Applications of FCS include measuring diffusion constants (Magde et al. 1974), aggregation states (Berland et al. 1995) hybridization of oligonucleotides (Kinjo and Rigler 1995; Oehlenschläger et al. 1996) and receptor-ligand interactions (Rauer et al. 1996). So far the use of FCS in combination with digital microscopy (FCM) (Brock et al. 1998; Brock and Jovin 1998) to study molecular diffusion and concentration in living cells has been limited. Still, FCM can be applied successfully to obtain quantitative information on concentration and diffusional behavior of fluorescent molecules in root hairs, as will be described below.

5.3.1 Study of Nod-Factor Diffusional Behavior on Living Root Hairs by FCM

In Fig. 9, we show the results of an FCM analysis of a growing *Vicia sativa* root hair in the presence of 10 nM initially extracellularly applied fluorescent Nod-factor. We used an orange fluorescing BODIPY 558/568 tagged Nod-factor in combination with the 543 nm He/Ne laser line. If the 488 nm or 514 nm laser line were used for excitation, the autofluorescence of the root hairs was too high. However, at 543 nm excitation the autofluorescence, was low enough to allow FCM experiments with orange or red fluorescing dyes. By moving the microscopy stage, the FCM confocal volume element can be positioned anywhere in the slide. We positioned the confocal volume just before the tip of a growing root hair after which several autocorrelation curves (60 s each) were acquired sequentially. The first three correlation curves (see Fig. 9), taken extracellularly, show fast Nod-factor diffusion ($D = 2.3\ 10^{-10}$ $m^2 \cdot s^{-1}$), but then suddenly a shift of the autocorrelation curves to slower diffusion is observed upon entering (growth) of the root hair into the confocal volume element. Also the fluorescence intensity is markedly increased when the cell wall enters the confocal volume. As soon as the cell wall exits and the cytosol moves into the volume element the fluorescence intensity drops, and autocorrelation curves of faster diffusing particles are obtained. From analysis of the correlation curve 5 in Fig. 9 (representing diffusion monitored in the cell wall) an average diffusion constant of approximately 2.3 10^{-13} $m^2 \cdot s^{-1}$ is calculated, indicating a 1000-fold decreased diffusional mobility as compared to Nod-factor diffusion in water. Under the employed labeling conditions, the average number of fluorescing particles inside the confocal volume element is usually around 50. Given the confocal volume of approximately 1.4 femtoliter ($1.4 \cdot 10^{-15}$ L) this implies a concentration of 60 nM. The actual concentration of Nod-factors in the cell wall will be several fold higher because the thickness of the cell wall is less than the diameter of the excited volume element. Considering the initial extracellularly applied concentration of Nod-

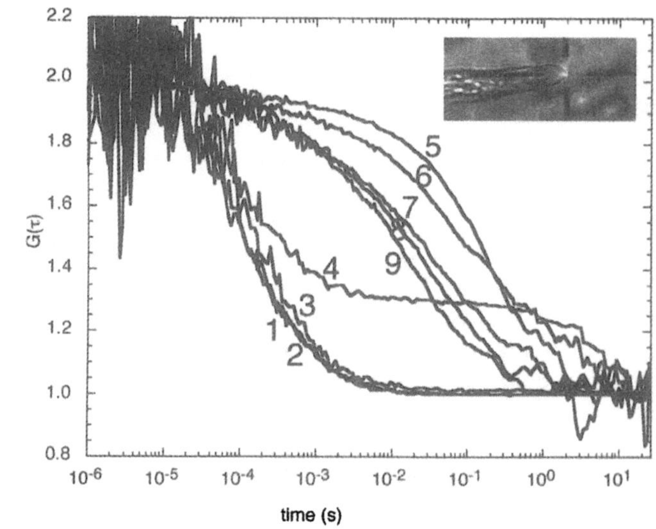

Fig. 9. Autocorrelation curves acquired with FCM after addition of 10 nM of NodRlv-IV (BODIPY 558/568-C_{12}) to the root system of *Vicia sativa*. Seedlings were grown a Fåhreus slide (Heidstra et al. 1994), which was inverted in such way that the coverglass was facing the microscope objective. The confocal volume element was positioned just before the tip of a growing root hair. The resulting autocorrelation curves of the fluorescent Nod-factor in medium (*1-4*), cell wall (*4-6*) and cytosol (*7-9*) of a root hair are shown. The *inset* shows a mixed fluorescence/phase contrast image of a root hair. The bright spot at the tip of the root hair is fluorescence originating from the excited confocal volume

factor of 10 nM, these results demonstrate a marked accumulation of the Nod-factor at the root hair cell wall, and a large reduction in mobility.

Even inside the root hair, fluorescent particles could be detected in the confocal volume above the autofluorescence background. It is remarkable that on average as few as ten fluorescent particles could be detected upon positioning the confocal volume element inside a *living* root hair and that the diffusional behavior of these particles could be studied quantitatively. These measurements clearly demonstrate that FCS provides close to single molecule-detection sensitivity even in living root hairs. Hence, FCM is a promising technique to obtain detailed information on biomolecular diffusion and quantity at very low probe concentrations with subcellular resolution.

6 Concluding Remarks

Implementation of spectroscopic modalities into a fluorescence microscope allows the study of detailed in vivo behavior of fluorescent molecules with high spatial and temporal resolution. In this chapter we have described how various spectroscopic parameters of fluorescent molecules, including excitation- (FRIM) or emission spectral properties (FSPIM), fluorescence lifetimes (FLIM) and molecular diffusion (FCM), can be monitored inside single living root hairs. There are a number of microspectroscopic methods which have been described elsewhere but still await their application for the study of root hairs. Examples are i) photobleaching microscopy, to measure the photochemical stability of fluorophores thereby reporting on their (subcellular) microenvironment (Arndt-Jovin et al. 1979; Gadella and Jovin 1995, 1997), ii) fluorescence anisotropy microscopy to measure rotational mobility of molecules in cells (Dix and Verkman 1990) and iii) fluorescence polarization microscopy to measure orientational distributions of molecules in cells (Axelrod 1989). Furthermore, two or more spectroscopic modalities can be integrated (e.g., combined temporally and spectrally resolved imaging (Vereb et al. 1998)) to obtain more detailed spectroscopic information in a single measurement.

A potentially interesting new development is the combination of two-photon excitation (TPE) with microspectroscopic methods, e.g., two-photon FLIM (So et al. 1995) and two-photon FCS (Berland et al. 1995). The main advantage of TPE is that the excitation is restricted to the focal point, which minimizes photobleaching and produces data lacking out-of-focus fluorescence, avoiding the need for insertion of a pinhole in the detection light path of the microscope. However, for thin samples such as root hairs, TPE might not provide significant improvement in signal-to-noise ratio over conventional confocal microscopy (Piston 1999). Still, TPE could be advantageous for exciting blue or UV-absorbing probes that are less compatible with common single photon CW laser systems.

The FCM experiments demonstrate that it is possible to attain close to single molecule-detection sensitivity even in the cytosol of living root hair cells. This is not only because of the high sensitivity of this technique but also because of the very low autofluorescence levels of growing root hairs upon excitation at wavelengths higher than 540 nm. The background fluorescence can be effectively reduced by yet another new microspectroscopic technique: cross-correlation FCS, which cross-correlates intensity fluctuations of molecules carrying two different fluorophores. As a result, only molecules emitting at the two corresponding wavelengths are observed (dual color fluorescence cross-correlation spectroscopy (see Schwille et al. 1997b; Rigler et al. 1998). By this technical advancement, in vivo detection of dual labeled single molecules might be accomplished, possibly revealing in vivo molecular conformation fluctuations, on- and off-rates for molecular association and single molecule enzyme kinetics (Weiss 1999). All of these potential and novel applications demonstrate that advanced microspectroscopy applied to the study of single plant cells (e.g., root hairs) is a rapidly emerging scientific field, and undoubtedly will shed new light on a variety of molecular processes inside living root hairs.

ACKNOWLEDGMENTS

We are grateful to Mark Hink for his contribution to the experiments shown in Fig. 9. J.G. is supported by the Netherlands Council of Earth and Life Sciences (ALW) and T.W.J.G. by the Royal Netherlands Academy of Arts and Sciences (KNAW).

References

Arndt-Jovin DJ, Latt SA, Striker G, Jovin TM (1979) Fluorescence decay analysis in solution and in a microscope of DNAs and chromosomes stained with quinacrine. J Histochem Cytochem 27:87-95

Axelrod D (1989) Fluorescence polarization microscopy. Meth Cell Biol 30: 333-352

Balaban RS, Kurtz I, Cascio HE, Smith PD (1986) Microscopic spectral imaging using a video camera. J Microscopy 141:31-9

Bastiaens PIH, Jovin TM (1996) Microspectroscopic imaging tracks the intracellular processing of a signal transduction protein: fluorescent-labeled protein kinase C. Proc Natl Acad Sci USA 93:8407-8412

Berland KM, So PTC, Gratton E (1995) Two-photon fluorescence correlation spectroscopy: method and application to the intracellular environment. Biophys J 68: 694-701

Bibikova TN, Jacob T, Dahse I, Gilroy S (1998) Localized changes in apoplastic and cytoplasmic pH are associated with root hair development in *Arabidopsis thaliana.* Development 125: 2925-2934

Brauer D, Otto J, Tu S (1995) Selective accumulation of the fluorescent pH indicator BCECF, in vacuoles of maize root-hair cells. J Plant Physiol 145: 57-61

Brauer D, Uknalis J, Triana R, Tu S (1996) Subcellular compartmentation of different lipophilic fluorescein derivatives in maize root epidermal cells. Protoplasma 192: 70-79

Bright GR (1993) Fluorescence ratio imaging: issues and artifacts. In: Herman B, Lemasters JJ (Eds) Optical microscopy - Emerging methods and applications. Academic Press, San Diego, pp 87-114

Bright GR, Fisher GW, Rogowska J, Taylor DL (1989) Fluorescence ratio imaging microscopy. Meth Cell Biol 30: 157-192

Brock R, Hink MA, Jovin TM (1998) Fluorescence Correlation Microscopy (FCM) of cells in the presence of autofluorescence. Biophys J : 2547-2557

Brock R, Jovin TM (1998) Fluorescence correlation microscopy (FCM) - FCS taken into the cell. Cell Mol Biol 44: 847-856

Bullen A, Saggau P (1999) High-speed, random-access fluorescence microscopy: II. Fast quantitative measurements with voltage-sensitive dyes. Biophys J 76: 2272-2287

Clegg RM (1995) Fluorescence resonance energy transfer. Curr Biol 6: 103-110

Clegg RM (1996) Fluorescence resonance energy transfer. In: Wang X-F, Herman B (eds) Fluorescence imaging spectroscopy and microscopy. John Wiley & Sons, New York, pp 179-252

Dix JA, Verkman AS (1990) Mapping of fluorescence anisotropy in living cells by ratio imaging. Biophys J 57: 231-240

Edward JT (1970) Molecular volumes and the Stokes-Einstein equation. J Chem Educ 47: 261-270

Ehrhardt DW, Atkinson EM, Long SR (1992) Depolarization of alfalfa root hair membrane potential by *Rhizobium meliloti* nod-factors. Science 256: 998-1000

Ehrhardt DW, Wais R, Long SR (1996) Calcium spiking in plant root hairs responding to *Rhizobium* nodulation signals. Cell 85: 673-681

Ellenberg J, Lippincott-Schwartz J, Presley JF (1999) Dual-colour imaging with GFP variants. Trends Cell Biol 9: 52-56

Felle HH, Kondorosi E, Kondorosi A, Schultze M (1995) Nod signal-induced plasma membrane potential changes in alfalfa root hairs are differentially sensitive to structural modifications of the lipochitooligosaccharide. Plant J 7: 939-947

Felle HH, Kondorosi E, Kondorosi A, Schultze M (1996) Rapid alkalinization in alfalfa root hairs in response to rhizobial lipochitooligosaccharide signals. Plant J 10: 295-301

Förster T (1948) Zwischenmolekulare Energiewanderung und Fluorescenz. Ann Phys 2: 55-75

Fricker MD, Plieth C, Knight H, Blancaflor E, Knight MR, White NS, Gilroy S (1999) Fluorescence and luminescence techniques to probe ion activities in living plant cells. In: Mason WT (ed) Fluorescent and luminescent probes for biological activity. Academic Press, London. pp 569-596

Gadella TWJ (1999) Fluorescence lifetime imaging microscopy (FLIM): Instrumentation and Applications. In: Mason WT (Ed) Fluorescent and Luminescent Probes for biological activity. 2nd edition, Academic Press, London. pp 467-479

Gadella TWJ, Arndt-Jovin DJ, Jovin TM (1994a) Visualization of lipid-receptor interactions on single cells by time-resolved imaging fluorescence microscopy. J Fluorescence 4: 295-298

Gadella TWJ, Clegg RM, Jovin TM (1994b) Fluorescence lifetime imaging micrscopy: pixel-by-pixel analysis of phase-modulation data. Bioimaging 2: 139-159

Gadella TWJ, Jovin TM (1995) Oligomerization of epidermal growth factor receptors on A431 cells studied by time-resolved fluorescence imaging microscopy. A stereochemical model for tyrosine kinase receptor activation. J Cell Biol 129: 1543-1558

Gadella TWJ, Jovin TM (1997) Fast algorithms for the analysis of single and double exponential decay curves with a background term. Application to time-resolved imaging microscopy. Bioimaging 5: 19-39

Gadella TWJ, Jovin TM, Clegg RM (1993) Fluorescence lifetime imaging microscopy (FLIM): Spatial resolution on the nanosecond time scale. Biophys Chem 48: 221-239

Gadella TWJ, van der Krogt GMN, Bisseling T (1999) GFP-based FRET microscopy in single living plant cells. Trends Plant Sci 4: 287-291

Gadella TWJ, van Hoek A, Visser AJWG (1997a) Construction and characterization of a frequency-domain fluorescence lifetime imaging microscopy system. J Fluorescence 7: 35-43

Gadella TWJ, Vereb G, Hadri A-E, Röhrig H, Schmidt J, John M, Schell J, Bisseling T (1997b) Microspectroscopic imaging of nodulation factor-binding sites on living *Vicia sativa* roots using a novel bioactive fluorescent nodulation factor. Biophys J 72: 1986-1996

Gerritsen HC, Sanders R, Draaijer A, Ince C, Levine YK (1997) Fluorescence lifetime imaging of oxygen in living cells. J Fluorescence 7: 11-15

Gilroy S (1997) Fluorescence microscopy of living plant cells. Ann Rev Plant Phys Plant Mol Biol 48: 165-190

Goedhart J, Röhrig H, Van Hoek A, Hink MA, Visser AJWG, Bisseling T, Gadella Jr TWJ (1999) Nod-factors integrate spontaneously in biomembranes, transfer rapidly between membranes and to root hairs, but transbilayer flip-flop does not occur. Biochemistry 38: 10898-10907

Goedhart J, Hink MA, Visser AJWG, Bisseling T and Gadella TWJ (2000) In vivo fluorescence correlation microscopy (FCM) reveals accumulation and immobilization of Nod-factors in root hair cell walls. Plant J. 21: 109-119.

Gordon GW, Berry G, Liang XH, Levine B, Herman B (1998) Quantitative fluorescence resonance energy transfer measurements using fluorescence microscopy. Biophys J 74: 2702-2713

Gross D, Loew LM (1989) Fluorescent indicators of membrane potential: microspectrofluorometry and imaging. Meth Cell Biol 30: 193-218

Grynkiewicz G, Poenie M, Tsien RY (1985) A new generation of Ca^{2+} indicators with greatly improved fluorescence properties. J Biol Chem 260: 3440-3450

Haugland RP (1996) Handbook of fluorescent probes and research chemicals. Molecular Probes Inc., Eugene

Heidstra R, Geurts R, Franssen H, Spaink H, van Kammen A, Bisseling T (1994) Root hair deformation activity of nodulation factors and the fate on *Vicia sativa.* Plant Physiol 105: 787-797

Hink MA, Van Hoek A, Visser AJWG (1999) Dynamics of phospholipid molecules in micelles: characterization with fluorescence correlation spectroscopy and time-resolved fluorescence anisotropy. Langmuir 15: 992-997

Hink MA, Visser AJWG (1998) Characterization of membrane mimetic systems with fluorescence correlation spectroscopy. In: Rettig W, Strehmel B, Schrader S, Seitert H (eds) Applied fluorescence in chemistry, biology and medicine. Springer Verlag, Berlin, pp 101-118

Jameson DM, Gratton E, Hall RD (1984) The measurement and analysis of heterogenous emissions by multifrequency phase and modulation fluorometry. Applied Spectroscopy Reviews 20: 55-106

Jovin TM, Arndt-Jovin DJ (1989a) FRET Microscopy: Digital imaging of fluorescence resonance energy transfer. Application in cell biology. In: Khen E, Hirschberg JG (eds) Cell structure and function by microspectrofluorometry. Academic Press, New York, pp 99-117

Jovin TM, Arndt-Jovin DJ (1989b) Luminescence digital imaging microscopy. Annu Rev Biophys Biophys Chem 18: 271-308

Jovin TM, Arndt-Jovin DJ, Marriott G, Clegg RM, Robert-Nicoud M, Schormann T (1990) Distance, wavelength and time: the versatile 3rd dimensions in light emission microscopy. In: Herman B, Jacobson K (eds) Optical Microscopy for Biology. Wiley-Liss, New York, pp 575-602

Kinjo M, Rigler R (1995) Ultrasensitive hybridization analysis using fluorescence correlation spectroscopy. Nucl Acids Res 23: 1795-1799

Lakowicz JR, Berndt KW (1991) Lifetime-selective fluorescence imaging using an rf phase-sensitive camera. Rev Sci Instrum 62: 1727-1734

Lakowicz JR, Szmacinski H, Johnson ML (1992) Calcium imaging using fluorescence lifetimes and long-wavelength probes. J Fluorescence 2: 47-62

Lerouge F, Roche P, Faucher C, Maillet F, Truchet G, Promé JC, Dénarié J (1990) Symbiotic host-specificity of *Rhizobium meliloti* is determined by a sulphated and acylated glucosamine oligosaccharide signal. Nature 344: 781-784

Magde D, Elson EL, Webb WW (1974) Fluorescence correlation spectroscopy. II. An experimental realization. Biopolymers 13: 29-61

Maiti S, Haupts U, Webb WW (1997) Fluorescence correlation spectrocopy: diagnostics for sparse molecules. Proc Natl Acad Sci USA 94: 11753-11757

Martínez-Zaguilán R, Gurulé MW, Lynch RM (1996) Simultaneous measurement of intracellular pH and Ca^{2+} in insulin-secreting cells by spectral imaging microscopy. Am J Physiol 270: 1438-1446

Martínez-Zaguilán R, Tompkins LS, Lynch RM (1994) Simultaneous analysis of multiple fluorescent probes in single cells by microspectroscopic imaging. Proc SPIE 2137: 17-28

Meseth U, Wohland T, Rigler R, Vogel H (1999) Resolution of fluorescence correlation measurements. Biophys J 76: 1619-1631

Miesenböck G, De Angelis DA, Rothman JE (1998) Visualizing secretion and synaptic transmission with pH-sensitive green fluorescent proteins. Nature 394: 192-195

Miyawaki A, Griesbeck O, Heim R, Tsien RY (1999) Dynamic and quantitative Ca^{2+} measurements using improved cameleons. Proc Natl Acad Sci USA 96: 2135-2140

Miyawaki A, Llopis J, Heim R, McCaffery JM, Adams JA, Ikura M, Tsien RY (1997) Fluorescent indicators for Ca^{2+} based on green fluorescent protein and calmodulin. Nature 388: 882-887

Oehlenschläger F, Schwille P, Eigen M (1996) Detection of HIV-1 RNA by nucleic acid sequence-based amplification combined with fluorescence correlation spectroscopy. Proc Natl Acad Sci USA 93: 12811-12816

Oida T, Sako Y, Kusumi A (1993) Fluorescence lifetime imaging microscopy (flimscopy). Methodology development and application to studies of endosome fusion in single cells. Biophys J 64: 676-685

Pawley JB (ed) (1995) Handbook of biological confocal microscopy. Plenum Press, New York

Periasamy A, Wodnicki P, Wang XF, Kwon S, Gordon GW, Herman B (1996) Time-resolved fluorescence lifetime imaging microscopy using a picosecond pulsed tunable dye laser system. Rev Sci Instrum 67: 3722-3731

Pingret J-L, Journet E-P, Barker DG (1998) *Rhizobium* Nod-factor signalling: evidence for a G protein-mediated transduction mechanism. Plant Cell 10: 659-671

Piston W (1999) Imaging living cells and tissues by two-photon excitation microscopy. Trends Cell Biol 9: 66-69

Pollok BA, Heim R (1999) Using GFP in FRET-based applications. Trends Cell Biol 9: 57-60

Rauer B, Neumann E, Widengren J, Rigler R (1996) Fluorescence correlation spectrometry of the interaction kinetics of tetramethylrhodamin α-bungarotoxin with *Torpedo calnifornica* acetylcholine receptor. Biophys Chem 58: 3-12

Read ND, Allan WTG, Knight H, Malhot R, Russell A, Shacklock PS, Trewavas AJ (1992) Imaging and measurement of cytosolic free calcium in plant and fungal cells. J Microscopy 166: 57-86

Rigler R (1995) Fluorescence correlations, single molecule detection and large number screening. Applications in biotechnology. J Biotechnol 41: 177-186

Rigler R, Foldes-Papp Z, Meyer-Almes F-J, Sammet C, Volcker M, Schnetz A (1998) Fluorescence cross-correlation: A new concept for polymerase chain reaction. J Biotechnol 63: 97-107

Rigler RJ, Widengren J, Mets Ü (1992) Interaction and kinetics of single molecules as observed by fluorescence correlation spectroscopy. In: Wolfbeis OS (ed) Fluorescence spectroscopy. Springer Verlag, Berlin, pp 13-24

Sanders R, Gerritsen HC, Draaier A, Houpt PM, Levine YK (1994) Fluorescence lifetime imaging of free calcium in single cells. Bioimaging 2: 131-138

Schwille P, Bieschke J, Oehlenschläger F (1997a) Kinetic investigations by fluorescence correlation spectroscopy: the analytical and diagnostic potential of diffusion studies. Biophys Chem 66: 211-228

Schwille P, Meyer-Almes F-J, Rigler R (1997b) Dual-color fluorescence cross-correlation spectroscopy for multicomponent diffusional analysis in solution. Biophys J 72: 1878-1886

So PTC, French T, Gratton E (1994) A frequency domain time-resolved microscope using a fast-scan CCD camera. Proc SPIE 2137: 83-92

So PTC, French T, Yu WM, Berland KM, Dong CY, Gratton E (1995) Time-resolved fluorescence microscopy using two-photon excitation. Bioimaging 3: 49-63

Spaink HP, Sheely DM, van Brussel AAN, Glushka J, York WS, Tak T, Geiger O, Kennedy EP, Reinhold VN, Lugtenberg BJJ (1991) A novel highly unsaturated fatty acid moiety of lipo-oligosaccharide signals determines host specificity in *Rhizobium*. Nature 354: 125-130

Stryer L (1978) Fluorescence energy transfer as a spectroscopic ruler. Ann Rev Biochem 47: 819-846

Sytsma J, Vroom JM, de Grauw CJ, Gerritsen HC (1998) Time gated fluorescence lifetime imaging and micro-volume spectroscopy using two-photon excitation. J Microscopy 191: 39-51

Szmacinski H, Lakowicz JR (1993) Optical measurements of pH using fluorescence lifetimes and phase-modulation fluorometry. Anal Chem 65: 1668-1674

Szmacinski H, Lakowicz JR, Johnson ML (1994) Fluorescence lifetime imaging microscopy: homodyne technique using high-speed gated image intensifier. Meth Enzymol 240: 723-748

Tsien RY (1998) The green fluorescent protein. Annu Rev Biochem 67: 509-544

Tsien RY, Backsai J, Adams R (1993) FRET for studying intracellular signalling. Trends Cell Biol 3: 242-245

Vereb G, Jares Erijman E, Selvin PR, Jovin TM (1998) Temporally and spectrally resolved imaging microscopy of lanthanide chelates. Biophys J 74: 2210-22

Wang XF, Periasamy A, Herman B (1992) Fluorescence lifetime imaging microscopy (FLIM): instrumentation and applications. Crit Rev Anal Chem 23: 369-395

Weiss S (1999) Fluorescence spectroscopy of single biomolecules. Science 283: 1676-83

Whitaker JE, Haugland RP, Prendergast F (1991) Spectral and photophysical studies of benzo[c]xanthene dyes: dual emission pH sensors. Anal Biochem 194: 330-344

Wu P, Brand L (1994) Resonance energy transfer: methods and applications. Anal Biochem 218: 1-13

Chapter 6
Molecules at the Tips of Root Hairs

Robert W. Ridge and Charlotta A. Fredriksson

Biology Department, International Christian University, 3-10-2 Osawa, Mitaka, Tokyo 181-8585, Japan

1 Introduction

Most of the work on this subject is associated with the *Rhizobium*/legume symbiosis, because the earliest interactions between rhizobia and legume host, at least for many important legume crops, are at the root hair-rhizosphere interface. Despite this, it is surprising that so few laboratories have published in this area of research: the presumed receptor(s) for communication molecules emanating from rhizobia would almost certainly be at the surface of the root hair.

This short chapter will discuss molecules at the tips of root hairs, either inside the apical dome or expressed on the external surface of the cell wall, but excluding molecules such as actin and tubulin, cell wall precursors in vesicles, ribosomes, organelles, calcium and other ions, that are normally found elsewhere in the cell and/or that are covered in other chapters of this book. Root hair specific proteins are also discussed elsewhere in this book and will not be discussed here.

Despite the very limited amount of information available, this is an important area of research for root hairs, because distinct molecules at the tips suggest many possibilities: control of polarity, communication to the rhizosphere through receptor activity, passive and active plant pathogen interactions. Indeed, the very nature of the growing tip of this cell, where the cell wall is expanding in direct exposure to the rhizosphere, makes it a likely site for many kinds of interaction. In this connection, it has been shown that in a nodulation-minus soybean, the stimulation of cell division in the cortex of the root by the synthetic auxin 2,4-D (2,4-dichlorophenoxyacetic acid), and the consequent cell wall expansion, allowed rhizobia-induced infection thread development directly in these cells rather than root hairs (Akao et al. 1991; Akao et al. 1999) with a subsequent successful symbiosis. This means that for the wild type soybean, invasion of root hairs by rhizobia requires an expanding cell wall rather than tip growth per se, and that symbiotic-specific communication molecules or receptors, if they exist, can also be expressed in other cells.

Key words. 2,4-D, cell-to-cell communication, hapten sugar, lectin, legume, infection, rhizobia, root hair, tip-growth

The first reports of distinct molecules at the tips of root hairs were in the mid to late 1980's (Diaz et al. 1986; Ridge and Rolfe 1986; Diaz 1989), although there were also reports of root hair specific molecules at that time (Werner and Kuhlmann 1985). Since then, there have been only a few publications directly addressing the role of known tip molecules or announcing the discovery of new ones. Indeed, the few molecules known to be specific for the tips of root hairs are lectins, sugars and glycoproteins, a spectrin-like molecule; and probably extensin, though not visualised at the tip (Arsenijevic-Maksimovic et al. 1997).

2 Lectins at the Tips of Root Hairs

The role of lectins in legume nodulation has recently been reviewed (Hirsch 1999). Lectin on the surface of pea root hairs was detected by Diaz et al. (1986) using an antibody against pea seed lectin. Labelling of root hair tips was found to be distributed in dense small patches rather than uniformly over the tip. Diaz et al. (1986) also showed that lectins are not present on all hairs, and gave evidence that lectin labelling of tips was detected predominantly opposite protoxylem poles. The significance of this finding is that early infection in the development of symbiotic nodules is usually opposite the protoxylem poles of the young root, and that infectivity of root hairs (as shown by the presence of lectin) is enhanced by unknown factors from within the root (a hormone?). Diaz et al. (1986) also showed that in the zone of mature hairs where lectins are rarely found in hairs, inoculation in the same zone with rhizobia produced few nodules compared to the immature root hair zone, which had high levels of lectin labelling and produced many nodules. Their work showed clearly that such root lectins are likely to play a role in the early stages of nodulation. Further work in this area showed that lectins were involved in host specificity of nodulation (Diaz et al. 1989a,b; 1995). The pea lectin gene was transformed into white clover plants, which were then able to be nodulated by *R. leguminosarum* bv. *viciae*. These workers also altered the carbohydrate-binding domain of the pea lectin by site-directed mutagenesis and when this mutant gene was introduced into white clover plants, the vicia rhizobia failed to nodulate them.

More recently, van Rhijn et al. (1998) showed similar results by transforming *Lotus* with soybean lectin gene. Soybean lectin was detected at the tips of young growing root hairs of the transgenic *Lotus* plants. Inoculation with *Bradyhizobium japonicum* resulted in nodulation of *Lotus*, which is normally not nodulated by *B. japonicum*. An interesting additional result of their work showed that exopolysaccharide of the bacteria is likely to be involved in early interactions with the root hair (see also Hirsch 1999). This suggests that lectins in the root hair tip may directly bind the bacteria to the root hair tip at the onset of the earliest interactions between plant and rhizobia. Certainly it is known that binding is an essential pre-requisite for the colonisation and infection of many kinds of tissue (Beuth et al. 1996), and also that lectins mediate many biological recognition processes (Sharon and Lis 1995).

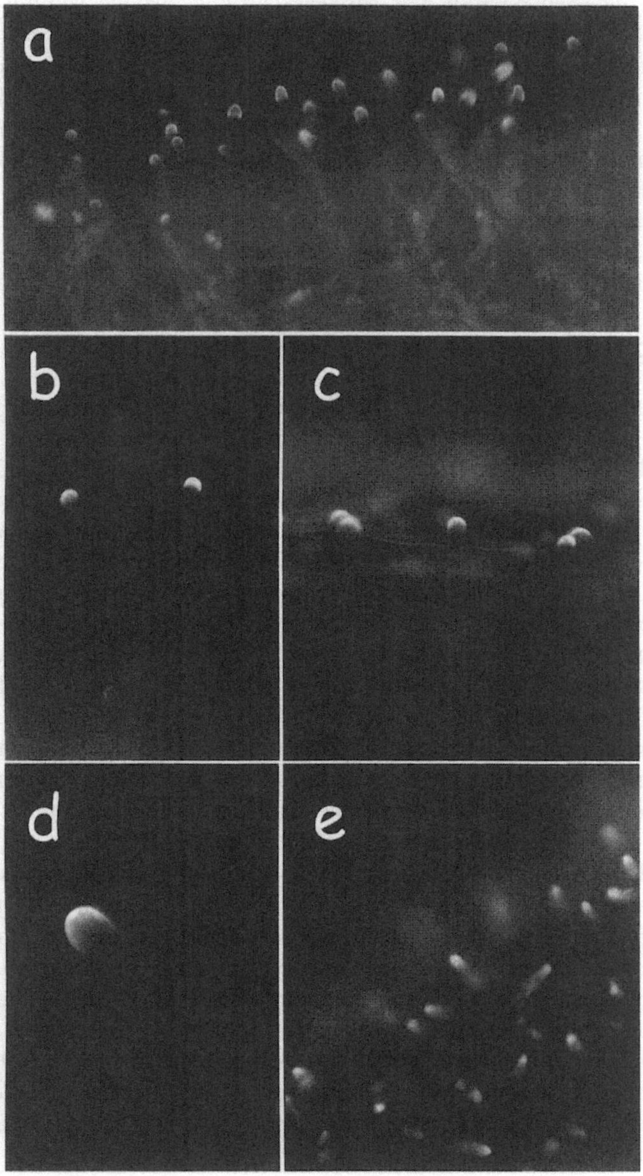

Fig. 1. Lectin binding to root hair tips demonstrating the presence of different sugar moities at the tip surface. Emerging (*c*) and actively growing hairs are shown in these micrographs. *a* and *b* siratro (*Macroptilium atropurpureum*) labelled with MPA-FITC (from *Maclura pomifera*, haptens: N-acetyl galactosamine, α-D-galactose); *c* siratro labelled with RCA-1-FITC (from *Ricinus communis*, hapten: β-D-galactose); *d Rafanus* labelled with RCA-1-FITC; *e* white clover labelled with VFA-FITC, which labels the whole root hair surface but prominently at the tip. From Ridge et al. (1998)

A lectin detected on root hair surfaces carrying enzyme activity similar to apyrase (ATP diphosphohydrolase) and that binds Nod-factors has also been reported, although it was not detected solely at root hair tips (Etzler et al. 1999). The lectin is called LNP by the authors (lectin nucleotide phosphohydrolase). Etzler et al. (1999) reported that this lectin binds Nod-factor, and that an antibody to LNP inhibits the ability of rhizobia to deform root hairs and to form nodules. However, their results are limited by single determination points in their Nod-factor-binding experiments, difficult-to-interpret micrographs, and lack of explanation of some of their techniques. This is a potentially exciting result, but we must await confirmation from other laboratories.

3 Sugar Molecules at the Tips of Root Hairs

First reported by Ridge and Rolfe (1986), using commercially-available lectins as probes (not to be confused with the lectins mentioned above; lectins are widely used as tools for detecting specific sugar molecules, particularly on cell surfaces), Ridge and Rolfe found that RCA-I lectin (from *Ricinus communis*, specific for β-D-galactose) bound to the root hair tips of siratro, a cowpea type legume known to be nodulated by several groups of rhizobia. Pre-incubation of roots, using a spot inoculation technique in the most susceptible zone (smallest emerging root hair zone) of the root for infection, showed that lectin bound to root hair tips prevented infection and nodulation by rhizobia. The authors concluded that carbohydrate molecules at the surface of the root hair tips could be directly involved in the initiation of the symbiosis. More recent work (Ridge et al. 1998) has shown that the broad host range siratro has a multiplicity of sugars at its root hair tips, whereas narrow host range plants such as clover and alfalfa have usually one (RCA-I-binding), sometimes two kinds of sugars at their tips (examples are shown in Fig. 1). The authors also showed that some non-legumes have sugars at their tips (RCA-I, VFA [*Vicia faba*] lectin binding), though their study was limited to the Brassicaceae. Such a defined area of sugar molecules at the tips of root hairs (Fig. 1) suggests that the molecules are probably the sugar moieties of membrane-bound glycoproteins, and that they are recycled at the tip, where there is known to be a large presence of clathrin at the base of the apical dome. Otherwise they would be detectable below the apical dome. Indeed in the case of one lectin, VFA, which is specific for α-D-mannose, this is true. In addition the results show that there may be a conserved sugar expression (or default expression) of one sugar, namely β-D-galactose, present on narrow host range legumes and on one member of the Brassicaceae.

More recently Samaj et al. (1999) have shown specific localisation of arabinogalactan-protein epitopes at the surface of maize root hairs, where strongest expression was at the bulge of root hair initiation (c.f. Ridge and Rolfe 1986) and at the tips of growing hairs (but depleted at mature, non-growing hairs). Samaj et al. (1999) also agree with previous suggestions that such glycoproteins may be involved in plant-microbe interactions (by signalling events).

What is the significance of sugar molecules at the tips of root hairs? In the case of root hairs that apparently have several kinds of sugars, are they part of the same glycoprotein molecule or is each on different molecules, or variations thereof? Only recently has it begun to be understood how plants use the structural diversity of oligosaccharides to regulate important cellular processes such as growth, development and defence (reviewed in John et al. 1997). These oligosaccharides are known to be biologically active at extremely low concentrations. In the case of sugar molecules at the tips of root hairs, especially those in legumes, it is imperative to understand if they are involved in early interactions with microorganisms in the rhizosphere. Indeed, the earlier work of Ridge and Rolfe (1986) showed that by blocking the sugar sites of siratro root hair tips, nodulation was prevented. Etzler et al. (1999) have recently reported a similar result whereby an anti-serum to a lectin at the *Dolichos* root hair surface prevented nodulation by rhizobia.

Thus sugars at the tips of root hairs need to be isolated and characterised, and further experimentation is required to understand their role in tip growth and/or interactions with the rhizosphere. It is the hope of the authors of this short chapter that interest will be stimulated in the area.

References

Akao S, Nakata S, Yoneyama T (1991) Formation of nodules on non-nodulating soybean T201 with treatment of 2,4-D. Plant Soil 138: 207-212

Akao S, Minakawa Y, Taki H, Khan MK, Yuhashi K-I, Nakayama Y, Asis CA Jr, Chebotar V, Kang U-G, Minamisawa K, Ridge RW (1999) Use of lacZ and gusA reporter genes to trace the infection process of nitrogen-fixing bacteria. Jap. Agric. Res. Quart. 33: 77-84

Arsenijevic-Maksimovic I, Broughton WJ, Krause A (1997) Rhizobia modulate root-hair-specific expression of extensin genes. MPMI 10: 95-101

Beuth J, Stoffel B and Pulverer G (1996) Inhibition of bacterial adhesion and infections by lectin blocking. Advances in Experimental Medicine and Biology 408:51

de Ruijter NCA, Rook MB, Bisseling T, Emons AMC (1998) Lipochito-oligosaccharides re-initiate root hair tip growth in *Vicia sativa* with high calcium and spectrin-like antigen at the tip. Plant Journal 13: 341-350

Diaz C (1989). Root lectin as a determinant of host-plant specificity in the *Rhizobium*-legume symbiosis. PhD thesis, University of Leiden, Netherlands

Diaz CL, van Spronsen PC, Bakhuizen R, Logman GJJ, Lugtenberg BJJ, Kijne JW (1986) Correlation between infection by *Rhizobium leguminosarum* and lectin on the surface of *Pisum sativum* L. roots. Planta 168:350-359

Diaz CL, Logman TJJ, Stam HC, Kijne JW (1995) Sugar-binding activity of pea lectin expressed in white clover hairy roots. Plant Physiol 109:1167-1177

Diaz C, Melchers L, Hooykaas P, Lugtenberg B and Kijne J (1989a). Root lectin as a determinant of host-plant specificity in the *Rhizobium*-legume symbiosis. Nature 338: 579-581

Diaz C, van Spronsen P, Bakhuizen R, Longman G, Lugtenberg B and Kijne J (1989b). Correlation between infection by *Rhizobium leguminosarum* and lectin on the surface of *Pisum sativum* L. roots. Planta 168: 530-539

Etzler ME, Kalsi G, Ewing NN, Roberts NJ, Day RB, Murphy JB (1999) A nod factor binding lectin with apyrase activity from legume roots. Proc. Nat. Acad. Sci. USA 96: 5856-5861

Hirsch AM (1999) Role of lectins (and rhizobial exopolysaccharides) in legume nodulation. Curr Opinions Plant Biol. 2:320-326

John M, Rohrig H, Schmidt J, Walden R, Schell J (1997) Cell signalling by oligosaccharides. Trends in Plant Sci. 2: 111-115.

Ridge RW, Rolfe BG (1986) Lectin binding to the root and root hair tips of the tropical legume *Macroptilium atropurpureum* Urb. Appl Envtal Micro 51: 328-332

Ridge RW, Kim R, Yoshida F (1998) The diversity of lectin-detectable sugar residues on root hair tips of selected legumes correlates with the diversity of their host ranges for rhizobia. Protoplasma 202: 84-90

Samaj J, Braun M, Baluska F, Ensikat H-J, Tsumuraya Y, Volkmann D (1999) Specific localisation of arabinogalactan-protein epitopes at the surface of maize root hairs. Plant Cell Physiology 40:874-883

Sharon N and Lis H (1995) Lectins - proteins with a sweet tooth: function in cell recognition. Essays in Biochemistry 30:59

van Rhijn P, Goldberg RB, Hirsch AM (1998) *Lotus corniculatus* nodulation specificity is changes by the presence of a soybean lectin gene. Plant Cell 10: 1233-1249

Werner D, Kuhlmann KP (1985) Legume root hair proteins. In: E Magnien and D de Nettancourt (Eds) Genetic engineering of plants and microorganisms important for agriculture. Nijhoff and Junk Publ., Dordrecht. pp 66-67

Chapter 7
Hormonal Control of Root Hair Growth and Development

Masayuki Katsumi, Mariko Izumo, and Robert W. Ridge

Biology Department, International Christian University, 3-10-2 Osawa, Mitaka, Tokyo 181-8585, Japan

1 Introduction

Physiological and morphological studies on root hairs trace back to the 18th Century (see Farr 1927a,b,c). The pattern of root hair growth and development is under the influence of various factors including genetic, physiological and environmental factors (see Ridge 1995b; Peterson and Farquhar 1996). Involvement of plant hormones in the control of root hair development and elongation has been the topic of much study. Whether or not a plant hormone is required for normal progress of the developmental process of root hairs, there are at least three criteria to be considered: (1) Production of the hormone by root hairs or supply of the hormone from other sources to root hairs. For this, though it is not easy to demonstrate root hairs per se produce plant hormones, root tissues can supply them to root hairs. In addition, under natural conditions, root hairs are exposed to microorganisms which produce plant hormones (Wang et al. 1982; Pegg 1985). (2) Stimulation of the normal process by exogenous supply of the hormone and/or suppression of the process by application of inhibitors of the hormone and its recovery by the hormone. For this, supportive evidence has been reported. (3) Suppression of the process in mutants related to the hormone such as those deficient in the hormone or hormone response.

Recent findings of many kinds of *Arabidopsis* mutants associated with root hair formation, morphology, and growth implicate plant hormones, especially auxin and ethylene in these processes (Schiefelbein and Somerville 1990; Baskin et al. 1992; Su and Howell 1992; Schiefelbein et al. 1993; Galway et al. 1994; Masucci and Schiefelbein 1994; 1996; Cernac et al. 1997; Grierson et al. 1997; Schneider et al. 1998; see also Aeschbacher et al. 1994). In addition, plant hormones have been reported to stimulate the growth of other tip growing cells such as pollen tubes by gibberellin (Kato 1955; Chandler 1957; Bose 1959),

Key words. 2,4-D, ABA, ACC, auxin, brassinolide, Ca^{2+}, cytokinin, cytoplasmic streaming, ethylene, gibberellin, IAA, kinetin, root-hair development, root-hair elongation auxin, fusicoccin and brassinolide (Hewitt et al. 1985). These facts suggest that plant hormones may be commonly involved in the control of tip growth

In this chapter, a comprehensive review of studies concerning the involvement of plant hormones in the control of root hair growth and development is presented, together with some of our experimental results using *Trifolium repens* L.

2 Root Hair Formation

Auxin and ethylene affect root hair formation. Stimulation of root hair formation by ethylene was probably first described by Borgström (1939) in *Pisum sativum, Vicia faba* and *Lupinus luteus* (see Abeles et al. 1992). Early studies also report stimulation of root hair production in tulips (de Munk and de Rooy 1971) and in cereals (Smith and Robertson 1971; Crossett and Campbell 1975).

Masucci and Schiefelbein (1994) have demonstrated that an *Arabidopsis* mutant, *rhd6* shows three distinct defects in root hair formation, namely, a reduction in the number, an overall basal shift in the site of formation, and a relatively high frequency of epidermal cells with multiple root hairs. These phenotypes can be rescued by application of exogenous indole-3-acetic acid (IAA) or 1-aminocyclopropane-1-carboxylic acid (ACC), an ethylene precursor. An ethylene-induced root hair formation in *Arabidopsis* has also been reported (Baskin and Williamson 1992). Dark-grown wild-type *Arabidopsis* seedlings which are largely root hairless and produce little ethylene can develop root hairs by exogenous treatment with ethylene and ACC (Cao et al. 1999). Treatment of wild-type *Arabidopsis* with aminovinylglycine (AVG), an inhibitor of ACC synthase in the pathway of ethylene biosynthesis, can phenocopy the *rhd6* phenotype. Abolishment of root hairs by treatments with AVG (Tanimoto et al. 1995; Masucci and Schiefelbein 1996) and with silver nitrate, an ethylene action inhibitor (Baskin and Williamson 1992) have also been reported in *Arabidopsis*. Additional supportive evidence for the involvement of ethylene/ACC in root hair production are ACC-induced ectopic root hairs formation in wild-type seedlings (Tanimoto et al. 1995; Masucci and Schiefelbein 1996), and presence of ectopic root hairs in an ethylene-response mutant (constitutive triple responses), *ctr1* (Dolan et al. 1994; Tanimoto et al. 1995). In *Arabidopsis* root hairs develop preferentially on trichoblasts overlying the cortical anticlinal cell walls, whereas non-hair cells develop on atrichoblasts overlying the cortical periclinal cell walls. Since in etiolated *Arabidopsis*, ethylene and ACC can induce root hairs on trichoblasts, and ethylene overproducing mutants (*eto1, eto2, eto3, abd, eto4*) preferentially develop root hairs on trichoblasts, these cells should be more sensitive to ethylene than atrichoblasts (Cao et al. 1999). The fact that exogenous auxins such as IAA and 2,4-dichlorophenoxyacetic acid (2,4-D), which can induce ethylene but do not cause ectopic root hair formation, and that the number of root hairs in the ethylene response mutants *etr1* and *ein2* are normal, suggest that ethylene may not be necessary for initiation of root hair growth (Masucci and Schiefelbein 1996). Lack of ectopic root hair formation due to ACC treatment has also been reported in wild-type and *etr1* seedlings (Pitts et al. 1998). Masucci and Schiefelbein (1996) have suggested that ethylene acts independently of *TTG* and GL2, positive regulators of trichome development (Hülskamp et al. 1994), and possibly after these genes. Cao et al. (1999) have also come to the same conclusion by a study with *CTR1* gene, a negative regulator of the ethylene response (Kieber et al. 1993, Roman et al. 1995).

An increase in root hair number by kinetin treatment has also been reported in *Raphanus sativus* L. (Bittner and Buschmann 1983).

It is concluded that ethylene is necessary for the initiation of root hairs, and that auxin acts through ethylene formation, though the mechanism is not clear yet.

3 Root Hair Elongation

3.1 Auxins/Ethylene

Until the discovery of auxin most of physiological studies on the control of root hair growth focussed on the effect of pH and inorganic ions (Seidel 1924; Farr 1927b,c; von Strugger 1926; 1928; Lundegårdh 1946). Earliest studies on the effect of auxin on root hair elongation have been reported by Czaja (1935) using *Lepidium sativum* L seedlings and by Meesters (1936) using *Agrostemma githago* seedlings. They both showed that IAA slightly inhibited normal growth of root hairs. Lundegårdh (1946) reported that IAA (0.01 mM–10 mM) had no effect on wheat root hair growth at pH 6.0, whereas at pH 5.0 IAA showed stimulation. He attributed this positive effect to a counteraction of IAA to H^+ which inhibits root hair growth. Neither was the chemotropism of root hairs affected by IAA (Meesters 1936). In general, auxin is inhibitory to the growth of the root system under ordinary conditions and a promotive effect, if any, should be at very low concentrations (Burström 1957). In fact, elongation of root hairs of *Agrostis alba* is stimulated by a low concentration of 0.1pM IAA and inhibited at high concentrations (0.1 mM – 1 mM) (Jackson 1960). Bates and Lynch (1996) reported that root hair elongation of the *axr2* mutant of *Arabidopsis* is stimulated by low phosphorus availability, and an auxin antagonist, 2-(p-chlorophenoxy)-2-methylpropionic acid inhibits root hair elongation in low phosphorus plants. They have also shown that exogenous IAA stimulates root hair elongation in the wild-type high phosphorus plants. On the other hand, Ekdahl (1957) reported that IAA at a concentration range from 0.1 mM–100 mM, which is relatively high for normal root growth, has a growth promoting effect in wheat root hairs. Since he measured root hair growth several hours after IAA treatment, this promotive effect is probably due to adaptation of root hairs to higher concentrations of IAA. These results suggest that auxin can stimulate root hair growth only at very low concentrations. Therefore, if root hairs require auxin for normal growth, the level of endogenous auxin supplied from the root tissues must also be very low. Pitts et al. (1998) obtained promotion of the root hair growth of *Arabidopsis* with a high concentration of 10 nM–10 mM 2,4-D. However, since they measured the effect 7 days after treatment, the root hairs might have adapted to the high concentration. If they had measured soon after the treatment, the effect could have been

Table 1. Effect of plant hormones on the elongation of root hairs in *Trifolium repens* L.

Plant hormones	Concentrations						
	0	0.1 nM	1 nM	10 nM	100 nM	1 μM	10 μM
2,4-D	313 ±6	327 ±6	320 ±6	368 ±6	551 ±8	520 ±8	410 ±9
GA	304 ±6		309 ±6	339 ±6	324 ±5	298 ±7	329 ±6
Kinetin	319 ±7		324 ±6	306 ±6	369 ±6	472 ±7	421 ±7
ABA	319 ±7		305 ±5	341 ±6	360 ±6	382 ±7	383 ±6
BR	313 ±6		383 ±6	364 ±6	384 ±7	400 ±7	396 ±6

Measurements were made on about 10 mature, randomly selected straight root hairs per seedling three days after treatments, and lengths (μm) of 237-424 root hairs were measured for each hormone concentration. For all measurements, data for each plant were averaged, and these averages were accumulated to achieve a mean of all the data for each treatment. Duncan's statistical test for group differences, P=5%. For 2,4-D and BR significant difference above 0.1nM; for kinetin and ABA above 10nM; for GA_3, no significant difference: ± standard error (Izumo et al. 1995)

inhibitory. We have also observed stimulation of the root hair growth of *Trifolium repens* by 3 days treatment of 10 nM – 10 μM 2,4-D (Table 1).

The mechanism of auxin-induced elongation of root hairs is still unknown. In ordinary cells, one of the major effects of auxin at subcellular levels which leads to cell elongation is cell wall loosening due to the acidification of the cell wall matrix via proton secretion from the cytosol into the apoplast (see Taiz 1984; Cleland 1987; Cosgrove 1986,1993). Though root hairs elongate by tip growth (see Ridge 1993), it is possible that auxin controls root hair elongation through a similar mechanism.

While calcium generally inhibits the growth of plant cells (see Hepler and Wayne 1985), it is an important element for the elongation of root hairs (Tanaka and Woods 1972; Ewens and Leigh 1985; also see Cormack 1962). It has been demonstrated that tip-growing cells like pollen tubes and root hairs require a continuous influx of Ca^{2+} at the growing tip and an internal Ca^{2+} gradient with higher levels at the tip for their continued elongation (Reiss and Herth 1979; Clarkson et al. 1988; Schiefelbein et al. 1992; de Ruijter et al. 1998). Schiefelbein et al. (1992) have demonstrated that root hair growth in *Arabidopsis* is dependent on the Ca^{2+} concentration in the growing medium, and that an extracellular Ca^{2+} gradient exists at the growing tip, suggesting the growing root hair tip as a sink for external Ca^{2+}. Ca^{2+} is thought to have a regulatory role in cytoplasmic streaming (see Seagull 1989), and by cytoplasmic streaming new cell wall materials packed in vesicles are transported to the expanding tip (Ridge 1995a). For cytoplasmic streaming to occur steadily, the presence of a minimum concentration of cytosolic Ca^{2+} seems necessary, though there is no clear correlation between cytoplasmic streaming and Ca^{2+} concentration (Clarkson et al. 1988). Auxin is known to affect cytoplasmic streaming in various types of plant cells including root hairs (see Ayling et al. 1994). In *Avena* root hairs, auxin stimulates cytoplasmic streaming at low concentrations, though it inhibits at high concentrations (Sweeney 1944). A similar effect of IAA has also been shown by Ayling et al. (1994). IAA between 10 nM–100 nM stimulates cytosolic streaming in tomato root hairs within 1-2 min after application, while concentrations of IAA above 1μM are inhibitory. Inhibition of streaming by a high concentration (10mM) of naphthyl acetic acid (NAA), a synthetic auxin, has also been reported in root hairs of *Hydrocharis dubia* (Tominaga et al. 1998). This inhibition was shown to be a result of acidification of the cytoplasm which disturbed the orientation of actin filaments and disrupted cortical microtubules. On the other hand, Tretyn et al. (1991) reported that auxin regulates cytoplasmic Ca^{2+} levels and pH in *Sinapis alba* root hair cells through changes in membrane potential, and suggest that a high concentration (100 nM) of IAA inhibits root hair growth by depolarizing root hairs to diminish Ca^{2+} gradient. However, auxin-induced changes in Ca^{2+} levels and pH is too fast to correlate with auxin-induced stimulation of cytoplasmic streaming (Ayling et al. 1994). It has been reported that cytoplasmic streaming is not indispensable for ordinary cell elongation, because cytochalasin B which inhibited cytoplasmic streaming completely did not entirely inhibit auxin-induced growth of *Avena* and maize coleoptiles (Cande et al. 1973).

It is very likely that auxin is required for root hair growth. At present, however, how auxin regulates root hair growth involving Ca^{2+} and cytoplasmic streaming is still unclear. Since auxin induces ethylene production, some of the results presented so far, especially those by high concentrations of auxin, might reflect the effect of ethylene rather than auxin. Baskin and Williamson (1992) have reported that ethylene increases root hair growth in *Arabidopsis* seedlings.

Recent studies on the molecular biology of *Arabidopsis* have revealed auxin/ethylene related genes associated with root hair growth and development. Many auxin mutants of *Arabidopsis* have shorter root hairs than the wild-type. The *aux1* (Okada and Shimura 1995; Pitts et al. 1998), *axr1* (Lincoln et al. 1990; Pitts et al. 1998), *axr2* and *axr3* (Leyser et al. 1996) genes all affect root hair growth. Although Cernac et al. (1997) have reported that the *axr1* mutants of *Arabidopsis* had reduced numbers of root hairs, Masucci and Schiefelbein (1996) found that the weak *axr1-3* allele had a normal number of root hairs. Pitts et al. (1998) examined this discrepancy in detail, and found that the discrepancy is due to the methods of measurements they employed. Pitts et al. found that the lengths of the root hairs of *axr1-12* and *axr1-7* were clearly shorter than wild-type hairs, while *axr1* mutations had little effect on root hair initiation. In the double mutants of *sar1* and *axr1*-12 the length of root hairs was the same as that of the wild-type (Cernac et al. 1997; Pitts et al. 1998). *SAR1* has been shown to act after *AXR1* in the pathway of auxin response (Cernac et al. 1997).

Pitts et al. (1998) have found that both 2,4-D, and ACC can stimulate root hair elongation of the wild-type, and mutants deficient in auxin response (*aux1*) and in ethylene response (*etr1*) of *Arabidopsis thaliana,* although these mutations have no significant effect on root hair initiation (Masucci and Schiefelbein 1996).

Since the *AUX1* gene product has been suggested to be an auxin flux carrier in *Arabidopsis* roots (Bennett et al. 1996), shorter root hairs of *axr1* mutant might be attributed to a defect in auxin supply to trichoblasts which is indispensable to root hair elongation. On the other hand, the *AXR1* gene product is similar to the ubiquitin-activating enzyme, E1, which is necessary for the activation of Rub1 protein (del Pozo et al. 1998) in *Arabidopsis*. For root hair growth, modification of nuclear proteins in the root hair by Rub is necessary, and auxin probably stimulates root hair elongation via this *AXR1* pathway.

Further analysis in this field will lead to the elucidation of the control mechanism of auxin/ethylene action in these processes.

Whatever the mechanism is, the presence of auxin-binding protein in root hairs of *Zea mays* (Radermacher and Klämbt 1993) might indicate that auxin has some controlling role in root hair growth and development.

3.2 Gibberellin

Stimulation of the growth of pollen tubes by gibberellin (Kato 1955; Chandler 1957; Bose 1958) implies that this hormone can also stimulate root hair growth. In our study using clover root hairs (Izumo et al. 1995), GA_3 had practically no growth stimulating effect in a concentration range from 1 nM to 10 μM (Table 1). However, uniconazole (10 μM), an inhibitor of gibberellin biosynthesis, markedly inhibited root hair growth, suggesting endogenous gibberellin was involved in the normal growth of clover root hairs. In fact, exogenous GA_3 (0.1 μM) could completely overcome the uniconazole-induced inhibition (Fig. 1). These results suggest that normal root hairs are dose saturated with endogenous gibberellin, and that the effective gibberellin concentration is very low. The results are also consistent with the finding of Tanimoto (1987; 1991) obtained by similar experiments using another gibberellin biosynthesis inhibitor, ancymidol, that the root requires a very low level of endogenous gibberellin for normal growth. Gibberellin-induced stimulation of root hair elongation has also been reported in *Pisum sativum* (Pecket 1960) and *Agrostis alba* (Devlin and Brown 1969). It is very likely that gibberellin is also required for normal growth of root hairs.

Gibberellin has been demonstrated to control cortical microtubule orientation, thus affecting the arrangement of cellulose microfibrils (Ishida and Katsumi 1991; 1992; also see Shibaoka 1994). This effect correlates with GA-induced cell elongation (Ishida and Katsumi1992). Since in root hairs microtubules are associated with tip growth (Emons and Wolters-Arts 1983, Emons et al. 1990; Ridge 1995a; Bibikova et al. 1999), GA-involved control of root hair growth might be related to this cytoskeleton protein.

3.3 Cytokinin/Abscisic Acid/Brassinosteroid

Cytokinin generally inhibits root growth (see Goodwin 1978; Evance 1984). However, it stimulates root hair elongation. Abutalybov and Akhundova (1982) reported that the growth of the root hairs of *Trianea bogotensis*, an aquatic plant, was stimulated by kinetin. Our results show that 0.1 μM–10 μM kinetin can

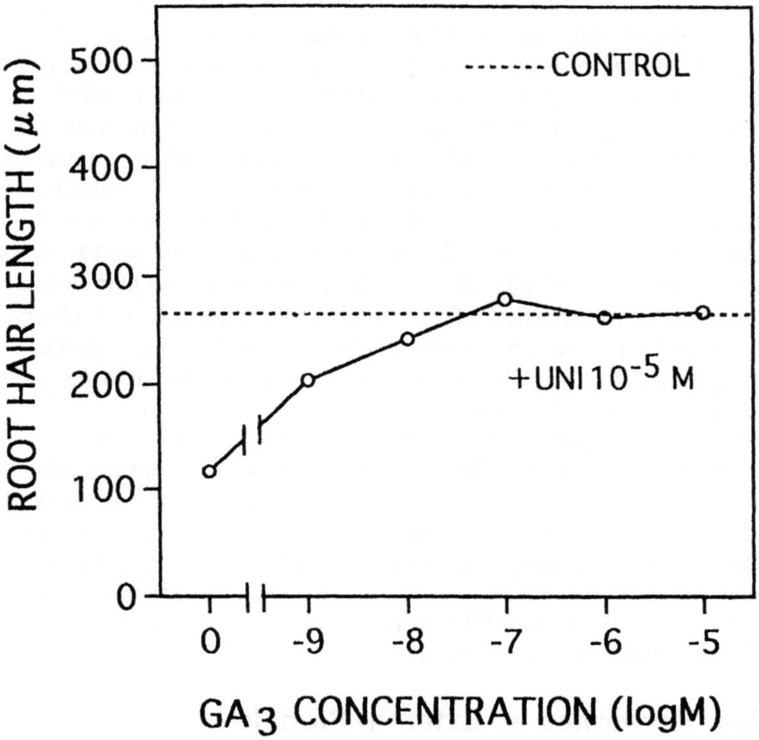

Fig. 1. Recovery by exogenous GA_3 of root hair growth inhibited by 10μM uniconazole Each point represents the mean of 287–511 root hairs, standard errors are too small to show. Measurements made 3 days after treatment

stimulate root hair growth in clover (Table 1). Roots of *Arabidopsis* seedlings are very sensitive to cytokinins. By treatment with 5 nM–10 μM benzyladenine (BA), the growth of primary roots was reduced, while that of root hairs was stimulated (Su and Howell 1992). Su and Howell (1992) have isolated cytokinin response mutants (*ckr1*) of *Arabidopsis* which have longer primary roots and shorter root hairs than wild-type in the absence of BA. The growth of the root hairs of *ckr1* was slightly stimulated by BA and could reach to the same length as that of the control wild-type in the presence of 50–500 nM BA. In normal *Arabidopsis* seedlings, primary roots and root hairs may counter-response to endogenous cytokinins (Su and Howell 1992). It is likely that cytokinin is involved in root hair growth. Since cytokinin can induce ethylene production (see Abeles et al. 1992), some of the cytokinin effects observed might be ascribed to ethylene.

The role of ABA in the growth and development of root hairs has been studied little. Endogenous ABA is known to mediate water stress given to plants (see Davies and Mansfield 1983, Hartung and Davies 1991). The effect is a significant reduction in overall growth. Worrall and Roughley (1976) and Vartanian (1981) reported that drought stress caused existing root hairs to become fragile and thin-walled, and produced short and swollen new root hairs in *Sinapsis alba* and *Trifolium subterraneum*, respectively. This fact suggests that ABA is involved in the formation of deformed root hairs. Schnall and Quatrano (1992) have demonstrated that treatment of *Arabidopsis* with ABA resulted in the formation of similar deformed root hairs. But ABA had no significant effect on the total number of root hairs. They have also shown that ABA insensitive mutants, *abi1* and *abi2* did not respond to exogenous ABA (no deformed root hairs), whereas an ABA sensitive mutant *abi3* displayed a response similar to the wild-type. Our results show that 0.1 μM – 10 μM ABA can slightly stimulate root hair elongation of clover (Table 1). Deformation of root hair morphology was also observed. There is a possibility that ethylene production induced by ABA (see Abeles et al. 1992) stimulated root hair growth. It is unlikely that ABA is required for normal growth and development of root hairs.

Since the growth of pollen tubes is stimulated by brassinolide (Hewitt et al. 1985), it may also stimulate root hair growth. However, there are no published reports on this subject available at present. Our results (Table 1) show that brassinolide can stimulate root hair elongation in clover. However, there was no concentration effect in the range examined from 0.1nM to 1μM. Simulation of stem cell elongation by this hormone occurs at very low concentrations compared with other hormones (Katsumi 1991). In the root system, a concentration effect could be observed below 0.1 nM of brassinolide.

4 Plant Hormones as External Signals

The involvement of plant hormones in the induction of the deformation of root hairs after attachment of *Rhizobium* is still ambiguous. Circumstantial supportive evidence is based on demonstration of the production of plant hormones by *Rhizobium* (see Mascarenhas and Canary 1985) The production of IAA (Badenoch-Jones et al. 1982; Wang et al. 1982; Wheeler et al. 1984; Ernstsen et al. 1987; Atzorn et al. 1988), gibberellins (Katznelson and Cole 1965; Atzorn et al. 1988) and GA-like substances (Williams and Sicardi de Mallorca 1982), and cytokinins (Phillips and Torrey 1970; 1972) have been reported. Badenoch-Jones et al. (1982) have conclusively identified by GC-MS the presence of IAA in the culture filtrate of several strains of *Rhizobium*. The major gibberellins excreted into the culture medium by *Rhizobium phaseoli* are GA_1 and GA_4, but this bacterium does not contribute to the endogenous content of the infected tissue (Atzorn et al. 1988) Mascarenhas and Canary (1985) have suggested that exogenous auxins supplied by neighbouring microorganisms may play a role in the induction of root hair curling. However, IAA alone may not be enough to induce curling of root hairs because IAA-producing strains of *Rhizobium* which attach to the surface of root hairs cannot induce significant curling of root hairs (Badenoch-Jones et al. 1982). It has also been reported that the expression of nodulation genes associated with root hair curling by *R. meliloti* infection requires the presence of root exudates in addition to IAA excreted by this bacterium (Mulligan and Long 1985).

5 Concluding Remarks

Cumulative data in the literature which have been reported until now provide strong evidence that plant hormones are involved in the process of root hair development and elongation. Findings of various auxin/ethylene-associated mutants in *Arabidopsis* and their molecular biological studies have demonstrated that auxin and ethylene in particular are important endogenous factors which control not only the elongation but also the development of root hairs. Cytokinin, gibberellin and brassinolide may also influence endogenously these processes, though evidence is still circumstantial.

The mechanism of the hormonal control of root hair elongation is not clearly understood yet. Whatever it is, it must be different from the mechanism of ordinary cell elongation which depends on cell wall loosening, because root hair elongation proceeds by tip growth.

Root hairs are the target cell of rhizobia, and by infection of these bacteria they show curling. Although it is known that Nod factors produced by these bacteria cause the curling, it is quite likely that plant hormones, especially auxin, also play a joint role in this process. Indeed, we have observed that exogenous auxin and kinetin induce deformation of clover root hairs, including curling (unpublished 1996).

References

Abeles FB, Morgan PW and Salveit ME (1992) Ethylene in plant biology. Academic Press, New York

Abutalybov MG and Akhundova, TS (1982) Cytokinin participation in regulation of potassium ion activity in root epidermal cells. Soviet Plant Physiol. 29: 395-402

Aeschbacher RA, Schiefelbein JW and Benfey PN (1994) The genetic and molecular basis of root development. Ann. Rev. Plant Physiol. Plant Mol. Biol. 45: 25-45

Atzorn R, Crozier A, Wheeler CT, and Sanberg G (1988) Production of gibberellins and indole-3-acetic acid by *Rhizobium phaseoli* in relation to nodulation of *Phaseolus vulgaris* roots. Planta 175:532-538

Ayling SM, Brownlee, C and Clarkson, DT (1994) The cytoplasmic streaming response of tomato root hairs to auxin; observations of cytosolic calcium levels. J. Plant Physiol. 143:184-188

Badenoch-Jones J, Summons RE, Djordevic MA, Shine J, Letham S and Rolfe BG (1982) Mass spectrometeric quantification of indole-3-acetic acid in *Rhizobium* culture supernatants, relation to root hair curling and nodule initiation. Appl. Environ. Microbiol. 44:275-280

Baskin TI, Betzner AS, Hoggart R, Cork A and Williamson RE (1992) Root morphology mutants in *Arabidopsis thaliana.* Aust. J. Plant Phsyiol. 19:427-437

Baskin TI and Williamson R (1992) Ethylene, microtubules and root morphology in wild-type and mutant *Arabidopsis* seedlings. Curr. Topics in Plant Biochem. Physiol. 11:118-130

Bates JR and Lynch JP (1996) Stimulation of root hair elongation in *Arabidopisis thaliana* by low phosphorous availability. Plant Cell Environ. 19:525-538

Bennett MJ, Marchant A, Green HG, May ST, Ward SP, Millner PA, Walker AR, Schulz B and Feldmann KA (1996) *Arabidopsis AUX1* gene: a permease-like regulator of root gravitropism. Science 273:948-950

Bibikova TN, Blancaflor FB and Gilroy S (1999) Microtubules regulate tip growth and orientation in root hairs of *Arabidopsis thaliana.* Plant J. 17:657-665

Bittner A and Buschmann C (1983) Uptake and translocation of K^+, Ca^{2+} and Mg^{2+} by seedlings of *Raphanus sativus* L. treated with kinetin. Z. Pflanzenphysiol. 109:181-189

Borgström G (1939) Theoretical suggestions regarding the ethylene responses of plants and observations on the influence of apple-emanations. Kungl Fisiografiska Sallskapets I Lund Forhandlingar 9:135-174

Bose N (1959) Effect of gibberellin on the growth of pollen tubes. Nature 184:1577

Burström H (1957) Auxin and the mechanism of root growth. Symp. Soc. Exp.Biol. 11:44-62

Cande WZ, Goldsmith MHM. and Ray PM (1973) Polar transport and auxin-induced elongation in the absence of cytoplasmic streaming. Planta 111:279-296

Cao XF, Linstead P, Berger F, Kieber J and Dolan L (1999) Differential ethylene sensitivity of epidermal cells is involved in the establishment of cell pattern in the *Arabidopsis* root. Physiol Plant 106:311-317

Cernac A, Lincoln C, Lammer D and Estelle M (1997) The *SAR1*gene of *Arabidopsis* acts downstream of the *AXR1* gene in auxin response. Development 124:1583-1591

Chandler C (1957) The effect of gibberellic acid on germination and pollen tube growth. Contrib Boyce Thompson Inst 19:215-224

Clarkson DT, Brownlee C and Ayling SM (1988) Cytoplasmic calcium measurments in intact higher plant cells: Results from fluorescence imaging of fura-2. J Cell Sci 91:71-80

Cleland R (1987) The mechanism of wall loosening and cell extension. In: Cosgrove DJ and Knievel DP (Eds) Physiology of cell expansion during plant growth. American Society of Plant Physiology. pp.18-27

Cormack RGH (1962) The development of root hairs in angiosperms II. Bot Rev 28:446-464

Cosgrove DJ (1986) Biophysical control of plant cell growth. Ann Rev Plant Physiol 37:377-405

Cosgrove DJ (1993) How do plant cell walls extend? Plant Physiol 102:1-6

Crossett AN and Campbell DJ (1975) The effects of ethylene in the root environment upon the development of barley. Plant and Soil 42:453-464

Czaja ATh (1935) Wurzelwachustum, Wuchusstoff und die Theorie der Wuchusstoffwirkung. Ber Dtsch Bot Ges 53: 221-245

Davies WJ and Mansfield TA (1983) The role of ABA in drought avoidance. In: Addicott, FT (Ed.) Abscisic acid. Praeger, New York. pp.589-614

de Munk WJ and de Rooy M (1971) The influence of ethylene on the development of 5°C-precooled apeldon tulips during forcing. Hortscience 6:40-41

de Ruijter NCA, Rook MB, Bisseling T, Emons AMC (1998) Lipochito-oligosaccharides re-initiate root hair tip growth in *Vicia sativa* with high calcium and spectrin-like antigen at the tip. Plant J 13:341-350

del Pozo JC, Timte C, Tan S, Callis J and Estelle M (1998) The ubiquitin-related protein RUB1 and auxin response in *Arabidopsis.* Science 280:1760-1763

Devlin RM and Brown DP (1969) Effect of gibberellic acid on the elongation rate of *Agrostis alba* root hairs. Physiol Plant 22:759-763

Dolan L, Duckett C, Grierson C Linstead P, Schneider K, Lawson E, Dean C, Poethig S and Roberts K (1994) Clonal relationships and cell patterning in the root epidermis of *Arabidopsis*. Development 120:2465-2474

Ekdahl L (1957) The effects of indole-3-acetic acid and 2,4-dichlorophenoxyacetic acid on the elongation rate of root hairs and roots of intact wheat seedlings. Physiol. Plant. 10:112-126

Emons AMC and Wolters-Arts A (1983) Cortical microtubules and microfibril deposition in the cell wall of root hairs of *Equisetum hyemale*. Protoplasma 117:68-81

Emons AMC and Wolters-Arts A, Traas J and Derksen J (1990) The effect of colchicine on microtubules and microfibrils in root hairs. Acta Bot. Neerl. 39:19-27

Ernstsen A, Sandberg G, Crozier A, and Wheeler GT (1987) Endogenous indoles and the biosynthesis of indole-3-acetic acid in cultures of *Rhizobium phaseoli*. Planta 177:422-428

Ewens M and Leigh RA (1985) The effect of nutrient solution composition on the length of root hairs of wheat (*Triticum aestivum* L.). J. Exp. Bot. 36:713-724

Farr CH (1927a) Studies on the growth of root hairs in solutions I. The problem, previous work, and procedure. Amer. J. Bot. 14:446-456

Farr CH (1927b) Studies on the growth of root hairs in solutions II. The effects of concentrations of calcium nitrate. Amer. J. Bot. 14:497-515

Farr CH (1927c) Studies on the growth of root hairs in solutions III. The effects of concentrations of $CaCl_2$ and $Ca(OH)_2$. Amer. J. Bot. 14:553-564

Galway M, Masucci J, Lloyd A, Walbot V, Davis R and Schiefelbein J (1994) The *TTG* gene is required to specify epidermal cell fate and cell patterning in the *Arabidopsis* root. Dev. Biol. 166:740-754

Goodwin PB (1978) Phytohormones and growth and development of organs of the vegetative plant. In: Letham DS, Goodwin PB and Higgins TJV (Eds.) Phytohormones and related compounds - A comprehensive treatise, Vol.II, Elsevier/Northolland Biochem Press, Amsterdam, pp.1 – 29

Grierson GS, Roberts K, Feldmann KA and Dolan L (1997) The *COW1* locus of *Arabidopsis* acts after *RHD2*, and in parallel with *RHD3* and *TIP1*, to determine the shape, rate of elongation and number of root hairs produced from each site of hair-formation. Plant Physiol 115:981-990

Hartung W and Davies WJ (1992) Drought-induced changes in physiology and ABA. In: Davies WJ and Jones HG (Eds.) Abscisic acid physiology and biochemistry. BIOS Sci Publ Ltd, Oxford, pp. 63-79

Hepler IB and Wayne RO (1985) Calcium and plant development. Ann Rev Plant 36:397-439

Hewitt FR, Hongh T, O'Neill P, Sasse JM, William EG and Rowan KS (1985) Effect of brassinolide and other growth regulators on the germination and growth of pollen tubes of *Prunus ovium* using a mutiple hanging-drop assay. Aust J Plant Physiol 12:201-211

Hülskamp M, Misera S and Jürgens G (1994) Genetic dissection of trichome cell development in *Arabidopsis*. Cell 76:555-566

Ishida K and Katsumi M (1991) Immunofluorescence microscopical observation of cortical microtubule arrangement as affected by gibberellin in d_5 mutant of *Zea mays* L. Plant Cell Physiol 32:409-417

Ishida K and Katsumi M (1992) Effect of gibberellin and abscisic acid on the cortical microtubule orientation in hypocotyl cells of light-grown cucumber seedlings. Internatl J Plant Sci 153:155-163

Izumo M, Ridge RW and Katsumi M (1995) Hormonal control of root hair growth in clover. In: Abstracts of the 15th International Conference on Plant Growth Substances, Minneapolis, MN, p.452

Jackson WT (1960) Effect of indoleacetic acid on rate of elongation of root hairs of *Agrostis alba* L. Physiol Plant. 13:36-45
Kato Y (1955) Responses of plant cells to gibberellin. Bot Gaz 117:16-24
Katsumi M (1991) Physiological modes of brassinolide action in cucumber hypocotyl growth. In: Cutler HG, Yokota T and Adam G (Eds.) Brassinosteroids - chemistry, bioactivity, and applications. Amer. Chem. Soc., Washington, D.C., pp.246-254
Katznelson H and Cole SE (1965) Production of gibberellin-like substances by bacteria and actinomycetes. Can J Microbiol 11:733-741
Kauss H (1987) Some aspects of calcium-dependent regulation in plant metabolism. Ann.Rev. Plant Physiol 38:47-72
Kieber JJ, Rothenberg M, Roman G, Feldman KA, Ecker JR (1993) *CTR1*, a negative regulator of the ethylene response pathway in *Arabidopsis*, encodes a member of the Raf family of protein kinases. Cell 72: 427-441
Leyser HMO, Pickett FB, Dharmaseri S and Estelle M (1996) Mutations in the *AXR3* gene of *Arabidopsis* result in altered auxin responses including ectopic expression of the *SAUR-ACI* promoter. Plant J 10:403-414
Lincoln C, Britton JH and Estelle M (1990) Growth and development of the *AXR1* mutants of *Arabidopsis*. Plant Cell 2:1071-1080
Lundegårdh H (1946) The growth of root hairs. Ark fur Bot Ser A33: No.5:1-19
Mascarenhas JP and Canary D (1985) Pollen. Symbionts and symbiont induced structure. In: Pharis RP and Reid DM (Eds.) Encycl Plant Physiol New Series, Vol.11 Hormonal Regulation of Development III, Springer-Verlag, Berlin, pp.579-598
Masucci JD and Schieflbein JW (1994) The *rhd6* mutation of *Arabidopsis thaliana* alters root hair initiation through an auxin-and ethylene-associated process. Plant Physiol 106:1335-1346
Masucci JD and Schieflbein JW (1996) Hormones act downstream of *TTG* and *GL2* to promote root hair outgrowth during epidermis development in the *Arabidopsis* root. Plant Cell 8:1505-1517
Meesters A (1936) The influence of hetero-auxin on the growth of root hairs and roots of *Agrostemma githago* L. Pro. Kon. Akad. Wetensch. Amsterdam 39:91-97
Mulligan JL and Long SR (1985) Induction of *Rhizobium meliloti nod* C expression by plant exudate requires *nod* D. Proc Natl Acad Sci USA 28:6609-6613
Okada K and Shimura Y (1995) Modulation of root growth by physical stimuli. In: Cold Spring Harbor laboratories (Eds.) *Arabidopsis*. Cold Spring Harbor Laboratory Press, pp.665-684
Pecket RC (1960) Effects of gibberellic acid on excised pea roots. Nature 185:114-115
Pegg GF (1985) Pathogenic and nonpathogenic microorganisms and and insects. In: Pharis RP and Reid DM Eds., Encycl. Plant Physiol., New Series, *11*: Hormonal regulation of development III, pp.599-624, Springer-Verlag, Berlin
Peterson L and Farquhar ML (1996) Root hairs: Specialized tubular cells extending root surfaces. Bot Rev 62:1-40
Phillips DA and Torrey JG (1970) Cytokinin production by *Rhizobium japonicum*. Physiol Plant 23:1057-1063
Phillips DA and Torrey JG (1972) Studies on cytokinin production by *Rhizobium*. Plant Physiol 49:11-15
Pitts RJ, Cernac A and Estelle M (1998) Auxin and ethylene promote root hair elongation in *Arabidopsis*. Plant J 16: 553-560
Radermacher E and Klämbt D (1993) Auxin-dependent growth and auxin-binding proteins in primary roots and root hairs of corn (*Zea mays* L.). Plant Physiol 141:698-703
Reiss H and Herth W (1979) Calcium gradients in tip growing plant cells visualized by chlorotetracycline fluorescence. Planta 146:615-621

Ridge RW (1993) A model of legume root hair growth and *Rhizobium* infection. Symbiosis 14:359-373.

Ridge RW (1995a) Microvesicles, pyriform vesicles and macro-vesicles associated with the plasma membrane in the root hair of *Vicia hirsuta* after freeze-substitution. J.Plant Res. 108:363-368

Ridge RW (1995b) Recent developments in the cell and molecular biology of root hairs. J.Plant Res.108:399-405

Roman G, Lubarsky B, Kieber JJ, Rothenberg M and Ecker JR (1995) Genetic analysis of ethylene signal transduction in *Arabidopsis thaliana*: Five novel mutant loci integrated into a stress response pathway. Genetics 139:1393-1409

Schiefelbein J, Galway M, Masucci J and Ford S (1993) Pollen tube and root-hair tip growth is disrupted in a mutant of *Arabidopsis thaliana*. Plant Physiol 103:979-985

Schiefelbein JW, Shipley A and Rowse P (1992) Calcium influx at the tip of growing root-hair cells of *Arabidopsis thaliana*. Planta 187:455-459

Schiefelbein JW and Somerveille C (1990) Genetic control of root hair development in *Arabidopsis thaliana*. Plant Cell 2:235-243

Schnall JA and Quatrano RS (1992) Abscisic acid elicits the water-stress response in root hairs of *Arabidopsis thaliana*. Plant Physiol 100:216-218

Schneider K, Mathur J, Boudonck K, Wells B, Dolan L and Roberts K (1998) The *ROOT HAIRLESS 1*gene encodes a nuclear protein required for root hair initiation in *Arabidopsis*. Genes Dev 12: 2013-2021

Seagull RW (1989) The plant cytoskeleton. CRC Crit Rev Plant Sci 8:131-167

Seidel K (1924) Untersuchungen uber das Wachstum und die Reizbarkeit der Wurzelhaare. Jb. Wiss. Bot. 63:501-552

Shibaoka H (1994) Plant hormone-induced changes in the orientation of cortical microtubules: alterations in the cross-linking between microtubules and the plasma membrane. Ann Rev Plant Physiol Plant Mol Biol 45:527-544

Smith KA and Robertson PD (1971) Effect of ethylene on root extension of cereals. Nature 234:148-149

Su W and Howell S (1992) A single genetic locus, *Ckr1*, defines *Arabidopsis* mutants in which root growth is resistant to low concentrations of cytokinin. Plant Physiol 99:1569-1574

Sweeney BM (1944) The effect of auxin on protoplasmic streaming in root hairs of *Avena*. Amer J Bot 31:78-80

Taiz L (1984) Plant cell expansion: regulation of cell wall, mechanical properties. Ann Rev Plant Physiol 35:585-675

Tanaka Y and Woods FW (1972) Root and root hair growth in relation to supply and internal mobility of calcium. Bot. Gaz. 133:29-34

Tanimoto E (1987) Gibberellin-dependent root elongation in *Lactuca sativa*: Recovery from growth retardant-suppressed elongation with thickening by low concentration of GA_3. Plant Cell Physiol. 28:963-973

Tanimoto E (1991) Gibberellin requirement for the normal growth of roots. In: Takahashi N, Phinney BO and MacMillan J (Eds.) Gibberellins. Springer-Verlag, New York, pp.229-240

Tanimoto M, Roberts K and Dolan L (1995) Ethylene is a positive regulator of root hair development in *Arabidopsis thaliana*. Plant J 8:943-948

Tominaga M, Sonobe S and Shimmen T (1998) Mechanism of inhibition of cytoplasmic streaming by auxin in root hair cells of *Hydrocharis*. Plant Cell Physiol 39:1342-1349

Tretyn A, Gottfried W and Felle HH (1991) Signal transduction in *Sinapis alba* root hairs: Auxins as external messengers. J Plant Physiol 139:187-193

Vartanian N (1981) Some aspects of structural and functional modifications induced by drought in root systems. In: Brouwer R, Gasparikova O, Kolek J and Loughman BC (Eds.) Structure and function of plant roots. Maritinus/Nijhoff/Dr. W Junk, The Hague, pp.309-318

von Strugger S (1926) Untersuchungen über den Einfluss der Wasserstoffionen auf das Protoplasma der Wurzelhaare von *Hordeum vulgare* L.(I). Sitz ber. Akad. Wiss. Wien. Abt. 1. 135:453-478

von Strugger S (1928) Untersuchungen über den Einfluss der Wasserstoffionen auf das Protoplasma der Wurzelhaare von *Hordeum vulgare* L.(II). Sitz ber. Akad. Wiss. Wien. Abt. 1. 143-169

Wang TL, Wood EA and Brewin NJ (1982) Growth regulators, *Rhizobium* and nodulation in pea. Indole-3-acetic acid from the cultured medium of nodulating and non-nodulating strains of *R. leguminosarum*. Planta 155:345-349

Wheeler CT, Crozier A and Sandberg G (1984) The biosynthesis of indole-3-acetic acid by *Frankia*. Plant and Soil 78:99-104

Williams PM and Sicardi de Malorca M (1982) Abscisic acid and gibberellin-like substances in roots and root nodules of *Glycine max*. Plant and Soil 65:19-26

Worrall VS and Roughley RJ (1976) The effect of moisture stress on infection of *Trifolium subterraneum* L. by *Rhizobium trifolii* Dong. J Exp Bot 27:1233-1241

Chapter 8
Electrobiology of Root Hairs

Roger R. Lew

York University, Biology Department, 4700 Keele Street, Toronto, Ontario M3J 1P3, Canada

1 Introduction

During growth and development of the plant root, root hairs often play a key role in nodulation, and water and ion uptake. Root hairs are not an obligatory cell type on the root surface. However, when present, they have highly active respiration (Connoly and Berlyn 1996), and are known to preferentially express a number of transport proteins, such as the plasma membrane H^+ ATPase (Samuels et al. 1992), ammonium and nitrate transporters (Lauter et al. 1996), and a phosphate transporter (Daram et al. 1998). In fact, root hair length increases when phosphorus levels are low (Bates and Lynch 1996). The preferential expression of ion transporters and morphological responses to nutrient conditions both suggest that root hairs have special and unique functions in ion transport. In addition, they are an ideal system for examining ion transport mechanisms in situ because they have anatomical and cytological characteristics that simplify the micro-manipulation, impalement, and micro-injection necessary to characterize ion transport. They are readily accessible and microscopic imaging is straightforward because of the lack of interference by surrounding tissue. This even results in an ability to manipulate specific intracellular targets, useful not only in studies of ion transport, but also cellular signal transduction in general.

For example, it is possible to impale the root hair in well defined cellular zones. Figures 1 and 2 show this explicitly. In young growing root hairs, the cytoplasm often extends back to the epidermal part of the cell, although tubular vacuolar structures may extend from the epidermal region into the hair. The nuclei is normally located at the base of the root hair (migration into the root hair occurs in a later developmental phase, when the root hair length exceeds about 100 μm). Cytoplasm-rich regions of the root hair are easily identified, and, impalement with a micropipette into these regions results in the micropipette tip localization in the cytoplasm. This is confirmed by ionophoretic injection of a tracer dye (lucifer

Key words. electrophysiology, membrane potential, resistance, conductance, micromanipulation, microinjection, cell-to-cell coupling, proton (H^+) ATPase, potassium (K^+) channels, anion cotransport, tip growth, calcium, extracellular ionic flux

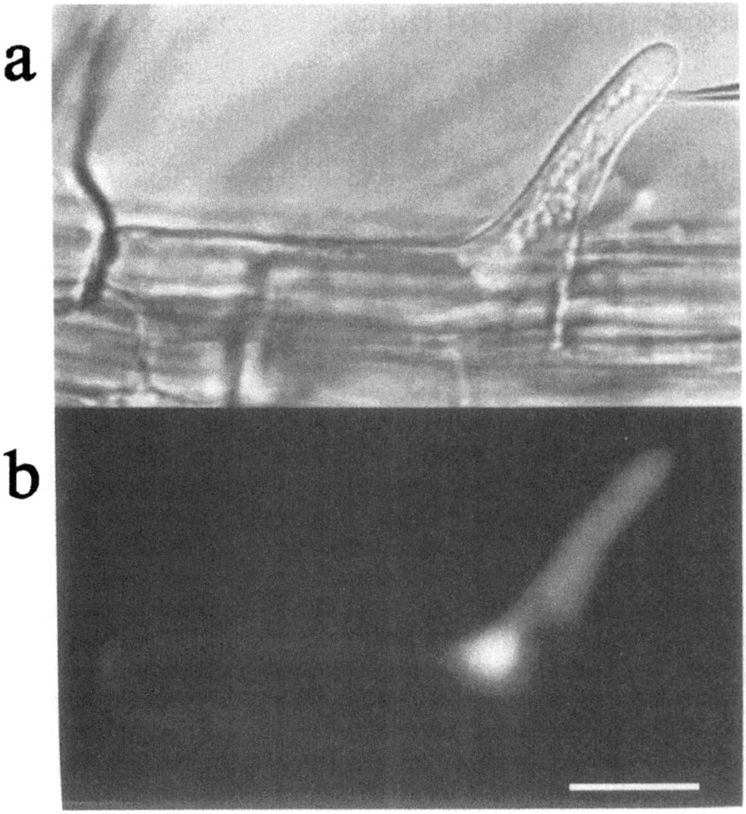

Fig. 1. *a,b*. Deliberate cytoplasmic impalement of a young growing root hair. *a* Bright field image of the impaled root hair. *b* Ionophoretic injection of lucifer yellow was used to demonstrate that the micropipette tip is located in the cytoplasm. Bar = 20 μm

yellow in the examples shown here) which diffuses throughout the cytoplasm (Fig. 1). The use of lucifer yellow as a tracer was purely circumstantial; other fluorescent dyes (including ion-sensitive dyes) could also be used. Interestingly, the lucifer yellow selectively stains the nuclei. In these and other dye injections, the fluorescent dye is retained in the injected cell. It does not diffuse into adjacent cells, even after 15 to 20 minutes. This confirms the observations of Duckett et al. (1994) who reported that dye movement between cells only occurs in developmentally "younger" expanding epidermal cells, prior to the initiation and extension of the root hair component of the cell. However, in developing, growing root hairs, there is significant ion movement between adjacent cells, based on measurements of electrical cell-to-cell coupling (Lew 1994), indicating a clear delineation between dye movement and ion movement through the intercellular plasmodesmata. Deliberate impalements into the vacuole are also feasible (Fig. 2). In other experiments not shown, it was also possible to initially impale the cell in

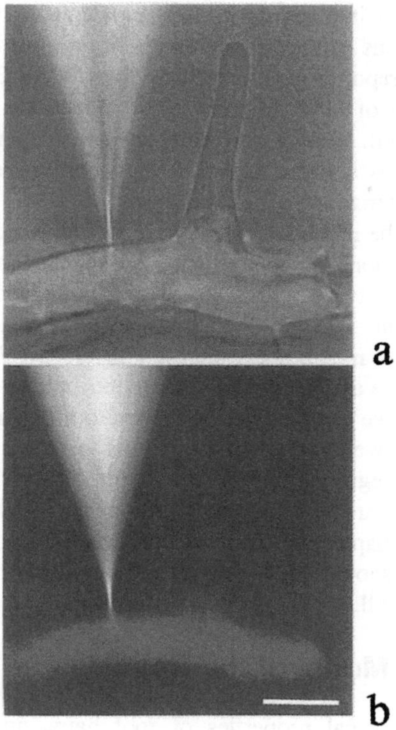

Fig. 2. *a,b*. Deliberate vacuolar impalement of a young growing root hair. *a* Bright field image of the impaled cell. The impalement was into the well-defined vacuole-rich epidermal part of the root hair cell, and was deliberately performed rapidly with intent to "place" the micropipette tip deep inside the cell, in the vacuole. *b* Confirmation of micropipette tip location in the vacuole using ionophoretic injection of lucifer yellow. Although tubular vacuolar structures extend into the root hair part of the cell, they do not appear to be directly connected to the large, central vacuole of the epidermal part of the cell. Bar = 20 μm

the cytoplasm, then move the micropipette tip into the vacuole. Confirmation of tip location is essential: That is, the use of a dye tracer is necessary to unequivocally establish the cellular location of the micropipette tip (Holdaway-Clarke et al. 1996). For example, in some experiments, impalement of the micropipette tip into the vacuole based on bright-field images was in fact cytoplasmic based on dye injection: the fluorescence image showed high fluorescence along the rim of the micropipette tip (suggesting the vacuole membrane had been "wrapped" around the microipette tip), and typical diffuse fluorescence solely in the cytoplasm. In addition to cytoplasmic and vacuolar targetting, one can also expect that impalements specifically targeted for the nucleus should be possible. In addition, because the root hairs can be impaled without causing damage to surrounding cells, wound responses are minimized and the impaled root hair commonly continues growth (Lew 1991; Lew unpublished 1994). This implies that the full consortia of physiological requirements for cellular "health" are maintained during the impalements. One consequence is that single cell transfections are relatively easy: Microinjection of a pBI121 vector containing the GUS reporter gene driven by the CAMV 35S promoter resulted in successful expression of GUS in 8.6% of 70 injections (V. Slowinski and R.R. Lew unpublished 1996). Clearly, root hairs are an excellent single cell system that can be used to address a full range of research objectives of which one area of research is the characterization of ion transport.

Ion transport can be involved in many physiological functions: 1) Maintenance of optimal cellular ion concentrations; 2) Maintenance and modulation of osmolarity, and thus turgor; and 3) Signal transduction controlling cellular responses to external stimuli. In all of these physiological roles, the ion movements across the membrane can result in electrical responses - changes in transmembrane voltage and/or current - that contribute to the electrobiology of the root hair. Some of these changes can be related to the dynamic growing process of the root hair itself as well as the uptake of ions to be transported to other parts of the root and the aboveground parts of the plant. All of the changes rely upon an examination of the electrobiological properties of the root hair.

The rest of this chapter will focus on the electrobiological aspect of root hair research, although it should be clear that root hairs offer many advantages to other areas of research as well.

2 Electrical Model of the Root Hair

To examine the electrical properties of root hairs, an excellent initial tool is electrical network analysis (Horowitz and Hill 1989; Weiss 1996; Fig. 3). This essentially creates an abstract conceptual framework that describes the movement of ions into and out of the cell on the basis of the fundamental electrical properties that underly the movement of electrically charged particles in any system. For root hairs, the plasma membrane, vacuolar membrane and cell-to-cell coupling through plasmodesmata are the three major factors which need to be accounted for. There are four components to the electrical network (Fig. 3): For both membranes, the voltage difference (the membrane potential) and resistance are in parallel with the membrane capacitance. Capacitance is the ability of the cell to hold charge: Q=CV, where Q is the net charge (in coulombs), C is the capacitance (in farads), and V is the voltage difference. Typical values for these three components are -190 mV voltage difference, 20-40 Mohm resistance, and *ca* 1 nanofarad (Lew unpublished 1996). The voltage differences and resistances are related through Ohm's law: V (voltage) = I (current) R (resistance). In root hairs, the plasma membrane H^+ ATPase generates a H^+ current that is the major contributor to the voltage difference between the cytoplasm and the outside of the cell (Felle 1982; Lew 1991), and may in fact be considered an additional fourth component, a constant current source since it exhibits little dependence on voltage at normal membrane potentials.

The other major contributor to the voltage component is the transmembrane gradient of $[K^+]$. The plasma membrane of higher plant cells typically have a relatively high permeability for K^+ compared to other ions. Because the

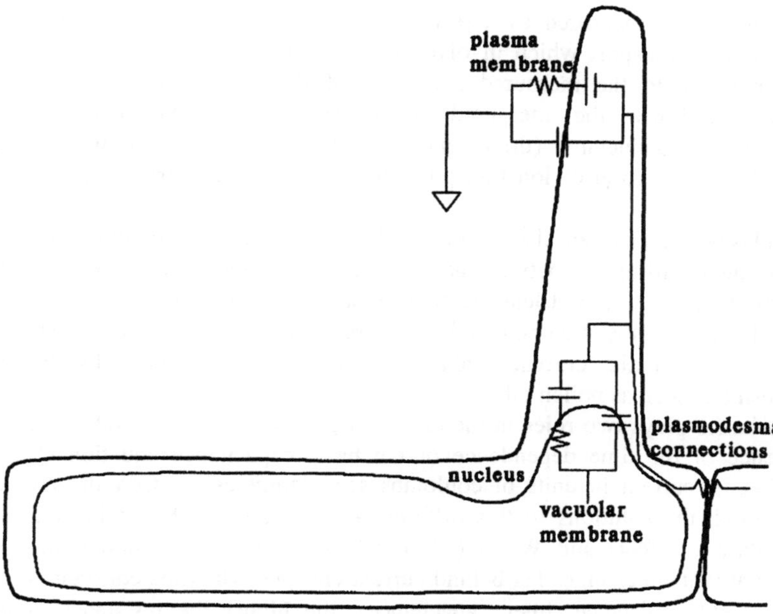

Fig. 3. An electrical model of the root hair. The symbolic nomenclature of electrical circuitry is overlaid on a diagram of a young growing root hair. The major membranes - plasma membrane and vacuolar membrane - are represented as electrical voltages in series with the membrane resistance. The capacitances are shown as parallel components. In root hairs, the major contributions to the electrical voltage at the plasma membrane are two: the "passive" potential due to K^+ and the "active" extrusion of H^+ due to the plasma membrane H^+-ATPase. The relatively high intracellular $[K^+]$ compared to the normally low extracellular $[K^+]$ causes a passive driving force for K^+ which is directed out of the cell. The tendency for the positively charged K^+ to move outward causes an inside negative electrical potential, described by the Nernst potential for K^+: $E_K = 55 \log([K^+]_{extracellular}/[K^+]_{intracellular})$ (mV units). Typical Nernst potentials for K^+ are about -50 to -100 mV. Summed with this "passive" component of the membrane potential is the activity of the H^+-ATPase, pumping positive charges out of the cell, causing a negative inside voltage difference. The parallel capacitance affects the response time of electrical changes, but this is generally much more rapid than typical response times for cellular transduction, and can usually be ignored. For the vacuole, information is not available from root hairs, but it is generally positive inside (about 20 mV) with a resistance similar to or less than the plasma membrane resistance

concentration of K^+ inside the cell is normally higher than the external concentration, there is a natural tendency for K^+ to diffuse out of the cell, driven by its concentration gradient. The outward positive charge movement results in a negative inside potential.

The resistance associated with the plasma membrane (and vacuolar membrane) is a compilation of all electrogenic ion transport acting at the membrane: K^+

channels would be measured as a resistance, as would H^+ coupled tranporters. Essentially, any transport which involved net ion movement across a membrane would contribute to the measured resistance of the membrane: the more ion transport, the lower the measured resistance. Because this tends to be counterintuitive, conductance (the inverse of resistance 1/R) is commonly used to describe the contribution of ion transport: the more ion transport, the higher the conductance.

The plasmodesmata can also be modeled as a resistance, although there may also be a voltage difference between adjacent cells. This additional component is a major source of ionic current leakage between adjacent cells (Lew 1994 1996), so much so that adjacent root hairs, could almost be considered an "ionic syncytium". This source of ionic current leakage would be absent in cells lacking plasmodesmata, such as pollen tubes.

Capacitance plays two roles in the electrical properties of plant cells. One is a consequence of the time dependence of the basic capacitance equation Q=CV: dQ/dt (that is, current in units of coulombs sec^{-1} [amperes]) = CdV/dt. Without going through the solutions to this differential equation (see Horowitz and Hill (1989), Starzak (1984) and Weiss (1996) for thorough presentations), the end result is that voltage changes lag behind current changes. The time constant (τ) for the lag is RC (resistance times capacitance): the higher the resistance and/or capacitance, the slower the voltage responds to changes in current. RC filtering, as this slow response is called, is primarily an important issue in instrumentation. The second consequence of capacitance has to do with the net charge required to develop a voltage difference. Nobel (1991) presents some realistic calculations which show that low μM charge imbalences between the inside and outside of the cell are required to develop normal membrane potentials. Typical ionic fluxes usually result in considerable concentration changes (in the range of mM min^{-1}), but counterion movement must also occur to minimize the net charge imbalence. Otherwise, the potential would exceed the ability of the membrane to withstand the resulting voltage field.

2.1 The Act of Measurement

There are a number of books which describe the methodology of electrophysiological measurements. Thomas (1978) and Purves (1981) detail the construction and use of microelectrodes. Ogden (1994) includes contributed chapters describing a variety of experimental techniques and analysis, while Sakmann and Neher (1995) focus exclusively on the patch clamp technique. Here, I will emphasize issues pertinent to intracellular measurements on root hairs (and most other higher plant cells as well).

When the cell is impaled, if the micropipette tip is located in the vacuole, the voltage difference, resistance and capacitance of both the vacuolar membrane and plasma membrane will be measured in series. This can result in significant errors. The actual error introduced by impalement through the vacuolar membrane was examined by Beilby (1984) in the coenocytic alga *Chara,* where the vacuole had a +20 mV potential difference relative to the cytoplasm and a 4-fold lower resistance compared to the plasma membrane at the normal resting potential of the cell. From an electrical network perspective, the plasma membrane and vacuolar resistances are additive because they are in series ($R_{total} = R_{vacuole} + R_{plasma\ membrane}$). Therefore, measurement errors would be relatively small (ca 1.25-fold overestimate). However, Holdaway-Clarke et al. (1996) found that the vacuolar and plasma membrane resistances were similar in corn suspension cells (which would result in *ca* 2-fold overestimate). The issue of whether impalements are vacuolar or cytoplasmic is a major reason why root hairs are so useful in measuring cellular electrical properties, because of the ease of assuring the impalement location is cytoplasmic, thus avoiding potential errors interpreting results.

However, there is no doubt that the act of impalement may affect the root hair. Immediately after impalement, a transient cessation of cytoplasmic streaming is commonly observed (Lew 1994; Lew unpublished 1992), possibly due to an influx of Ca^{2+} at the impalement site. Over a period of about 5 minutes, a plug of optically dense material, refractile when using phase contrast microscopy, commonly forms in the area surrounding the micropipette tip. One additional potential artefact that would affect electrical measurements is a region of ion leakage at the impalement site. This represents a shunt resistance which would cause underestimation of both the resistance and the membrane potential of the root hair. This is because the shunt resistance (R_{shunt}) is in parallel with the plasma membrane resistance ($R_{plasma\ membrane}$). In electrical networks, for parallel resistances the total resistance (R_{total}) is described by: $1/R_{total} = 1/R_{shunt} + 1/R_{plasma\ membrane}$. So, the presence of a low shunt resistance, equal in magnitude or lower than the plasma membrane resistance would result in a *measured* resistance (R_{total}) 0.5-fold or lower than the *actual* resistance ($R_{plasma\ membrane}$). In addition, because the shunt resistance would cause current leakage out of the cell, the *measured* membrane potential would be lower than the *actual* membrane potential, at least 0.5-fold less for the example given above. This potential problem has led to controversies. In the case of potassium uptake into the cell, Gassmann and Schroeder (1994) pointed out that shunt resistances would have a significant effect upon estimates of the electrophoretic driving force available for high affinity uptake of K^+ via K^+ selective inward ion channels, because of an underestimate of the *actual* membrane potential. Thankfully, shunt resistances are not a potential source of error for electrical measurements of *Arabidopsis thaliana* root hairs. This is best demonstrated by presenting data from dual impalements into a single root hair cell (Fig. 4). During an experiment, after testing the microelectrode resistance, it is first pressed against the cell wall and measures a characteristic negative voltage, normally -30 to -60 mV which is due to the Donnan potential created by negative charges in the cell wall. Upon entry into the cell, the microelectrode registers an almost instantaneous negative voltage, which hyperpolarizes to a more negative stable value over 20-40 seconds. Subsequent impalement into the same cell with a second micropipette causes transient depolarization of the voltage measured by the first microelectrode, followed by a return to the original voltage. If there was a significant shunt resistance at the impalement sites, the second impalement would have caused attenuation of the voltage measured by the first microelectrode. Since

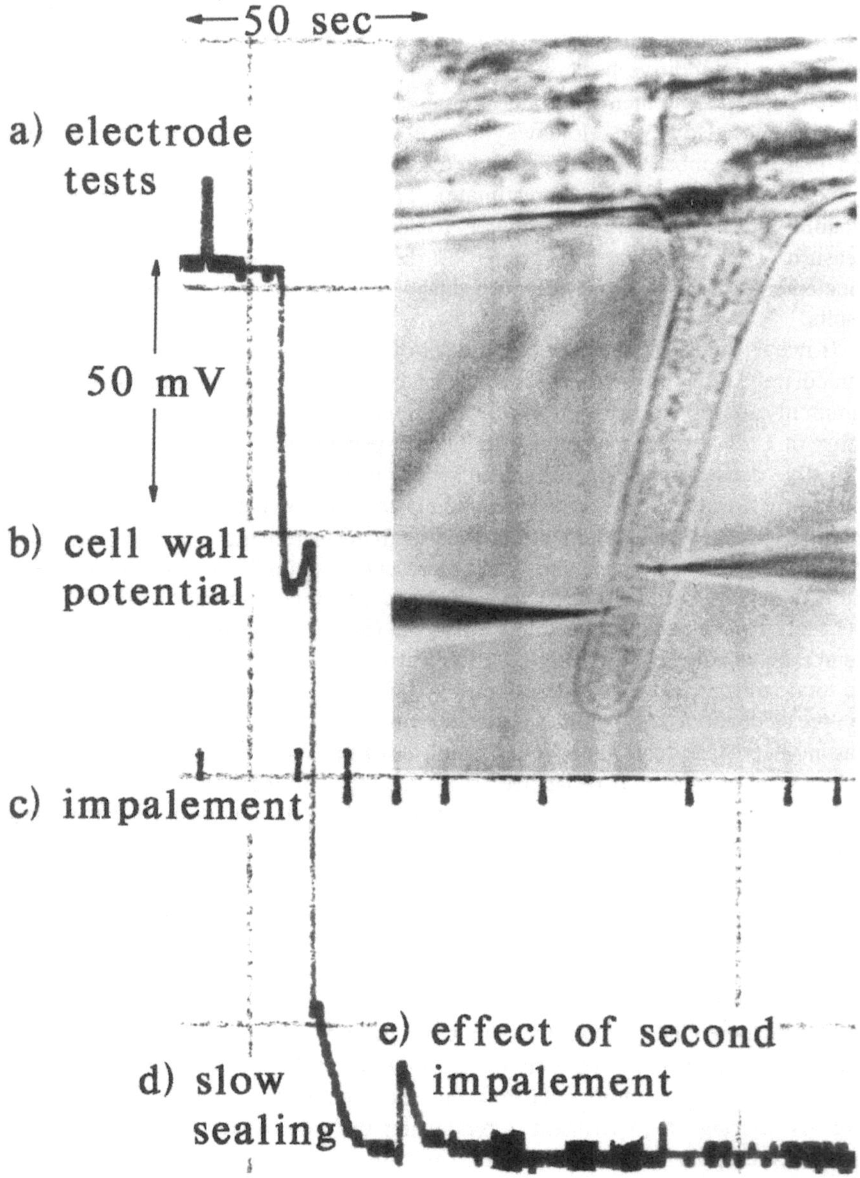
50 sec
a) electrode
tests
50 mV
b) cell wall
potential
c) impalement
d) slow
sealing
e) effect of second
impalement

this did not occur, we can conclude that shunt resistances (R_{shunt}) are not a significant source of error. In fact, it is likely that the slow hyperpolarization of the voltage observed after the initial impalement is due to the formation of a seal between the plasma membrane and the micropipette glass at the impalement site. Dual impalements indicate the lack of shunt resistances around the impalement site, but more importantly, the ability to make multiple impalements into root hairs extends the range of electrical measurements and intracellular manipulations possible in a higher plant cell.

Over the years, most measurements of the electrical properties of higher plant cells relied upon impalement with one micropipette. This effectively limited the types of measurements to only one: voltage. This is because measurements of resistance with a single microelectrode are complicated by the fact that the micropipette itself can have a resistance (10-20 Mohm) similar to typical cell resistances. Since changes in microelectrode resistance after impalement are not measureable, any estimate of membrane resistance is inherently uncertain, leaving voltage as the sole measureable component. One alternative is the use of the discontinuous single-electrode voltage clamp (Forestier et al. 1998), sucessfully used in guard cells. However, the best method to measure the full range of a higher plant cell's electrical properties is to use two microelectrodes. This may require the impalement of multiple micropipettes into a single cell, something that is relatively easy in accessible cells such as root hairs, where dual impalements are easily performed (Fig. 4). Alternatively, double-barrel micropipettes (Blatt and Slayman 1983; Lew 1991) (one of the micropipettes impaling the cell in Fig. 4 is in fact double-barreled) may be used. With two microelectrodes, one may be dedicated to voltage measurements, while the other microelectrode is used to inject current into the cell. The deflection of the voltage

Fig. 4. (*opposite page*). Experimental excerpt showing the act of impalement and effect of multiple impalements. The data is taken directly from a recording in which the root hair was impaled in the cytoplasm with a double-barrel micropipette and a single barrel micropipette (the impalements are shown in the *inset*). The potential measurement is from one of the barrels in the double-barrel micropipette. *a* Initial electrode tests (1 nA current injection) were used to ascertain the miropipette tip resistance and the absence of crosstalk between the two barrels of the double-barrel micropipette. *b* Initially, the micropipette tip is pressed against the root hair wall, and "reports" a potential of about -60 mV, due to negative charges within the cell wall matrix. *c* When the tip penetrates the wall and enters the cytoplasm, it measures an immediate negative-inside potential of about -150 mV. *d* Then, a slow hyperpolarization to about -180 mV occurs; the potential remains stable at this value for considerable periods of time. This slow hyperpolarization probably reflects "sealing" between the micropipette glass and the plasma membrane. *e* Subsequent impalement with a second micropipette causes a only a transient depolarization; after which, the measured potential returns to its orginal, stable value. The absence of a depolarized potential after impalement with the second micropipette indicates that shunt resistances at the sites of impalements are not a significant source of ion leakage from the cell. The experimental excerpt is from a set of experiments exploring the effect of modulating $[Mg^{2+}]_{cytoplasm}$ to assess Mg^{2+} effects on the electrical properties of the plasma membrane (Lew, unpublished 1996) (measured with the first microelectrode) caused by the current injection (through the second microelectrode) can be used to calculate the membrane resistance. Two types of measuring techniques are used: current clamping, in which a known amount of current is injected into the cell and the voltage measured, and voltage clamping, in which the current required to hold the membrane potential at a specified value is measured.

2.2 Cable Properties and Electrical Coupling

Having explored the "act of measurement" in the context of a simple electrical model, it should be clear that root hairs are an excellent system for electrical measurements because of cytoplasm-localized impalements and the lack of shunt resistances at the impalement site. There is one major drawback which must be acknowledged, due to the cable properties of root hairs (a consequence of their long, narrow shape) (Meharg et al. 1994) and the presence of electrical cell-to-cell coupling between adjacent cells (Lew 1994). The effect of both cable and coupling properties is to attenuate current flow at points distant from the point of impalement where the actual measurements are performed (the term cable is traditionally used since it was the laying of the first trans-Atlantic telegraph cables which inspired the initial analysis of current attenuation) (Weiss 1996; Rall 1977). The problem is most severe when normalizing current flow and resistance to cell area. As well, changes in resistance due to some treatment can be attenuated. Meharg et al. (1994) used dual impalements to measure the attenuation of current flow along mature root hairs (about 1000 μm length), and measured a 37% attenuation (1/e) of injected current about 100 μm from the site of current injection. This distance (known as the length constant) is the same in younger root hairs (less than 120 μm length) (Lew 1996). Besides the current attenuation due to the cable-like shape of the root hair, cell-to-cell coupling causes significant current flow into adjacent cells (Lew 1994). There is no simple way to correct for errors introduced by the dual effects of cable properties and cell-to-cell coupling. So current density estimates (in units of amperes per cell area) are inherently uncertain. Nevertheless, the qualitative changes in resistance that can be measured with dual impalements when one uses either two separate micropipettes or double-barrel micropipettes are very important in interpreting the ion transport events that underlie changes in the electrobiological properties of the root hair. As one example, treatment of *Arabidopsis* root hairs with a K^+ channel blocker, tetraethylammonium (TEA), causes the membrane potential to become more negative, and results in a slight increase in the resistance of the membrane (Lew 1991). A more negative potential is commonly interpreted as activation of the plasma membrane H^+ ATPase, but an increase in resistance suggests an alternative: that the K^+ channel blocker is inhibiting an inward flux of K^+. The inward movement of positive charge would normally depolarize the potential; in its absence, the potential hyperpolarizes; the decrease in ion movement across the membrane is reflected by the increase in the membrane resistance.

2.3 Beyond Potential and Resistance

So far, the emphasis has been on the standard electrophysiological techniques normally used to characterize the electrical properties of ion transport across the plasma membrane. But there are other adjunct techniques which are extremely useful. These are described in the following sections.

2.3.1 Ion-Selective Microelectrodes

Besides measurements of voltage, resistance, and cell-to-cell electrical coupling, micropipette impalements offer the ability to measure other important properties of root hairs. For example, ion-selective microelectrodes can be used to probe the intracellular concentrations of various ions (or, more accurately, the *activity* of the ion [Castellan 1971], sometimes called the "free" ion concentration). Research measuring intracellular ion activities in root hairs comes primarily from the laboratory of Hubert Felle (e.g., Felle and Bertl 1986; Felle 1988; Felle et al. 1992; Felle 1994; Felle 1996; Hermann and Felle 1995; Felle and Hepler 1997). The use of ion-selective microelectrodes in higher plant cells is reviewed in Felle (1993). Normally, a short column of ion-sensitive liquid ion exchanger (Ammann 1986; Thomas 1978) is loaded into the micropipette tip. Poly vinyl chloride (PVC) is used to hold the ion exchanger in place; otherwise the cell turgor pressure can easily displace it, resulting in erroneous measurements. Calibrations with solutions containing known concentrations of the measured ion are performed before and after the actual impalements to assure the accuracy of the intracellular ion measurements. In general, ion-selective microelectrodes are probably the most accurate method for determining intracellular ion activities, but they are also technically challenging to fabricate and use. Alternatives, such as ion-sensitive fluorescent dyes (Clarkson et al. 1988; Chapter 9 this volume) offer the ability to "map" spatial gradients of ion activities, but the determination of the actual ion activity requires the use of dissociation concentrations (K_d); not only do published values of K_d's vary by orders of magnitude, but K_d's are quite sensitive to ionic strength, pH, and even temperature. Hence, intracellular ion activities determined with fluorescent dye techniques are inherently inaccurate compared to ion-selective microelectrodes.

Most of the research using ion-selective microelectrode has focussed on H^+ and Ca^{2+}, although intracellular Cl^- concentrations have also been measured in a characterization of Cl^- uptake by root hairs (Felle 1994) and intracellular K^+ was measured in a study of K^+ transport by *Arabidopsis* root cells (Maathuis and Sanders 1993).

2.3.2 Extracellular Measurements of Net Ion Fluxes

Another technique which offers the advantage of being non-invasive is the "vibrating-probe" technique. This method also utilizes ion-sensitive liquid ion exchangers, but the measurements are made outside of the cell. An example of a measurement on a root hair is shown in Fig. 5. The probe is "vibrated" between a point close to the cell and a "reference" position some distance (normally 10-20 μm) away from the root hair, parallel to the surface of the root to avoid spurious signals from the root. The voltages measured at the two positions are a function of the concentration of the ion measured by the ion-selective liquid ion exchanger, and any difference in the concentrations is expected to be due to a gradient of ion concentration generated by fluxes either into or out of the root hair. In addition to being non-invasive, the technique is a very effective way to measure *net* ion fluxes. There are some significant drawbacks: 1) Typically root hairs are small, resulting in small signals (in the range of nanovolts to microvolts) that can be "buried" in background electrical noise (fluorescent lights and static on sweaters are two examples of noise sources); 2) In cells which are growing, the deposition of new cell wall exposes negatively charged groups which would bind positively charged ions entering the cell wall matrix from the bulk solution – the net result is an *apparent* influx of either H^+ or Ca^{2+} (Ryan et al. 1992); and 3) Buffering of the extracellular solution, not just with added buffers but also as a consequence of the buffering capacity of H_2O and dissolved CO_2, may cause underestimates of H^+ fluxes (Arif et al. 1995). Nevertheless, the technique has been sucessfully used to "map" ion fluxes around growing root hairs, especially a calcium influx localized to the growing tip of the hair (Schiefelbein et al. 1992) which may function in the maintenance of an internal, tip-high [Ca^{2+}] apparently required for root hair tip growth (Hermann and Felle 1995; Felle and Hepler 1997).

2.3.3 Measurements of Turgor Pressure

A pressure probe coupled to a micropipette can be used to measure the turgor pressure of the root hair. This method has commonly been used in relatively large cells, such as *Chara* (Zimmerman and Steudle 1974); but has also been used successfully in higher plant root cells (e.g., Zimmerman et al. 1992; Pritchard et al. 1996 used a modified technique to sample contents of single cells), root hairs (Lew 1996) and even cells as small as guard cells (Franks et al. 1995).

3 Ion Transport in Root Hairs

Having described the techniques used to characterize ion transport at the level of a single root hair cell, it should be clear that there are numerous methods that can be used to address various aspects of ion transport. Indeed, there has been considerable research on ion uptake, the role of ions in various cellular processes, such as growth, the regulation of intracellular ion levels, and the regulation of cell-to-cell communication by intracellular ion levels. For ion uptake, a major role must be attributed to the H^+ ATPase, which generates the electrical driving force required for the uptake of charged ions and other nutrients.

3.1 The Plasma Membrane H^+ ATPase

In plants and fungi, the plasma membrane H^+ ATPase (Serrano 1989; Palmgren 1998) generates much of the negative inside electrical potential. The protein has

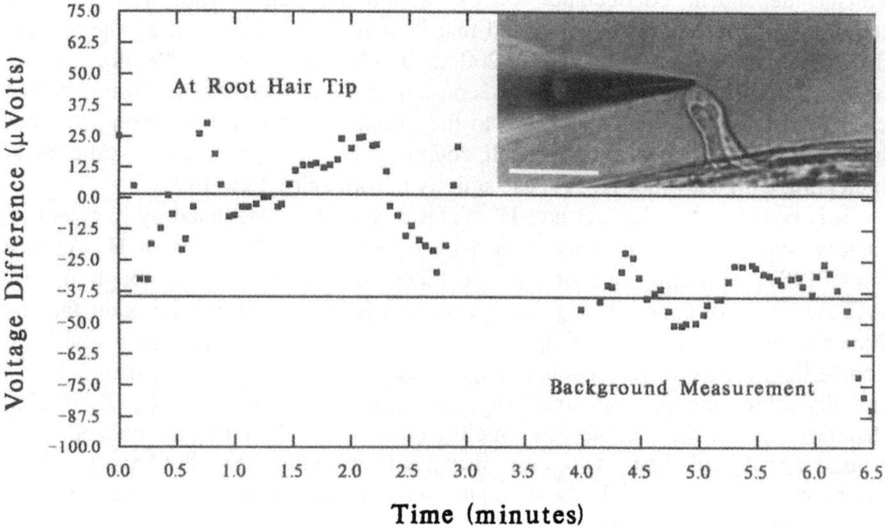

Fig. 5. Experimental excerpt showing the vibrating probe technique. The *inset* shows a microprobe containing a short (about 50 μm) column of a Cl^--sensitive liquid ion exchanger. The probe is first positioned close to the root hair tip, as shown, then moved 10 μm away, parallel to the root surface. The very slight differences in [Cl^-] between the two positions are recorded as a μVolt difference by the Cl^--sensitive liquid ion exchanger, which can subsequently be converted into a [Cl^-] difference. Multiple measurements must be made to account for the high noise level. The probe has a very high electrical resistance, which makes it act like an antenna, picking up natural background electrical noise (for example, from fluorescent lights, or even static on a sweater), an especial problem when μVolts are being measured. A background measurement is made some distance away from the root hair (and the root). The difference between the two measurements, at the root hair and background, is due to the [Cl^-] differences near the root hair. In this case, the [Cl^-] near the root hair is lower than 10 μm away, indicating that the root hair is taking up Cl^-. This experimental excerpt comes from a characterization of osmotic effects on H^+, Ca^{2+}, K^+, and Cl^- transport during root hair growth (Lew 1998)

been cloned from a variety of sources, and is known to exist as multiple isoforms in higher plants, which are often expressed in specific tissues (Michelet and Boutry 1995). It would be difficult to undertake biochemical characterization of root hair specific H^+ ATPase because of the difficulty of isolating sufficient protein. However, it is clearly present on the basis of the effect of inhibitors on the electrical potential of the root hair (Lew 1991). A relatively specific inhibitor, vanadate, depolarizes the potential about 100 mV from an initial potential of -170 mV, similar in magnitude to the effect of cyanide which would inhibit the pump by depleting intracellular ATP (Lew and Spanswick 1984). A comparison of the voltage dependent conductance of the membrane before and after vanadate inhibition of the H^+ ATPase revealed that its activity is unaffected by the voltage difference, in the range of about -200 to 0 mV (Lew 1991). Therefore, under normal circumstances, it exhibits a "constant current" activity. Estimates of the current density are problematic due to the problems discussed regarding the root hair cable properties and cell-to-cell coupling, but it appears to be quite high, indicating a high level of H^+ ATPase activity in root hairs (Lew 1991).

It is possible that the root hair H^+ ATPase activity is regulated by intracellular factors, such as calcium (known to activate the *Neurospora crassa* H^+ ATPase [Lew 1989]), protons, one of the substrates for the enzyme (although the H^+ ATPase does not appear to regulate cytosolic pH [Felle 1996]), and other factors. Experiments ionophoresing either Ca^{2+} or H^+ ions into the root hair revealed no compelling evidence for regulation by either ion. Increases in cytosolic [Ca^{2+}] initially inhibit cytoplasmic streaming, then at higher levels cause the potential to depolarize and inhibit cell-to-cell coupling (Lew 1994). Increases in cytosolic [H^+] caused cytological changes - a rapid increase in vacuolar area - but have no effect on the membrane potential (Lew, unpublished, 1993) and minimal effects on cell-to-cell coupling (Lew 1994).

Hormone effects on the electrical properties of root hairs have been reported which may be due to modulation of H^+ ATPase activity: both auxin (Tretyn et al. 1991; Ayling et al. 1994) and cytokinin (Silverman et al. 1998) cause changes in the root hair membrane potential For cytokinin, the effect - a shift to a more negative potential, consistent with activation of the H^+ ATPase - was not observed with injection of the hormone into the root hair, only when it was applied exogenously (Silverman et al. 1998). Interestingly, cytokinin also stimulates root hair growth (Silverman et al. 1998). For auxins, the responses are equivocal. Either depolarization (Ayling et al. 1994; Tretyn et al. 1991) or hyperpolarization of the membrane potential was observed which depended upon the exogenous concentration of auxin (Tretyn et al. 1991). In addition to an auxin-induced increase in cytoplasmic streaming rates in root hairs (Ayling and Butler 1993; Ayling et al. 1994), auxin is implicated as part of the transduction pathway regulating root hair initiation (Masucci and Schiefelbein 1994).

The negative inside electrical potential generated by the H^+ ATPase is part of a proton motive force (pmf), which includes both the potential and pH differences between the inside of the cell and the external environment: pmf (mV) = ΔVoltage +55 ΔpH. Either the potential and/or the pH difference can be used to drive the uptake of other ions into the cell, particularly K^+ and Cl^-.

3.2 K^+ Transport

Generally, K^+ is an essential nutrient for higher plants which is often accumulated to relatively high levels from the soil solution. The recent progress in characterizing K^+ transport genes is described by Fox and Guerinot (1998). Most ongoing research uses functional complementation of yeast potassium transport mutants to screen cDNA libraries and/or confirm a role in potassium transport for genes cloned by other methods. Putative K^+ transporters are being cloned at a remarkable pace, but it will be a while before the dust completely settles, revealing their physiological significance and transport properties. K^+ uptake by root cells apparently relies upon multiple mechanisms that depend upon the external concentrations of K^+. One mechanism is uptake through a K^+ channel, normally described as an inward rectifier because it is activated by negative voltages and would function to carry K^+ into the cell (Maathuis and Sanders 1995). Such a K^+ inward rectifying channel was directly characterized using patch clamp techniques on individual wheat root hair protoplasts (Gassmann and Schroeder 1994). One cloned gene that may be responsible for the inward channel is the AKT1 K^+ channel. It is an inward rectifier which is preferentially expressed in mature regions of roots in peripheral cell layers, although it is not induced by potassium deprivation (Legarde et al. 1996). An *Arabidopsis akt1* mutant exhibits reduced potassium (^{86}Rb) uptake, poor growth under potassium deprivation and apparent absence of the inward K^+ channel (Hirsch et al. 1998). For an inward K^+ channel, uptake relies solely upon the electrophoretic force generated by the negative inside potential of the cell. Especially at low external concentrations of $[K^+]$, there may be a need for a different transport mechanism using a co-transported ion (Maathuis and Sanders 1993), such as a H^+/K^+ symport (characterized in *Neurospora*: Rodriguez-Navarro et al. 1986; Blatt et al. 1987), which would increase the driving force and thus affinity for K^+ uptake. One candidate symport is a putative Na^+/K^+ symport, HKT1, isolated using functional complementation of *Saccharomyces cerevisiae* mutants deficient in high affinity K^+ uptake, and characterized in both *Saccharomyces* and *Xenopus laevis* oocytes (Gassmann et al. 1996). The gene is rapidly "up-regulated" after potassium depletion (Wang T-B et al. 1998), but Na^+ as the co-transported ion is unexpected. From the viewpoint of ion transport mechanisms, it is clear that H^+ and Na^+ ions share many physical similarities. At face value, this may seem a strange statement, but the most likely proton species to function in transport is the hydronium ion H_3O^+ (Boyer 1988) which would be much more similar to Na^+ then the "naked" hydrogen cation, H^+. In fact, Na^+ replaces H^+ as the transported ion in some bacterial F_1F_0 ATP synthetases (Dimroth 1995). At least one glucose co-transporter, the mammalian SGLT1, can use either H^+ or Na^+ ions to "drive" glucose uptake, although there is a higher affinity for glucose when the cotransported ion is Na^+ (Hirayama et al. 1994). This mechanistic complexity, plus potential modifications of ion transport regulation in heterologous expression systems, could easily affect the apparent transport mechanisms of HKT1. Hopefully, further research will resolve its physiological role.

Even in a single species, regulation of K^+ transport activity can obscure interpretation. For example, a mutation in the yeast H^+ ATPase (*pma1-105*) causes altered K^+ channel activation by cytosolic ATP (Ramirez et al. 1989), implying a linkage between the H^+ ATPase, ATP usage and channel activity. Such an "energy charge" linkage is known to exist in renal proximal tubules between the Na^+-K^+ ATPase and K^+ channel activity (Tsuchiya et al. 1992). Attempts to delineate the interrelationship between K^+ uptake activity and the H^+ ATPase in higher plants have unfortunately been few, the results so far indicate no direct biochemical linkage (Briskin and Gawienowski 1996), although a regulatory linkage between the H^+ ATPase and K^+ channel activity cannot be discounted.

One of the things that makes root hairs so special as a model system for examining issues related to transport of K^+ and other osmotically active ions is that they can be examined in the context of cellular growth. *Arabidopsis* root hair growth rates are normally about 1 μm min^{-1}. This corresponds to a volume increase of about 50 fL min^{-1} for a hair with a diameter of 8 μm (Lew 1991). Because the root hairs are expected to maintain their turgor during growth, the increase in volume requires uptake/production of osmotically active substances. For a cellular osmolarity of 500 mosmol, the rate of uptake/production would have to be about 25 fmol min^{-1}. One of the osmotically active substances taken up by the cell is expected to be K^+. In support of this, inhibition of the inward K^+ channel with TEA (tetraethylammonium), a "standard" K^+ channel inhibitor (Hille 1984), inhibits an inward K^+ current and *Arabidopsis* root hair growth (Lew 1991). The inhibition of growth is transient (growth resumes after about 4 minutes), so K^+ uptake cannot be considered an obligatory mechanism for maintaining the internal osmolarity at a level sufficient to "drive" cellular expansion. Using the vibrating probe technique, a net K^+ influx was found in growing *Limnobium stoloniferum* root hairs (Jones et al. 1995). *Limnobium* root hairs do have TEA-sensitive inward (and outward) K^+ currents, presumably due to K^+ channels (Grabov and Bottger 1994; described in detail below). The K^+ influx continued after inhibition of growth by 20 μM Al^{3+}. Aluminum does inhibit inward K^+ channels of wheat root hairs (half-maximal at about 8 μM; Gassmann and Schroeder 1994) at the same concentration inhibiting *Limnobium* root hair growth (half-maximal at about 7 μM; Jones et al. 1995). This leads to an apparent discrepancy: Al^{3+} levels sufficient to inhibit an inward K^+ channel measured on root hair protoplasts with the patch clamp technique do not affect K^+ influx in intact root hairs, even though growth inhibition is observed. Whether or not multiple K^+ influx mechanisms and/or differential sensitivity are the explanation for the discrepancy, the results do imply incomplete coupling between K^+ influx and cellular expansion (as well as multiple aluminum effects [cf Jones et al. 1998; Sivaguru et al. 1999]).

The electrical properties of *Limnobium* root hairs are easily measured because of their large size (cf Ullrich and Novacky 1990). This also makes it possible to examine them with a technique that is not commonly used in higher plant electrophysiology: the sucrose gap technique (Grabov and Bottger 1994). With this method, voltage dependence of ionic currents in the root hairs can be measured by "clamping" the voltage to specified levels. As noted above, Grabov and Bottger (1994) identified TEA-sensitive inward and outward K^+ channels which were active at potentials more negative and more postive than the normal membrane potential, respectively. Modification of the redox potential of the root hairs by adding the electron acceptor hexacyanoferrate III to the external solution activated the inward K^+ channels and inhibited outward K^+ channels. The mechanism modulating channel activity and its physiological role are not known, but may be related to a redox system located on the plasma membrane that mediates iron uptake (Moog et al. 1995; Robinson et al. 1999). Interestingly, external redox modification (with reductive agents) is known to control root hair morphogenesis (Sanchez-Fernandez et al. 1997).

Studies of K^+ transport extend well beyond a "simple" single cell system like a root hair, but even in a root hair, K^+ transport is clearly quite complex. Some of the complexities related to K^+ transport are probably due to variable K^+ supply and utilization rates during plant growth. The inward K^+ channel is the best characterized K^+ transporter, but it is highly likely that multiple K^+ transporters – including cotransporters and possibly K^+ ATPases – are present; their activity may vary depending upon intracellular and extracellular $[K^+]$ levels (cf Wang T-B et al. 1998), the energy status of the cell, and redox poise (Grabov and Bottger 1994). While they may not be an obligatory component of cellular expansion, K^+ transporters can contribute to maintenance of intracellular osmolarity.

Beside K^+, another osmotically active ion which is actively accumulated by plant cells is Cl^-.

3.3 Cl^- Transport

Growing *Arabidopsis* root hairs normally exhibit Cl^- uptake as measured using the extracellular vibrating probe (Lew 1998). The mechanism of Cl^- uptake into the root hair (or other plant cell) cannot be passive, say, through a Cl^- channel, because the negative inside voltage would electrophoretically "expel" the chloride anion from the cell. The best evidence elucidating the mechanism of Cl^- uptake by plant cells has been obtained using root hairs. The experiments involved the use of ion-selective microelectrodes in conjunction with membrane potential measurements: Ullrich and Novacky (1990) monitored the potential and both intracellular and extracellular pH during Cl^- uptake using *Limnobium* root hairs. After the addition of Cl^- extracellularly, they observed a depolarization of the membrane potential, 60-100 mV more positive than the normal membrane potential, concomitant with cytoplasmic acidification and alkalinization of the extracellular solution. These results were consistent with a nH^+/ Cl^- symport activity, actively accumulating Cl^- along with H^+ inside the cell. The stoichiometry could not be assessed, but more than 1 H^+ per Cl^- would be necessary to account for the observed depolarization of the membrane potential. Using *Sinapis alba* root hairs, Felle (1994) extended these results by utilizing both H^+- and Cl^--sensitive microelectrodes in conjunction with membrane potential and resistance measurements. Thus, the intracellular $[Cl^-]$ could be directly monitored during the presentation of extracellular Cl^-, explicitly demonstrating its uptake. The evidence supported the presence of a nH^+/ Cl^- symport, but interestingly, Felle (1994) found that a shift to an acid extracellular pH (from 9.5 to 4.5) was sufficient to increase intracellular $[Cl^-]$ about 2.5 fold, even though the membrane potential depolarizes under these conditions. This indicates a kinetic dependence of symport activity on the ΔpH component of the proton motive force, separate from the ΔVoltage component.

3.4 NH_4^+, NO_3^-, and Pi Transport

There are a variety of other ions which must be actively imported by the cell. For example, ammonium, nitrate, phosphate, sulphate. For some of these, putative transporters have been cloned. But in general, there is only limited knowledge regarding their mechanisms of uptake.

Ammonium uptake must involve multiple mechanisms which would depend upon pH and hence the relative contribution of ammonia (NH_3) and ammonium (NH_4^+) to uptake into the cell. In barley and tomato root hairs, the addition of NH_4^+ extracellularly cause a depolarization of the membrane potential consistent with positive charge movement into the hair (Ayling 1993). Because K^+ channels do exhibit some permeability to NH_4^+, the NH_4^+ uptake into the cell could utilize a K^+ inward rectifier channel. However, an ammonium transporter (LeAMT1) has been cloned from a tomato root hair specific cDNA library (Lauter et al. 1996). Its expression was root hair specific and constitutive. Whether this transporter is responsible for the observed electrical changes remains to be seen.

In their experiments on *Limnobium* root hairs, Ullrich and Novacky (1990) examined the effect of NO_3^- on the membrane potential and cytoplasmic pH. Nitrate addition caused a depolarization of the membrane potential, but, unexpectedly, it induced an alkalinization of both the cytoplasm and extracellular solution. This is not completely consistent with a nH^+/NO_3^- symporter, but the possibility of almost immediate cytoplasmic reduction of NO_3^-, consuming imported H^+, could explain the discrepancy. A more complete analysis of the electrical properties of NO_3^- transport into *Arabidopsis* root hairs by Meharg and Blatt (1995) used voltage clamping to assess the voltage dependence of the ionic currents induced by NO_3^- transport. They observed larger NO_3^--induced currents at voltages more negative than the normal membrane potential and at acid extracellular pH, supporting the presence of a voltage dependent nH^+/NO_3^- symporter. This nH^+/NO_3^- symport activity is probably due to recently cloned nitrate transporters (c.f. Wang and Crawford 1996; Wang R et al. 1998; Lauter et al. 1996), some preferentially expressed in root hairs (Lauter et al. 1996), since mutations in both the *Arabidopsis* CHL1 (NRT1) and NRT2 nitrate transporters minimize membrane potential changes observed after additions of low levels of NO_3^- (Wang and Crawford 1996; Wang R et al. 1998).

As is true for other anions, phosphate appears to be taken up via a nH^+/Pi symporter (Ullrich and Novacky 1990). Some of the cloned putative phosphate transporters do appear to be expressed in root epidermal cells, including root hairs (Daram et al. 1998; Liu et al. 1998).

4 Ion Transport and Root Hair Morphogenesis

Because of their characteristic tip growth habit, root hairs offer the ability to explore potential roles that ion transport may have during root hair initiation and elongation. Besides the role(s) that K^+ and Cl- may play as osmotically active agents maintaining intracellular osmolarity during growth, ion transport may also play a role in signal transduction. Modulation of hormones and/or intracellular components of signal transduction may regulate ion transport as part of their regulation of morphogenesis. Alternatively, ion transport itself may modify the concentrations of ions that act as second messengers, especially Ca^{2+}.

4.1 Ca^{2+} Transport and Root Hair Morphogenesis

Ca^{2+} is very important to the physiological function of plant cells, primarily because of its role as a second messenger. The importance of Ca^{2+} in signal transduction will be detailed elsewhere in this book (Chapter 9). However, one aspect of Ca^{2+} transport will be explored here, potential transport mechanisms associated with root hair morphogenesis.

Nutrient status, both phosphate nutrition and redox potential, affects root hair growth, but Ca^{2+} itself probably plays a key role in regulating the actual morphology of the root hair. There is a tip-high Ca^{2+} gradient within the root hair cytoplasm (Felle and Hepler 1997; Wymer et al. 1997) which appears to control the polarity and direction of tip growth (Bibikova et al. 1997). It may be maintained by tip-localized Ca^{2+} transport into the tip region during growth of the root hair. A number of researchers have documented a tip-localized Ca^{2+} influx in growing root hairs using the vibrating probe (Schiefelbein et al. 1992; Hermann and Felle 1995; Felle and Hepler 1997; Lew unpublished 1997)(although it may be due to Ca^{2+} binding to newly deposited cell wall). Evidence that suggests a possible transport mechanism to account for Ca^{2+} uptake at the root hair tip during growth comes from patch clamp measurements of another tip-growing organism, the oomycete *Saprolegnia ferax* (Garrill et al. 1992; Garrill et al. 1993; Levina et al. 1994; Lew 1998a). *Saprolegnia* has stretch-activated channels permeable to Ca^{2+} which are located solely at the growing hyphal tip. Inhibiting them with an inhibitor of stretch-activated channels, gadolinium, inhibits growth and dissipates a tip-high cytosolic Ca^{2+} gradient. The stretch-activated channels appear to function as a "growth-sensor": as the tip is stretched during growth, the channels open, letting Ca^{2+} into the tip, signalling for continued deposition at the tip of the materials necessary for continued growth. In root hairs, a similar scenario is certainly possible, especially because of the tip-specific Ca^{2+} influx commonly observed during growth. To examine this possibility in situ, a pressure probe was used to modify the root hair turgor. At increased internal pressure, the cell wall will "stretch" (particularly at the root hair tip which is relatively extensible). If channels are activated, the Ca^{2+} movement into the cell would result in the membrane potential becoming more positive. The actual result of such experiments was that direct pressure modulation had no effect on the potential, resistance, or cell-to-cell coupling, indicating that stretch-activated channels are not activated by direct pressure-induced tip swelling (Lew 1996), and may not regulate the tip growth process. This is supported by experiments using gadolinium: Its addition has no effect on the membrane potential, resistance, or growth rate of *Arabidopsis* root hairs (Lew unpublished 1993). Alternative sources of the tip-high Ca^{2+} gradient in root hairs could be tip-localized electroneutral transporters or release from some internal store.

4.2 Abscisic Acid, Phosphatases and Ion Transport

During water stress, abscisic acid is synthesized by roots (Griffiths et al. 1996). Both water stress and abscisic acid application induce short and swollen root hairs in *Arabidopsis* (Schnall and Quatrano 1992). Because abscisic acid induces stomatal closure (Anderson et al. 1994) and activates an outward K^+ current in guard cells (Blatt 1990), it was possible that abscisic acid would also affect the electrical properties of root hairs. However, application of 100 μM abscisic acid caused no change, or in 2 of 5 experiments a slight hyperpolarization (12 mV) of the membrane potential, and there was no indication of either activated or inactivated currents at voltages either more negative or positive than the normal membrane potential (Lew unpublished 1994). So, its immediate effects (within 50 s) on the electrical properties of the root hair are nil, in contrast to the dramatic effects observed in intact guard cells (Blatt 1990). It should be emphasized that even though abscisic acid is without effect, water stress, in the form of mannitol/sorbitol or various salts (NaCl, KCl, $MgCl_2$) does have significant effects on the electrical properties of growing root hairs. The non-charged osmotica cause a hyperpolarization of the membrane potential and an increase in membrane resistance (Lew 1996), while KCl and NaCl depolarize the potential without affecting the membrane resistance (Lew 1998), as does $MgCl_2$ and $CaCl_2$ (Lew unpublished 1999). K^+ fluxes measured with a vibrating probe do not respond to water stress in a manner consistent with a role in regulating intracellular osmolarity (Lew 1998).

Another morphogen is okadaic acid, an inhibitor of serine/threonine protein phosphatases, which effectively inhibits root hair elongation (Smith et al. 1994). Unfortunately, as was the case with abscisic acid, application of okadaic acid at 1 μM had no effect on the electrical properties of the root hair (or cytoplasmic streaming) for periods of time up to 300-400 seconds, although, the long-term effect (no root hairs) was observed after 3-4 days (Lew, unpublished 1995).

5 Concluding Remarks

As noted in the introduction, root hairs offer many advantages for intracellular manipulation and electrical characterization of ion transport. In many instances, the experimental sophistication possible using root hairs has not been approached in other cellular types in higher plants (with the exception of guard cells). Hopefully, the ability to use the root hair as a single cell "test tube", as indicated by the research described here and elsewhere in this book will encourage researchers to consider using root hairs as an experimental system of choice.

References

Ammann D (1986) Ion-selective microelectrodes. Principles, design and application. Springer-Verlag, Berlin, New York. pp 1-346

Anderson BE, Ward JM, Schroeder JI (1994) Evidence for an extracellular reception site for abscisic acid in *Commelina* guard cells. Plant Physiol 104:1177-1183

Arif I, Newman IA, Keenlyside N (1995) Proton flux measurements from tissues in buffered solution. Plant Cell Environ 18:1319-1324

Ayling SM (1993) The effect of ammonium ions on membrane potential and anion flux in roots of barley and tomato. Plant Cell Environ 16:297-303

Ayling SM, Butler RC (1993) Time-series analysis of measurements on living cells illustrated by analysis of particle movement in the cytoplasm of tomato root hairs. Protoplasma 172:124-131

Ayling SM, Brownlee C, Clarkson DT (1994) The cytoplasmic streaming response of tomato root hairs to auxin: Observations of cytosolic calcium levels. J Plant Physiol 143:184-188

Bates TR, Lynch JP (1996) Stimulation of root hair elongation in *Arabidopsis thaliana* by low phosphorus availability. Plant Cell Environ 19:529-538

Beilby MJ (1984) Current-voltage characteristics of the proton pump at *Chara* plasmalemma: I. pH dependence. J Memb Biol 81:113-125

Bibikova TN, Zhigilei A, Gilroy S (1997) Root hair growth in *Arabidopsis thaliana* is directed by calcium and an endogenous polarity. Planta 203:495-505

Blatt MR (1990) Potassium channel currents in intact stomatal guard cells: Rapid enhancement by abscisic acid. Planta 180:445-455

Blatt MR, Slayman CL (1983) KCl leakage from microelectrodes and its impact on the membrane parameters of a nonexcitable cell. J Memb Biol 72:223-234

Blatt MR, Rodriguez-Navarro A, Slayman CL (1987) Potassium-proton symport in *Neurospora*: Kinetic control by pH and membrane potential. J Membrane Biol 98:169-189

Boyer PD (1988) Bioenergetic coupling to proton motive force: Should we be considering hydronium ion coordination and not group protonation? Trends Biochem Science 13:5-7

Briskin DP, Gawienowski MC (1996) Role of the plasma membrane H^+-ATPase in K^+ transport. Plant Physiol 111: 1199-1207

Castellan GW (1971) Physical Chemistry. Addison-Wesley. Reading, MA. pp 357-381

Clarkson DT, Brownlee C, Ayling SM (1988) Cytoplasmic calcium measurements in intact higher plant cells: Results from fluorescent ratio imaging of fura-2. J Cell Science 91:71-80

Connolly JH, Berlyn GP (1996) Cytochemical assay for differential respiratory activity in roots and root hairs. Biotech Histochem 71:197-201

Daram P, Brunner S, Persson BL, Amrhein N, Bucher M (1998) Functional analysis and cell-specific expression of a phosphate transporter from tomato. Planta 206:225-233

Dimroth P (1995) On the way towards the Na^+-binding site within the F_1F_0 ATPase of *Propiongenium modestum*. Biochem Soc Trans 23:770-775

Duckett CM, Oparka KJ, Prior DAM, Dolan L, Roberts K (1994) Dye-coupling in the root epidermis of *Arabidopsis* is progressively reduced during development. Development 120:3247-3255

Felle H (1982) Effects of fusicoccin upon membrane potential, resistance and current-voltage characteristics in root hairs of *Sinapis alba*. Plant Sci Lett 25:219-225

Felle H (1988) Cytoplasmic free calcium in *Riccia fluitans* L. and *Zea mays* L.: Interaction of Ca^{2+} and H^+? Planta 176:248-255

Felle HH (1993) Ion-selective microelectrodes: Their use and importance in modern cell biology. Bot Acta 106:5-12

Felle HH (1994) The H^+/Cl- symporter in root hairs of *Sinapis alba*: An electrophysiological study using ion-selective microelectrodes. Plant Physiol 106:1131-1136

Felle HH (1996) Control of cytoplasmic pH under anoxic conditions and its implication for plasma membrane proton transport in *Medicago sativa* root hairs. J Exper Bot 47:967-973

Felle H, Bertl A (1986) The fabrication of H^+-selective liquid-membrane micro-electrode for use in plant cells. J Exp Bot 37:1416-1428

Felle H, Hepler PK (1997) The cytosolic Ca^{2+} concentration gradient of *Sinapsis alba* root hairs as revealed by Ca^{2+}-selective microelectrode tests and fura-dextran ratio imaging. Plant Physiol 114:39-45

Felle HH, Tretyn A. Wagner G (1992) The role of the plasma-membrane Ca^{2+}-ATPase in Ca^{2+} homeostasis in *Sinapis alba* root hairs. Planta 188:306-313

Forestier C, Bouteau F, Leonhardt N, Vavasseur A (1998) Pharmacological properties of slow anion currents in intact guard cells of *Arabidopsis*. Application of the discontinuous single-electrode voltage-clamp to different species. Eur J Physiol 436:920-927

Fox TC, Guerinot ML (1998) Molecular biology of cation transport in plants. Annu Rev Plant Physiol Plant Molec Biol 49:669-696

Franks PJ, Cowan IR, Tyerman SD, Cleary AL, Lloyd J, Farquhar JD (1995) Guard cell pressure/aperture characteristics measured with a pressure probe. Plant Cell Environ 18:795-800

Garrill A, Lew RR, Heath IB (1992) Stretch-activated Ca^{2+} and Ca^{2+}-activated K^+ channels in the hyphal tip plasma membrane of the oomycete *Saprolegnia ferax*. J Cell Sci 101:721-730

Garrill A, Jackson SL, Lew RR, Heath IB (1993) Ion channel activity and tip growth:tip-localized, stretch-activated channels generate a Ca^{2+} gradient that is required for tip growth in the oomycete *Saprolegnia ferax*. J Eur Cell Biol 60: 358-365

Gassmann W, Schroeder JI (1994) Inward-rectifying K^+ channels in root hairs of wheat. A mechanism for aluminum-sensitive low-affinity K^+ uptake and membrane potential control. Plant Physiol 105:1399-1408

Gassmann W, Rubio F, Schroeder JI (1996) Alkali cation selectivity of the wheat root high-affinity potassium transporter HKT1. Plant J 10:869-882

Grabov A, Bottger M (1994) Are redox reactions involved in regulation of K^+ channels in the plasma membrane of *Limnobium stoloniferum* root hairs? Plant Physiol 105:927-935

Griffiths A, Parry AD, Jones HG, Tomos AD (1996) Abscisic acid and turgor pressure regulation in tomato roots. J Plant Physiol 149:372-376

Hermann A, Felle HH (1995) Tip growth in root hair cells of *Sinapis alba* L.: Significance of internal and external Ca^{2+} and pH. New Phytol 129:523-533

Hille B (1984) Ionic channels of excitable membranes. Sinauer Associates, Sunderland, MA, ix and 426 pp

Hirayama BA, Loo DDF, Wright EM (1994) Protons drive sugar transport through the Na^+/glucose cotransporter (SGLT1). J Biol Chem 269:21407-21410

Hirsch RE, Lewis BD, Spalding EP, Sussman MR (1998) A role for the AKT1 potassium channel in plant nutrition. Science 280:918-921

Holdaway-Clarke TL, Walker NA, Overall RL (1996) Measurement of the electrical resistance of plasmodesmata and membranes of corn suspension-culture cells. Planta 199:537-544

Horowitz P and Hill W (1989) The art of electronics. Cambridge University Press, pp 1-59

Jones DL, Shaff JE, Kochian LV (1995) Role of calcium and other ions in directing root hair tip growth in *Limnobium stoloniferum*. I. Inhibition of tip growth by aluminum. Planta 197:672-680

Jones DL, Gilroy S, Larsen PB, Howell SH, Kochian LV (1998) Effect of aluminum on cytoplasmic Ca^{2+} homeostasis in root hairs of *Arabidopsis thaliana*. Planta 206:378-387

Lauter F-R, Ninnemann O, Bucher M, Riesmeier JW, Frommer WB (1996) Preferential expression of an ammonium transporter and of two putative nitrate transporters in root hairs of tomato. Proc Natl Acad Sci 93:8139-8144

Legarde D, Basset M, Lepetit M, Conejero G, Gaymard F, Astruc S, Grignon C (1996) Tissue-specific expression of *Arabidopsis* AKT1 gene is consistent with a role in K^+ nutrition. Plant J 9:195-203

Levina NN, Lew RR, Heath IB (1994) Cytoskeletal regulation of ion channel distribution in the tip-growing organism *Saprolegnia ferax*. J Cell Sci 107: 127-134

Lew RR (1989) Calcium activates an electrogenic proton pump in *Neurospora* plasma membrane. Plant Physiol 91:213-216

Lew RR (1991) Electrogenic transport properties of growing *Arabidopsis thaliana* root hairs. The plasma membrane proton pump and potassium channels. Plant Physiol 97:1527-1534

Lew RR (1994) Regulation of electrical coupling between *Arabidopsis* root hairs. Planta 193:67-73

Lew RR (1996) Pressure regulation of the electrical properties of growing *Arabidopsis thaliana* L. root hairs. Plant Physiol 112:1089-1100

Lew RR (1998) Immediate and steady state extracellular ionic fluxes of growing *Arabidopsis thaliana* root hairs under hyperosmotic and hypoosmotic conditions. Physiol Plant 104:397-404

Lew RR (1998a) Mapping fungal ion channel locations. Fung Genet Biol 24:69-76

Lew RR, Spanswick RM (1984) Characterization of the electrogenicity of soybean (*Glycine max* L.) Roots: ATP dependence and the effect of ATPase inhibitors. Plant Physiol 75:1-6

Liu C, Muchhal US, Uthappa M, Kononowicz AK, Raghothama KG (1998) Tomato phosphate transporter genes are differentially regulated in plant tissues by phosphorus. Plant Physiol 116:91-99

Maathuis FJM, Sanders D (1993) Energization of potassium uptake in *Arabidopsis thaliana*. Planta 191:302-307.

Maathuis FJM, Sanders D (1995) Contrasting roles of two K^+ -channel types in root cells of *Arabidopsis thaliana*. Planta 197:456-464

Masucci JD, Schiefelbein JW (1994) The *rhd6* mutation of *Arabidopsis thaliana* alters root-hair initiation through an auxin- and ethylene-associated process. Plant Physiol 106:1335-1346

Meharg AA, Blatt MR (1995) NO_3^- transport across the plasma membrane of *Arabidopsis thaliana* root hairs: Kinetic control by pH and membrane voltage. J Memb Biol 145:49-66

Meharg AA, Maurousset L, Blatt MR (1994) Cable correction of membrane currents recorded from root hairs of *Arabidopsis thaliana* L. J Exp Bot 45:1-6

Michelet B, Boutry M (1995) The plasma membrane H^+-ATPase. A highly regulated enzyme with multiple physiological functions. Plant Physiol 108:1-6

Moog PR, van der Kooij TAW, Bruggemann W, Schielfelbein JW, Kuiper PJC (1995) Responses to iron deficiency in *Arabidopsis thaliana*: The Turbo iron reductase does not depend on the formation of root hairs and transfer cells. Planta 195:505-513

Noble PS (1991) Physicochemical and environmental plant physiology. Academic Press, San Diego, xx and 635 pp

Ogden D (1994) Microelectrode techniques. The Plymouth workshop handbook. Second edition. The Company of Biologists Ltd, Cambridge, x and 448 pp

Palmgren MG (1998) Proton gradients and plant growth: role of the plasma membrane H^+-ATPase. Adv Bot Res 28:1-70

Pritchard J, Fricke W, Tomos D (1996) Turgor-regulation during extension growth and osmotic stress of maize roots. An example of single-cell mapping. Plant Soil 187:11-21

Purves RD (1981) Microelectrode methods of intracellular recording and ionophoresis. Academic Press, London, x and 146 pp

Rall W (1977) Core conductor theory and cable properties of neurons. In Kandel ER (ed) Handbook of Physiology, Vol 1, American Physiological Society, Bethesda, Maryland, pp 79-97

Ramirez JA, Vacata V, McCusker JH, Haber JE, Mortimer RK, Owen WG, Lecar H (1989) ATP-sensitive K^+ channels in a plasma membrane H^+-ATPase mutant of the yeast *Saccharomyces cerevisiae*. Proc Natl Acad Sci 86:7866-7870

Robinson NJ, Procter CM, Connolly EL, Guerinot ML (1999) A ferric-chelate reductase for iron uptake from soils. Nature 397:694-697

Rodriguez-Navarro A, Blatt MR, Slayman CL (1986) A potassium-proton symport in *Neurospora crassa*. J Gen Physiol 87:649-674

Ryan PR, Newman IA, Arif I (1992) Rapid calcium exchange for protons and potassium in cell walls of *Chara*. Plant Cell Environ 15:675-683

Sakmann B, Neher E (1995) Single channel recording. Second edition. Plenum Press, New York, London, xxii and 700pp

Samuels AL, Fernando M, Glass ADM (1992) Immunofluorescent localization of plasma membrane H^+-ATPase in barley roots and effect of K nutrition. Plant Physiol 99:1509-1514

Sanchez-Fernandez R, Fricker M, Corben LB, White NS, Sheard N, Leaver CJ, Van Montagu M, Inze D, May MJ (1997) Cell proliferation and hair tip growth in the *Arabidopsis* root are under mechanistically different forms of redox control. Proc Natl Acad Sci 94:2745-2750

Schiefelbein JW, Shipley A, Rowse P (1992) Calcium influx at the tip of growing root-hair cells of *Arabidopsis thaliana*. Planta 187:455-459

Schnall JA, Quatrano RS (1992) Abscisic acid elicits the water-stress response in root hairs of *Arabidopsis thaliana*. Plant Physiol 100:216-218

Serrano R (1989) Structure and function of plasma membrane ATPase. Annu Rev Plant Physiol Plant Mol Biol 40:61-94

Silverman FP, Assiamah AA, Bush DS (1998) Membrane transport and cytokinin action in root hairs of *Medicago sativa*. Planta 205:23-31

Sivaguru M, Baluska F, Volkmann D, Felle HH, Horst WJ (1999) Impacts of aluminum on the cytoskeleton of the maize root apex. Short-term effects on the distal part of the transition zone. Plant Physiol 119:1073-1082

Smith RD, Wilson JE, Walker JC, Baskin TI (1994) Protein-phosphatase inhibitors block root hair growth and alter cell shape of *Arabidopsis* roots. Planta 194:516-524

Starzak ME (1984) The physical chemistry of membranes. Academic Press, Orlando, x and 334 pp

Thomas RC (1978) Ion-sensitive intracellular microelectrodes. How to make and use them. Academic Press, London, New York, San Francisco. xiii and 110 pp

Tretyn A, Wagner G, Felle HH (1991) Signal transduction in *Sinapis alba* root hairs: Auxins as external messengers. J Plant Physiol 139:187-193

Tsuchiya K, Wang W, Giebisch G, Welling PA (1992) ATP is a coupling modulator of parallel Na-K-ATPase-K-channel activity in the renal proximal tubule. Proc Natl Acad Sci 89:6418-6422

Ullrich CI, Novacky AJ (1990) Extra- and intra-cellular pH and membrane potential changes induced by K^+, Cl^-, $H_2PO_4^-$, and NO_3^- uptake and fusicoccin in root hairs of *Limnobium stoloniferum*. Plant Physiol 94:1561-1567

Wang R, Crawford NM (1996) Genetic identification of a gene involved in constitutive, high-affinity nitrate transport in higher plants. Proc Natl Acad Sci 93:9297-9301

Wang R, Liu D, Crawford NM (1998) The *Arabidopsis* CHL1 protein plays a major role in high-affinity nitrate uptake. Proc Natl Acad Sci 95:15134-15139

Wang T-B, Gassmann W, Rubio F, Schroeder JI, Glass ADM (1998) Rapid up-regulation of *HKT1*, a high-affinity potassium transporter gene, in roots of barley and wheat following withdrawal of potassium. Plant Physiol 118:651-659

Weiss TF (1996) Cellular biophysics. Volume 2: Electrical properties. MIT Press, Cambridge MA, pp 1-162

Wymer CL, Bibikova TN, Gilroy S (1997) Cytoplasmic free calcium distributions during the development of root hairs of *Arabidopsis thaliana*. Plant J 12:427-439

Zimmerman U, Steudle E (1974) Hydraulic conductivity and volumetric elastic modulus in giant algal cells: Pressure- and volume-dependence. In: Zimmerman U, Dainty J (Eds) Membrane transport in plants. Springer-Verlag, New York, Heidelberg, Berlin. pp 64-71

Zimmerman U, Rygol J, Balling A, Klock G, Metzeler A, Haase A (1992) Radial turgor and osmotic pressure profiles in intact and excised roots of *Aster tripolium*. Plant Physiol 99:186-196

Chapter 9
Calcium in Root Hair Growth

Tatiana Bibikova[1] and Simon Gilroy[2]

[1]Plant Molecular Science Group, Division of Biochemistry and Molecular Biology, Institute of Biomedical and Life Sciences, Bower Building, University of Glasgow, Glasgow G12 8QQ, UK
[2]Biology Department, The Pennsylvania State University, 208 Mueller Lab, University Park, PA 16802, USA

1 Introduction

The growth of cells as diverse as fungal hyphae, pollen tubes, algal rhizoids, and root hairs is characterized by a highly localized control of cell expansion confined to the growing tip. The concentration of cytoplasmic calcium ($[Ca^{2+}]_c$) has been proposed to play an important role in both structuring the apical cytoplasm of these tip-growing cells as well as localizing the site of exocytosis to the expanding apex. Indeed, a tip-focused $[Ca^{2+}]_c$ gradient is associated with the growing point of these cells and dissipating the gradient with Ca^{2+} buffers and Ca^{2+}-channel blockers inhibits apical growth. In addition, responses to factors such as biotic and abiotic stresses and rhizobial nodulation factors (Nod factors) are all thought to be mediated, at least in part, by changes in root hair Ca^{2+} dynamics. In this chapter we will review two aspects of the involvement of Ca^{2+} in root hair development, the evidence for $[Ca^{2+}]_c$ as a central determinant of localized growth, and the role of cytoplasmic Ca^{2+} as a second messenger in Nod factor signaling. In both these cases the combined research of many laboratories has begun to reveal $[Ca^{2+}]_c$ as a multifunctional regulator of root hair responses. In addition, we will highlight elements of the root hair growth machinery, such as membrane trafficking and the cytoskeleton, that may be regulated by these changes in $[Ca^{2+}]_c$.

2 Root Hair Growth

The hair like growth pattern of root hairs is essential to their role in enhancing the water and nutrient uptake capacity of the plant. In addition this growth habit is thought to aid in anchoring the plant in the soil (Peterson and Farquhar 1996). The first indications of root hair formation become evident when the epidermal cell

Key words. root hair, calcium, tip growth, initiation, pollen tube, *Fucus*, rhizoid, CDPK, calmodulin, annexin, actin, tubulin, cytoskeleton, Nod factor, calcium channels that will give rise to the root hair (the trichoblast) undergoes a switch from diffuse growth to highly localized expansion from one side to form a bulge in the cell wall. This is the process of root hair initiation. The expanding region of the trichoblast then enters a second, and distinct phase of localized growth, tip growth, where polarized secretion limits the deposition of new membrane and cell wall material to the growing apex of the hair (Schnepf 1986).

These two phases of root hair formation, initiation and subsequent tip growth, both reflect a very tightly controlled localization of growth by the trichoblast. However, the mechanisms whereby this control is brought about appear to be different. Cytoplasmic Ca^{2+} has emerged as a central regulator of growth and signal/response systems of the tip-growing root hair. However, current data argue against an equivalent role for Ca^{2+} in the localized growth phenomena that occurs during root hair initiation.

2.1 Root Hair Initiation

During the process of root hair initiation, a localized bulge forms on the outer cell wall of the trichoblast. This expanded region will give rise to the root hair and its site on the lateral wall of the trichoblast is precisely regulated. In *Arabidopsis thaliana*, for example, root hairs are always formed at the apical end of the cell (i.e., that nearest the root apex; Schiefelbein and Somerville 1990). The process of root hair initiation therefore requires establishment of a new growth point within the trichoblast at a predefined site, as well as accumulation of the appropriate cell constituents at this site to support its subsequent expansion. How the trichoblast defines this site of new growth remains enigmatic.

Gradients of cytoplasmic pH and Ca^{2+}, are recognized as important determinants of polarized growth in a range of eukaryotic cells including protists and plants (e.g., Hyde and Heath 1995; Kropf 1997; Malho 1998a). Thus ionic gradients in the cytoplasm have represented attractive candidates to specify the root hair initiation site.

Perhaps the strongest evidence for a role of Ca^{2+} as a determinant of the site of growth initiation comes from work on the formation and fixation of a new cellular axis of polarity in the zygotes of fucoid algae. The *Fucus* zygote is initially spherical and unpolarized. It responds to a variety of environmental cues, such as the direction of light or gravity, by forming a polar axis which specifies the site of subsequent localized outgrowth of a rhizoid. Formation and fixation of this axis appears to be dependent on reorganization of the actin-based cytoskeleton and gradients of Ca^{2+} and pH (Kropf 1997; Kropf et al. 1998). The localized influx of Ca^{2+} has been proposed to be an important determinant of the site of future outgrowth of the rhizoid (e.g., Robinson and Jaffe 1973,1975; Speksnijder et al. 1989; Berger and Brownlee 1993; Robinson 1996; Love et al. 1997; Brownlee et al. 1998; Pu and Robinson 1998; although see also Kropf and Quatrano 1987). Gradients in $[Ca^{2+}]_c$ may be generated through asymmetrical redistribution of Ca^{2+} channels (Shaw and Quatrano 1996) and act through, for example, the localization of the calcium-dependent regulatory protein calmodulin to the site of new growth (Love et al. 1997; Pu and Robinson 1998). Thus for *Fucus* there is a wealth of information suggesting Ca^{2+} is involved in the polarization and its associated localized growth phenomenon.

Studies on the morphogenesis in the green alga *Micrasterias denticulata* highlight the alternate possibility that Ca^{2+} gradients are not a universal determinant specifying the site of future growth. As in *Fucus* zygotes (and root hairs), the shape of this organism is determined by localized patterns of secretion and cell wall deposition (Kiermayer 1981). Measurements of the $[Ca^{2+}]_c$ in growing cells of *M. denticulata* demonstrated an even distribution of this ion throughout the entire cell despite such localized growth. In addition, microinjection of different BAPTA-type buffers, which are known to disturb gradients in $[Ca^{2+}]_c$ (Speksnijder et al. 1989), did not have any measurable effect on *M. denticulata* morphogenesis (Holzinger et al. 1995). These results argue against the presence of intracellular Ca^{2+} gradients that define the localized growth points in *M. denticulata*, although it is important to note that growth in *M. denticulata* is very slow compared to that of root hairs.

Current evidence suggests *M. denticulata* most closely relates to the Ca^{2+} involvement (or lack thereof) in the bulge formation as the root hair initiates. Initial polarization of the trichoblast during root hair initiation is known to occur prior to the formation of the bulge, because the nucleus migrates to the initiation site well before any observable deformation of the trichoblast cell wall (Meekes 1985; Sato et al. 1995). These events set the time scale over which a localized change in $[Ca^{2+}]_c$ associated with specifying the initiation site might be observed. However, no changes in $[Ca^{2+}]_c$ were seen to precede or coincide with the initiation process of root hairs in *A. thaliana* despite monitoring for several hours prior to nuclear migration and root hair formation (Fig. 1; Wymer et al. 1997). In addition, application of Ca^{2+} channel-blocking drugs that inhibit tip growth in root hairs (Hermann and Felle 1995; Wymer et al. 1997), a process that we will see is well characterized as being calcium regulated, fail to arrest the initiation process (Wymer et al. 1997).

If root hair initiation is Ca^{2+}-independent, the question remains as to nature of the molecular events that specify the initiation site and mediate the wall bulging associated with the initiation process. Bulge formation is inhibited by protein phosphatase inhibitors (Smith et al. 1994), providing some very tentative indications that protein phosphorylation might play a role in determining localized growth. Similarly microtubule rearrangements are associated with the formation of the bulge (Emons and Derksen 1986). *Arabidopsis* mutants such as Rhd 1 and 6 are known to affect the initiation process (Schiefelbein and Somerville 1990; Masucci and Schiefelbein 1994) and ethylene and auxin appear to be involved in initiation site selection (Masucci and Schiefelbein 1994). Despite these clues to the events participating in the initiation process, we presently lack a clear molecular model for how the initiation site is selected. The subsequent wall bulging however appears intimately linked to changes in wall pH (Bibikova et al. 1998), a topic more fully covered in Chapter 10 this volume.

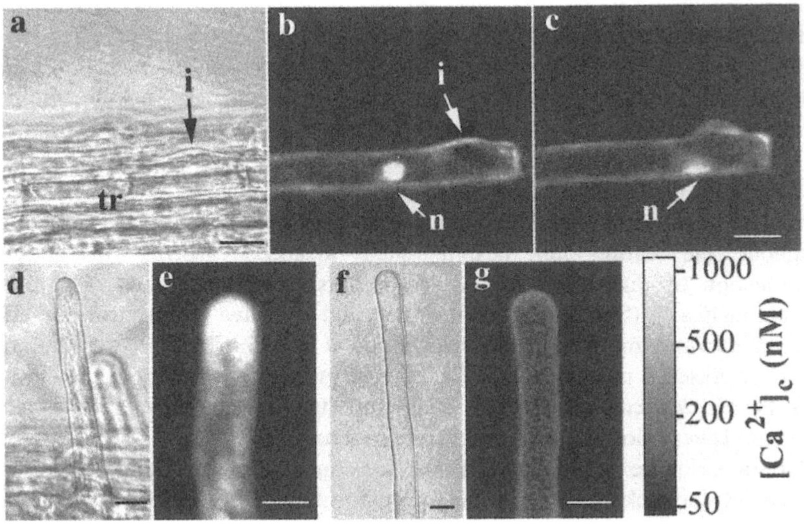

Fig. 1. *a-g*. Cytoplasmic $[Ca^{2+}]$ in initiating and tip-growing root hairs of *Arabidopsis*. The root hairs were microinjeted with the fluorescent Ca^{2+} indicating dye Indo-1 and $[Ca^{2+}]$ monitored by confocal ratio imaging. *a* Bright field and *b,c* $[Ca^{2+}]$ in an initiating root hair. *b* 5 min and *c* 15 min after the start of wall bulging. *d,f* Bright field and *e,g* $[Ca^{2+}]$ in a tip-growing and non-growing root hair. Note the development of a tip-focused gradient in $[Ca^{2+}]$ when tip growth commences and its loss as the hair stops growing. i, initiation site; *n*, nucleus; *tr*, trichoblast. Bar 10 μm

2.2 Tip Growth

Following initiation, the root hair commences a genetically (Schiefelbein and Somerville 1990; Masucci and Schiefelbein 1994) and physiologically (Wymer et al. 1997; Bibikova et al. 1998; and see below) distinct localized growth process, tip growth. In the tip-growing root hair, deposition of new plasma membrane and cell wall material is confined to the expanding tip, leading to the elongated hair-like morphology (Sievers and Schnepf 1981; Schnepf 1986). The ultrastructure of the growing apex, and general mechanisms underlying apical growth, seem to be similar in all known tip-growing cells. Thus, as we will see, tip-focused gradients of $[Ca^{2+}]_c$ and cytoskeletal rearrangements appear to be central determinants of the site and direction of apical growth in cells as varied as fungal hyphae, pollen tubes and root hairs.

2.2.1 Tip-Growing Systems Show Structural Zonation

The ultrastructure of the apex of the growing root hair provides a good description of the highly organized and polarized cytoplasmic zonation seen in most tip-growing plant and algal cells (Emons 1987; Rosen et al. 1964). The tip of these cells predominantly contains non-streaming cytoplasm enriched with Golgi derived vesicles (Rosen et al. 1964), although sparse microtubules and actin filaments may also be present (e.g., Lloyd and Wells 1985; Traas et al. 1985; Emons 1987; Ridge 1988). Further from the tip are found mitochondria, Golgi bodies, vesicles, endoplasmic reticulum, and plastids (Rosen et al. 1964; Ridge 1988; reviewed in Sievers and Schnepf 1981). The structural zonation appears to reflect the polarized secretory machinery necessary for tip growth. Thus when the apically growing cell stops elongating, the structural zonation of the apical cytoplasm disappears (Rosen et al. 1964; Lancelle and Hepler 1992; Bonnett and Newcomb 1966; Galway et al. 1997).

The regulation of the zonation and the direction in which the secretory apparatus operates seems to be intimately linked to the $[Ca^{2+}]_c$ at the apex. The resting level of $[Ca^{2+}]_c$ in eukaryotic cells is between 100 and 400 nM but $[Ca^{2+}]_c$ at the tip of apically growing cells is elevated to several micromolar (e.g., pollen tubes: Malho 1998a). This region of high $[Ca^{2+}]_c$ appears to be essential for the polarized growth of all kinds of tip-growing cells. Thus analysis of the Ca^{2+} dynamics in elongating fungal hyphae, algal rhizoids and especially pollen tubes, has provided a backdrop against which the unique aspects of Ca^{2+} regulated tip growth in root hairs has been investigated.

2.2.2 Calcium and Fungal Tip Growth

The $[Ca^{2+}]_c$ is highest at the very tip of growing hyphae of *Saprolegnia* and *Neurospora* (Garril et al. 1993; Knight et al. 1993; Levina et al. 1994; 1995). The steepness of this gradient correlates well with growth, and the gradient is negligible in non-growing hyphae (Garril et al. 1993; Levina et al. 1995). Stretch-activated, Ca^{2+}-permeable channels are concentrated at the growing tip of *Saprolegnia ferax* hyphae (Garril et al. 1993). Therefore it has been suggested that these Ca^{2+} channels maintain the elevated $[Ca^{2+}]_c$ at the tip by allowing polarized influx of Ca^{2+} (Garril et al. 1993). Treatment with cytochalasin disrupts this asymmetrical distribution of mechanosensitive channels and inhibits hyphal growth, implying that an actin network maintains the channel gradient in the tip (Levina et al. 1994). In contrast, the plasma membrane of *Neurospora* hyphae contain Ca^{2+}-activated Ca^{2+} channels (Levina et al. 1995). These channels however, are not concentrated at the apex but rather are evenly distributed throughout the growing hyphae. As a tip-focused $[Ca^{2+}]_c$ gradient is still essential for the elongation of these tip-growing cells, this gradient must be generated by either differential regulation of the Ca^{2+} channels at the apical and basal parts of the hyphae, or possibly by the release of Ca^{2+} from internal stores towards the tip (Levina et al. 1995). These data suggest a model whereby stretch-activated Ca^{2+} channels at the hyphal tip might trigger a small Ca^{2+} influx which is amplified and becomes self-reinforcing due to Ca^{2+} activation of the Ca^{2+} channels at the apex. Such a self-reinforcing, self-sustaining Ca^{2+} gradient bears striking similarity to that found at the tip of *Arabidopsis* root hairs (Bibikova et al. 1997; and see below). However, a note of caution must be added when applying models derived from studies of fungal hyphae to the mechanisms of tip growth in plant cells. In fungal cells, secretory vesicles accumulate in the hyphal tip to create an apical body called the Spitzenkorper (Girbardt 1969). Although the structure of the apex of tip-growing plant cells resembles fungi, a structure equivalent to a Spitzenkorper has not been identified in plant cells.

2.2.3 Algal Rhizoids

The spherical, single-celled zygote of the fucoid alga produces a tip-growing protrusion from its surface that forms the rhizoid. Outgrowth of the rhizoid is accompanied by the appearance of a positive current entering the rhizoid apex and leaving from subapical regions. The inward current at the rhizoid apex is carried primarily by Cl^- efflux and Ca^{2+} influx (Nuccitelli and Jaffe 1976). Growing *Fucus* rhizoids also possess a tip-focused $[Ca^{2+}]_c$ gradient (Brownlee and Wood 1986; Brownlee and Pulsford 1988; Berger and Brownlee 1993; Taylor et al. 1996; Pu and Robinson 1998). This gradient is essential for growth since microinjection of Ca^{2+} buffers that dissipate the gradient (a so called "shuttle buffering" experiment) blocks subsequent extension of the rhizoid (Speksnijder et al. 1989; Taylor et al. 1996). Mechanosensitive Ca^{2+} channels have been identified in the plasma membrane of *Fucus* rhizoids, and activation of these channels has been shown to elevate $[Ca^{2+}]_c$ (Taylor et al. 1996; Brownlee et al. 1998). These stretch-activated channels may therefore play an important role in the establishment and maintenance of the tip-focused Ca^{2+} gradient in the algal rhizoid (Taylor et al. 1997; Brownlee et al. 1998).

2.2.4 Pollen Tubes

Pollen tubes represent one of the most intensively studied tip-growing plant cells. The structure and activities observed at the growing pollen tube apex suggests close parallels to root hair elongation. These studies also indicate that the theme of a Ca^{2+}-dependent growth machinery may be a universal feature of tip-growing plant cells.

The presence of Ca^{2+} at between 10 μM and 10 mM in the culture media is essential for pollen tube growth, suggesting an important role of this ion in the regulation and/or maintenance of pollen tube growth (Picton and Steer 1983). Studies with the self-referencing (vibrating) microelectrode indicated that Ca^{2+} enters the pollen tube predominantly at the growing tip (Kuhteiber and Jaffe 1990). A number of studies further characterized a gradient in $[Ca^{2+}]_c$ in the growing pollen tube from above 3 μM at the apex, to 150-300 nM at the base of the pollen tube (Obermeyer and Weisenseel 1991; Rathore at al. 1991; Miller et al. 1992; Malho et al. 1994; Pierson et al. 1994; 1996; Messerli and Robinson 1997). Inhibition of pollen tube elongation by thermal shock, application of osmoticum, or caffeine is accompanied by the dissipation of the Ca^{2+} gradient and elimination of tip-directed Ca^{2+}-influx (e.g., Kuhteiber and Jaffe 1990; Pierson et al. 1994; 1996). Disruption of the $[Ca^{2+}]_c$ gradient by application of Ca^{2+} channel blockers or use of the shuttle buffering technique also leads to immediate growth arrest (Pierson et al. 1994). Furthermore, elevation of $[Ca^{2+}]_c$ at one side of the growing pollen tube leads to the reorientation of the growth direction towards the new $[Ca^{2+}]_c$ "hot spot", suggesting that localized elevation of $[Ca^{2+}]_c$ can drive growth (Malho and Trewavas 1996). These results all strongly indicate that a gradient in $[Ca^{2+}]_c$ is one of the factors that can direct the tip growth process in the pollen tube.

The source of the Ca^{2+} generating the gradient in $[Ca^{2+}]_c$ could be from influx across the plasma membrane and/or release from intracellular stores. The contribution of intracellular stores has been implied by the observation of inositol-1,4,5- trisphosphate induced transient increases and waves of elevated $[Ca^{2+}]$ in growing pollen tubes (Franklin-Tong et al. 1996; Malho 1998c). $[Ca^{2+}]$ waves are generally mediated by Ca^{2+} release from intracellular stores (Malho et al. 1998). However, inositol-1,4,5- trisphosphate-induced Ca^{2+} changes could not be elicited at the tip (Malho 1998c), implying they may be part of a signal transduction system that operates in parallel to the Ca^{2+} changes involved in the apical growth machinery. Indeed, current data suggests Ca^{2+}-influx via open Ca^{2+} channels preferentially localized to the apical plasma membrane of the tip is more closely correlated with growth. Thus the manganese quench technique, an indirect way of visualizing Ca^{2+}-channel activity, indicates a high level of open Ca^{2+} channels localized to the apical plasma membrane of the pollen tube (Malho et al. 1995). Similarly, measurements made with the self-referencing microelectrode have revealed pronounced Ca^{2+} influx at the apex of the growing pollen tube, implying influx across the plasma membrane at this point. Detailed measurements of these fluxes and the $[Ca^{2+}]$ gradient in Lily pollen have provided strong support for the association of Ca^{2+} and growth but have also revealed inconsistencies with the apical site of Ca^{2+} fluxes. Thus, the growth rate of the Lily pollen tube oscillates with a period of 20 to 30 s (Pierson et al. 1996). Growth rate and the $[Ca^{2+}]_c$ gradient oscillate in phase, with the highest $[Ca^{2+}]_c$ at the time of highest growth rate. However, Ca^{2+}-influx at the tip is delayed relative to changes in growth rate by 11 s and is significantly larger than that needed to support the observed magnitude of the $[Ca^{2+}]_c$ gradient (Pierson et al. 1996; Holdaway-Clarke et al. 1997; Messerli and Robinson 1997). The reason for these discrepancies in timing and magnitude of the flux are unknown, but if the current estimates of the peak $[Ca^{2+}]_c$ in the tip-focused gradient are underestimates (Messerli and Robinson 1997) then the influx may be close to that required to sustain the growth-related gradient. In addition, the influx at the pollen tube apex could reflect not only influx to the cytosol but also Ca^{2+} uptake to the newly synthesized cell wall produced during the pulse of growth. Such Ca^{2+} binding by the cell wall could explain the temporal shift in Ca^{2+} flux and growth. Calcium interaction with the wall may also be an important growth regulating step. Pectins in the cell wall may be cross-linked by Ca^{2+}, rigidifying the wall. Conversely if the pectins are esterified, the Ca^{2+} binding sites are blocked and the wall is less rigid. Thus extracellular Ca^{2+} and pectin esterification levels may play a role in determining the mechanical properties of the cell wall and so the ease of turgor driven tip growth. It is interesting to note that in pulsatile growth of pollen tubes, there are regions of the apical cell wall that show high degrees of methyl esterification, and these regions correlate to wall deposition during periods of accelerated growth (Li et al. 1996). A regulatory role for wall Ca^{2+} and the interactions between the root hair cell wall, plasma membrane and cytoskeleton (Fowler and Quatrano 1997; Miller et al. 1997a,b; see below) cannot be excluded.

2.2.5 Calcium and Root Hair Tip Growth

A tip-focused $[Ca^{2+}]_c$ gradient appears to be an important regulator of growth in all tip-growing systems known (Malho 1998a), and root hairs have proven no exception. Electrophysiological studies using the self-referencing microelectrode showed that Ca^{2+}-influx is higher at the tip than at the base or sides of growing root hairs (Schiefelbein et al. 1992; Pierson et al. 1994; Herrmann and Felle 1995; Jones et al. 1995). Disrupting gradients in $[Ca^{2+}]_c$ by the addition of Ca^{2+}-ionophores, Ca^{2+}-channel blockers, or the microinjection of Ca^{2+} shuttle-buffers also all inhibit root hair growth (Herrmann and Felle 1995; Wymer et al. 1997; de Ruijter et al. 1998; Jones et al. 1998).

Ayling et al. (1994) and Ehrhardt et al. (1996) were able to microinject fluorescent Ca^{2+}-indicators into root hairs of tomato and alfalfa, and established a basal $[Ca^{2+}]_c$ level of 10^{-7} M in these cells. Herrmann and Felle (1995), using Ca^{2+}-sensitive microelectrodes, then observed a 3-fold elevation in $[Ca^{2+}]_c$ towards the tip of root hairs of *Sinapis alba*. The presence of this tip-focused $[Ca^{2+}]_c$ gradient has now been confirmed by several groups using both ion selective microelectrodes and the imaging of Ca^{2+} sensitive fluorescent dyes (Bibikova et al. 1997; Felle and Hepler 1997; Wymer et al. 1997; de Ruijter et al. 1998; Gehring et al. 1998; Jones et al. 1998). For example, the growing root hairs of *Arabidopsis* maintain a $[Ca^{2+}]_c$ of >1 μM localized to the apical 1-5 μm of cytoplasm, with $[Ca^{2+}]_c$ of 100-200 nM elsewhere in the hair (Fig. 1). The steepness of the $[Ca^{2+}]_c$ gradient correlates well with growth rate of individual root hairs, being most pronounced in the rapidly elongating root hairs. Root hairs that have stopped elongating either by reaching their final mature length, or due to some experimental manipulation, show no elevation of $[Ca^{2+}]_c$ at the tip (Fig. 1; Wymer et al. 1997; Bibikova et al. 1997; de Ruijter et al. 1998; Jones et al. 1998). Similarly, the rhd2 mutant of *Arabidopsis* shows a general basal $[Ca^{2+}]_c$ that is higher throughout the cell than in the wild-type. This mutant is incapable of tip growth (Schiefelbein and Sommerville 1990) and its disrupted $[Ca^{2+}]_c$ only becomes evident after an apparently normal initiation process has occurred. Thus disruption of $[Ca^{2+}]_c$ homeostasis in rhd2 root hairs correlates with their inability to establish tip growth (Wymer et al. 1997). In addition, when root hairs are forced to produce multiple tips, for example by cytoskeletal disruption (Bibikova et al. 1999), these new tips always possess a tip-focused gradient in $[Ca^{2+}]_c$ if they are growing. Similarly, a gradient in $[Ca^{2+}]_c$ is always seen in the new tips that form on root hairs after Nod factor treatment (de Ruijter et al. 1998). These observations are all consistent with a model whereby a localized increase in $[Ca^{2+}]_c$ is specifically associated with growth.

2.2.6 Calcium and the Orientation of Growth

There seems little question that a tip-focused gradient in $[Ca^{2+}]_c$ is present in growing root hairs and is necessary for growth. The question that remains is at what level does the gradient act? Is it simply part of the machinery that supports growth, or could it be an important factor determining the direction of growth? Support for and against a role of $[Ca^{2+}]_c$ in controlling the direction of root hair growth has come from attempts to reorient the gradient, and so reorient growth. In *Arabidopsis,* root hairs grow at approximately 90 degrees from the surface of the root, maintain a single growth point, and deviate from straight growth by less than 2 degrees along their length (Bibikova et al. 1997). Redirecting growth by applying a touch stimulus causes a shift in the position of the tip-focused $[Ca^{2+}]_c$ gradient and an equivalent shift in subsequent growth direction. Likewise, localized Ca^{2+}-ionophore application can shift the orientation of the influx of Ca^{2+} and such treatment elicits a change in the orientation of root hair growth toward the new $[Ca^{2+}]_c$ gradient (Bibikova et al. 1997). A similar phenomenon has been observed in pollen tubes wherein UV photoactivation of caged calcium triggers pollen tubes to grow toward elevated $[Ca^{2+}]$ (Malho and Trewavas 1996; Bibikova et al. 1997). The difference between pollen tubes and root hairs seems to lie in the duration of the change in growth direction. In root hairs, the reorientation of growth, be it induced by Ca^{2-+}-ionophore application or by a touch stimulus (but interestingly not by Nod factors) is transient, with a return to the original growth direction within 15 to 20 min. In contrast pollen tubes grown in vitro can show sustained reorientation of growth in response to alterations in the Ca^{2+} gradient (Malho and Trewavas 1996; Bibikova et al. 1997). Why might pollen tubes behave differently to root hairs in showing a sustained reorientation of growth? One answer might be that root hairs are studied in their intact environment, i.e., attached to the root, where an orientation mechanism ensuring their growth away from the root is in operation. In contrast, pollen tubes grown in vitro are probably divorced from the physical and chemical cues they normally use to orient their growth through the style (Malho 1998b). It seems probable that in vivo the orientation of both these tip-growing systems will be very highly controlled by factors from the plant that ensure growth is directed to its correct path. The obvious question arising from this observation is what are these cues that orient the Ca^{2+} gradient and so control the direction of root hair growth? The answer to this question remains elusive but the spatial information might include extracellular gradients of ions or organic molecules or an electric current that flows around the growing root. This current might serve as a growth localizing mechanism by facilitating electrophoretic migration of organelles and membrane proteins (Jaffe and Nuccitelli 1977).

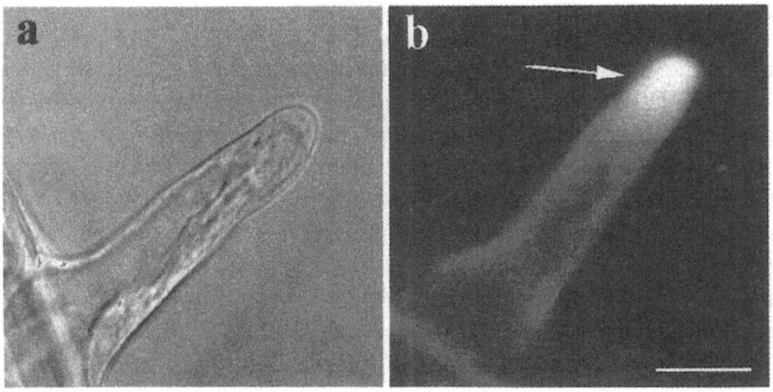

Fig. 2. *a,b.* Apical gradient in fluorescence from a root hair stained with 5μM fluorescent (BODIPY-labeled) dihydropyridine (DHP). *a* bright field image, *b* fluorescence image. DHP is thought to bind to Ca^{2+} channels. Images were taken using a Zeiss LSM 410 confocal microscope, 488 nm excitation and > 525 nm emission. Bar 10 μm

3 Regulation of Calcium Channels in the Root Hair Plasma Membrane

Pharmacological evidence suggests that the root hair plasma membrane contains at least three distinct classes of Ca^{2+} channels (White 1998). Unfortunately, we can only speculate as to which of these channels are important in transport of Ca^{2+} at the growing root hair tip as the electrophysiology of growing root hairs has proven technically demanding. However we do have some clues as to how these channels might be regulated. The reorientation of root hair growth in response to the asymmetrical application of Ca^{2+}-ionophore (Bibikova et al. 1997; 1999) suggests that a localized influx of Ca^{2+} can recruit and stabilize a Ca^{2+} gradient, i.e., recruit open Ca^{2+}-channels. This is because reorientation of the Ca^{2+} gradient and growth are sustained for at least 15 min despite asymmetrical application of the Ca^{2+}-ionophore lasting less than 1 s. Plant Ca^{2+}-activated Ca^{2+} channels (Ward and Schroeder 1994) provide an obvious mechanism for amplification of initial Ca^{2+} gradients as well as explaining the self sustaining nature of the gradient. The evidence from fungi and protists also suggests Ca^{2+}- or stretch-activated Ca^{2+} permeable channels might underlie the gradient formation (Garril et al. 1993; Taylor et al. 1997; Brownlee et al. 1998).

The evidence for clustering of open Ca^{2+} channels at the apex of the growing root hair is circumstantial. The self-referencing microelectrode has shown Ca^{2+} influx occurs in the apical region of the root hair, suggesting channel clustering or localized activation (Schiefelbein et al. 1992; Pierson et al. 1994; Herrmann and Felle 1995; Jones et al. 1995), and manganese quench data implies more open channels at the tip (Wymer et al. 1997). Localization of putative Ca^{2+} channels to the site of growth has been detected in *Fucus* using a fluorescent dihydropyridine, a drug that binds to animal Ca^{2+} channels (Shaw and Quatrano 1996). Clustering of fluorescence was interpreted as clustering of Ca^{2+} channels in the zygote plasma membrane at the site of rhizoid outgrowth. Similar labeling protocols applied to *Arabidopsis* root hairs yields a tantalizing gradient of fluorescence towards the apex of the elongating hair (Fig. 2). Confocal microscopy indicates this fluorescence comes from throughout the volume of the apical cytoplasm and so does not simply reflect Ca^{2+} channel accumulation to the root hair apical plasma membrane. Such staining patterns may well reflect Ca^{2+} channels on their way to be inserted into the apical plasma membrane as part of the secretory machinery.

It is also important to note that, as in the pollen tube, Ca^{2+} release from intracellular stores might play a role in regulating root hair Ca^{2+} dynamics. Although inositol-1,4,5-trisphosphate, cyclic ADP-ribose and Ca^{2+}-induced Ca^{2+}-release channels have been identified in plants, the best defined site(s) from which such intracellular Ca^{2+} release occurs is from the tonoplast (Ward and Schroeder 1994; Allen et al. 1995). In the growing root hair, the vacuole is excluded from the cytoplasm at the apex, i.e., it is excluded from the site where the $[Ca^{2+}]_c$ gradient forms. Thus the vacuole is an unlikely candidate for maintaining the tip-focused gradient. However, Ca^{2+}-release channels have been characterized on other intracellular membranes (notably the endoplasmic reticulum: Klusener et al. 1995; Muir and Sanders 1997) and the fluorescent-DHP staining in Fig. 2 implies that Ca^{2+} channels occur throughout the membrane-rich apical cytoplasm of the root hair. However, their participation in maintaining tip growth has yet to be determined.

4 Mechanism of Calcium-Dependent Regulation of Growth

It is clear that the tip-focused $[Ca^{2+}]_c$ gradient is required to maintain root hair growth. Presumably therefore it must regulate elements of the growth machinery at the growing tip. The candidates for such Ca^{2+}-regulated elements have been drawn from studies on many tip-growing systems (principally pollen tubes), although none has yet been firmly demonstrated as mediating Ca^{2+}-regulated tip growth events.

4.1 Interactions Between the Cytoskeleton and Calcium

The cytoskeleton is an obvious candidate for an element of the tip growth machinery regulated by $[Ca^{2+}]_c$. Treatment with actin filament antagonists arrests apical growth in most tip-growing cells, indicating that intact actin filaments are essential for the tip growth (Alfano et al. 1993; Kropf et al. 1998; Yamashita and May 1998; Bibikova et al. 1999; Miller et al. 1999).

In pollen tubes, actin filaments run along the side walls of the pollen tube and extend towards the tip. These actin filaments have been proposed to organize cytoplasmic structure at the tip (Derksen et al. 1995) as well as mediate delivery of secretory vesicles to the apical plasma membrane (Cai et al. 1997). However, these filaments do not seem to enter the clear zone at the tip of the tube (Doris and

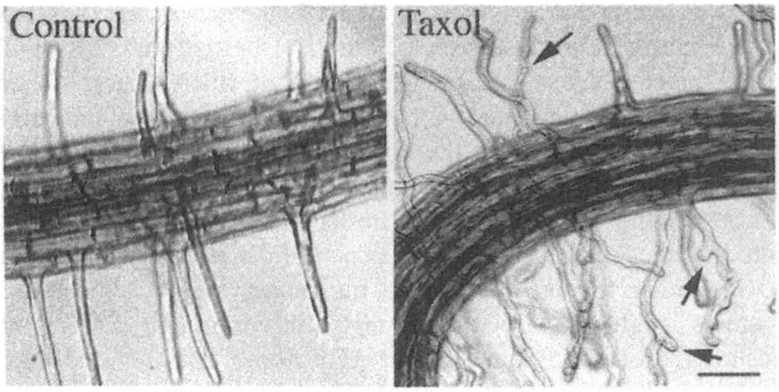

Fig. 3. Loss of straight root hair growth induced by treatment with 5 μM Taxol. Note that although tip growth is maintained, its directionality is impaired in the taxol treated plants. Taxol-treated root hairs show waving and/or branching patterns (*arrows*). Bar 50 μm

Steer 1996). In growing root hairs actin also runs along the length of the root hair and in all cytoplasmic strands. Actin then flares into fine bundles subapically but filaments are excluded from the vesicle rich apex (Miller et al. 1999). In non-growing root hairs the actin bundles extend throughout the tip. This structural zonation may indicate that actin is required to move secretory vesicles towards the apical clear zone, but the Ca^{2+} gradient both maintains this as a cytoskeleton free zone and facilitates the movement and fusion of vesicles once in this area. Consistent with this model is the observation that when the $[Ca^{2+}]_c$ gradient is lost, actin microfilaments protrude to the tube or root hair tip and streaming extends much further towards the apex (Miller et al. 1999). In addition, elevated $[Ca^{2+}]_c$ is known to promote secretory vesicle fusion with the plasma membrane (Homann and Tester 1997; Carrol et al. 1998), inhibit cytoplasmic streaming, fragment F-actin and depolymerize microtubules (Kohno and Shimmen 1987; 1988; Cyr 1994; see also Chapter 3 this volume).

In contrast to actin, an intact microtubule cytoskeleton does not appear to be required for tip growth to proceed. Thus, treatments with microtubule depolymerizing agents do not stop apical growth of fungal hyphae (Yamashita and May 1998), protist rhizoids (Kropf et al. 1998), pollen tubes (Heslop-Harrison et al. 1988) or root hairs (Bibikova et al. 1999) and, for example, the complex structure of the root hair wall does not appear to be controlled by microtubule orientation (van Amstel et al. 1993). These antagonists do however cause tip swelling and aberrant branching in many tip-growing cells, including yeast, moss protonemata, fungal hyphae, pollen tubes and root hairs (Fig. 3; Schmiedel and Schnepf 1979; Mizukami and Wada 1983; Doonan et. al. 1988; Heslop-Harrison et al. 1988; That et al. 1988; Hagen 1998; Emons et al. 1990; Bibikova et al. 1999). Similarly, in filamentous fungi, genetic deletion of the microtubule associated motor protein kinesin, does not inhibit growth, but does induce the formation of gnarled and contorted hyphae (Seiler et al. 1997; Wu et al. 1998). Genetic evidence also indicates that a calmodulin-regulated, kinesin-like protein is involved in the regulation of trichome branching (Oppenheimer et al. 1997). These observations suggest that while the microtubule-based cytoskeleton is not part of the apical growth machinery per se, it is involved in stabilizing the location and size of the site over which localized growth can occur. In root hairs where the microtubule cytoskeleton has been disrupted by drug treatment, the $[Ca^{2+}]_c$ gradient at the apex becomes more labile (Bibikova et al. 1999). Thus, in taxol treated root hairs of *Arabidopsis*, the $[Ca^{2+}]_c$ gradient is able to split into several, independent gradients that can migrate to the sides of the root hair and so initiate aberrant root hair branching. It is tempting to speculate that microtubules might be ordering actin (Tominaga et al. 1997) that might in turn be localizing Ca^{2+}-channels (Levina et al. 1994) to the apex of the elongating root hair.

Whilst the microtubule cytoskeleton may be stabilizing the Ca^{2+} gradient, Ca^{2+} has also been shown to be intimately involved in regulating microtubule dynamics. Calcium and calmodulin can promote microtubule assembly and disassembly (Fisher et al. 1996) and microtubule associated proteins such as elongation factor 1 alpha are known to elicit Ca^{2+}-dependent alterations in microtubule structure (Cyr 1994). Thus control of the root hair growth machinery may well involve both microtubule stabilization of the Ca^{2+} gradient as well as feedback control from this gradient on the microtubule dynamics. For further information on microtubules and root hairs see Chapter 2 this volume.

4.2 Annexins

Annexins are Ca^{2+}-dependent phospholipid binding proteins that have been proposed to play important roles in mediating secretory activity in animals (Gruenberg and Emans 1993). Plant annexins induce aggregation of liposomes or secretory vesicles, suggesting a similar role in plants (Delmer and Potikha 1997). High levels of annexins at the growing tips of pollen tubes and fern rhizoids have been reported (Blackbourn et al. 1992; Clark et al. 1995), implying they may be mediating Ca^{2+}-dependent secretory events in tip growth. However, the Ca^{2+} affinity of plant annexins ranges from 60 to several hundred micromolar depending on the system studied (reviewed in Delmer and Potikha 1997). Such high $[Ca^{2+}]_c$ has not been reported in a plant cell. In addition, plant annexins lack the highly conserved protein motifs that are important for Ca^{2+} binding to animal annexins, suggesting that in plant annexins Ca^{2+} binding is either unimportant or accomplished by a different mechanism (Delmer and Potikha 1997). In addition, several animal annexins exhibit voltage-gated Ca^{2+} channel activity when incorporated into lipid membranes in vitro (Pollard et al. 1992). The similarity between plant and animal annexins in hydrophobicity plots suggests that plant annexins might also exhibit such an activity (Delmer and Potikha 1997). Annexins also bind to actin in a Ca^{2+}-dependent manner (Ikebuchi and Wiseman 1990; Calvert et al. 1996) and so could be involved in regulating the structure of the apical actin cytoskeleton.

4.3 Calmodulin and CDPK

Two other elements of a Ca^{2+} -dependent tip growth regulatory system have been identified from pollen tubes. Calmodulin has been shown to localize to the tip of growing pollen tubes and Calmodulin antagonists inhibit tip growth (Moutinho et al. 1998a). In addition, in pollen tubes a calmodulin-domain protein kinase (CDPK) localizes to the actin cytoskeleton (Putnam-Evans et al. 1989) and growth-related relocalization of a putative Ca^{2+} -dependent protein kinase has been observed (Moutinho et al. 1998b). Such a tip-localized gradient of kinase activity could help regulate Ca^{2+}-dependent exocytosis at the pollen tube tip (Moutinho et al. 1998b). CDPKs are a group of conserved proteins found in plants and protists that exist as large multi-gene families (Roberts and Harmon 1992; Satterlee and Sussman 1998). Most of the putative CDPK substrates are membrane-associated proteins, and many CDPK family members are characterized by an N-terminal myristoylation site that may aid in their membrane associations (Satterlee and Sussman 1998). Considering the ubiquitous nature of protein phosphorylation as a regulator of eukaryotic cell function, Ca^{2+}-regulated kinase and phosphatases represent obvious candidates for Ca^{2+} -dependent regulators of membrane trafficking and apical growth. However, the role of these proteins in regulating growth phenomena in general, and apical growth in particular, remains to be determined.

5 Calcium and Nod Factor Signal Transduction

In addition to its role in growth regulation, Ca^{2+} is recognized as a ubiquitous second messenger and signal transduction element in plant cells (Bush 1995). Such a signaling role in root hairs is perhaps best exemplified in the studies on Nod factor action. Rhizobial lipochito-oligosaccharide Nod-factors are secreted by symbiotic bacteria as signals that lead to alterations in root hair growth and trigger root cortical cell division (Mylona et al. 1995). E.g., in an assay in which root hairs are grown between glass slides, the Nod factors induce swelling of the tip and subsequent new hair outgrowth from the swelling (de Ruijter et al. 1998). These changes in root architecture aid in the establishment of a successful interaction between root and bacterium and involve highly selective signal (Nod factor) recognition and signal transduction events. Challenging root hairs with a Nod-factor results in Ca^{2+} influx, membrane depolarization, cytoplasmic alkalization and a plateau-like increase in $[Ca^{2+}]_c$ (Ehrhardt et al. 1992; 1996; Kurkdjian 1995; Felle et al. 1995; 1996; 1998; Gehring et al. 1997). All these changes in cell physiology occur simultaneously with the cessation of root hair growth, and prior to root hair swelling (Ehrhardt et al. 1992; 1996; Ridge 1992; Felle et al. 1995; 1996; Kurkdjian 1995; Gehring et al. 1997; Cardenas et al. 1998). A general increase in $[Ca^{2+}]_c$ and cytoplasmic alkalization occur simultaneously with, or precede membrane depolarization. Such changes always precede the onset of periodic Ca^{2+} spiking in the nucleus (Gehring et al. 1997; Ehrhardt et al. 1996).

Cytoplasmic [Ca^{2+}] is acting at many levels in this system. Nod factor-related root hair growth is still associated with a [Ca^{2+}]$_c$ gradient. Thus 1-2 h after Nod factor addition to root hairs of *Vicia sativa* new growing tips form and their growth is accompanied by an increase in [Ca^{2+}]$_c$ (de Ruijter et al. 1998). However, the rapid plateau-like change in [Ca^{2+}]$_c$ (Gehring et al. 1997) and Ca^{2+} spiking around the nucleus (Erhardt et al. 1996) elicited by Nod factors may serve a signaling role rather than an element of the apical growth machinery. The oscillations in [Ca^{2+}] reported in the nucleus are particularly intriguing since they represent a unique response element in plant signaling events. They appear to be associated with successful Nod factor perception as they are not induced in nodulation deficient mutants of alfalfa (Erhardt et al. 1996). Signal induced changes in nuclear [Ca^{2+}] have been well documented in animals (Badminton et al. 1996). Malho et al. (1998) have also reported that in tobacco seedlings, cold shock-induced increases in [Ca^{2+}] in the nuclear region can be delayed relative to changes in the cytoplasm, implying some degree of independent control of [Ca^{2+}] and possibly signaling between these 2 compartments. The nuclear spiking seen in root hairs in response to Nod factors may therefore have a role in regulating nuclear events independent of bulk cytoplasmic changes, especially [Ca^{2+}]$_c$ changes associated with the tip growth process. What might these transient increases in the nucleus be regulating? Changes in the nuclear, but not cytoplasmic, [Ca^{2+}] environment have been shown to mediate cAMP-dependent gene expression in mammalian cells (Hardingham et al. 1997) and in tobacco, touch-related calmodulin gene expression may also be correlated to changes in nuclear [Ca^{2+}] (Malho et al. 1998). Thus the localized control of nuclear [Ca^{2+}] has the potential to selectively mediate Nod-factor-related changes in gene expression (Pingret et al. 1998). The oscillatory nature of the change may also relate to information about the Nod factor. Oscillations in [Ca^{2+}]$_c$ have been observed in other plant cells, such as guard cells and proposed to play a role in frequency encoding of information (e.g., [ABA], Staxen et al. 1999). Recent evidence supports the idea that each aspect of the kinetics (rise time, duration, maximum [Ca^{2+}]) of the transient can independently regulate different cellular responses (Dolmetsch et al. 1997). In addition, Ca^{2+}-regulated enzymes such as CAM-kinase II have been shown to be able to decode the different frequencies of [Ca^{2+}] oscillations into different levels of enzyme activity (deKoninck and Schulman 1998). Thus there is evidence for cellular systems that can both encode and decode information in [Ca^{2+}] transients. Whether such encoding underlies the Nod-factor induced changes in nuclear [Ca^{2+}] remains to be determined.

In the cytoplasm, [Ca^{2+}] also seems a candidate for carrying Nod factor information. Gehring et al. (1997) have reported a plateau-like increase in [Ca^{2+}]$_c$ in response to Nod factors, and artificial increases in Ca^{2+} levels, triggered by ionophore application, can induce the membrane depolarization (Felle et al. 1998) and gene expression (Pingret et al. 1998) associated with Nod factor action. Similarly application of a Ca^{2+}-channel blocker prevented Nod factor action, suggesting Ca^{2+} influx to the cytoplasm is critical for successful Nod factor signaling (Felle et al. 1998). What might these changes in [Ca^{2+}]$_c$ be regulating?

Actin may be rapidly fragmented in Nod factor treated root hairs (Cardenas et al. 1998) or show redistribution and formation of many fine bundles of filaments (de Ruijter et al. 1999). Actin severing, alterations in actin filament polymerization kinetics and unbundling of thick actin filaments to finer bundles are all activities readily accomplished by Ca^{2+}. In addition, Ca^{2+} may be coordinating the membrane transport activity associated with Nod factor action. Thus the depolarization of root hairs elicited by Nod factors is caused by Cl^- efflux (Felle et al. 1998) that may well be mediated by the action of Ca^{2+} regulated anion channels (Hedrich 1994).

6 Conclusions and Perspectives

It is perhaps most clear from the work on Nod factor signaling that Ca^{2+} is acting in many roles in the root hair. It is a signal transduction element, a coordinator of cellular activities and a component of the tip growth machinery. How these roles are separated is unclear, but the pronounced spatial (tip focused gradient) and temporal (nuclear Ca^{2+} spiking versus steady cytoplasmic change) separations within the cell are obvious mechanisms to maintain independent regulatory functions. It is also highly unlikely that Ca^{2+} acts in isolation in the root hair. For example, cytosolic pH is well characterized as a second messenger that controls cellular activities in plant (Felle 1989), animal (Epel and Dube 1987) and protist (Kropf et al. 1995) cells, and in particular it can trigger developmental (Sater et al. 1994; Friedman et al. 1990) and morphological (Roncal et al. 1993) transitions (see Chapter 10 this volume). Ca^{2+} and pH are known to be interacting regulatory systems in plants (Felle 1989; Plieth et al. 1997) and so understanding how Ca^{2+} regulates root hair function will be inextricably linked to our progress in understanding such parallel regulatory systems.

ACKNOWLEDGMENTS

The authors thank Elison Blancaflor and Sarah Swanson for critical reading of the manuscript. This work was supported by a grant from the National Science Foundation.

References

Alfano F, Russel A, Gambardella R, Duckett J (1993) The actin cytoskeleton of the liverwort *Riiccia fluitans*: effects of cytochalasin B and aluminum ions on rhizoid tip growth. J Plant Physiol 142:569-574

Allen GJ, Muir SR, Sanders D (1995) Release of Ca^{2+} from individual plant vacuoles by both $InsP_3$ and cyclic ADP-ribose. Science 268:735-737

Ayling SM, Brownlee C, Clarkson DT (1994) The cytoplasmic streaming response of tomato root hairs to auxin: observations of cytoplasmic calcium levels. J Plant Physiol 143:184-188

Badminton MN, Campbell AK, Rembold CM (1996) Differential regulation of nuclear and cytosolic Ca^{2+} in HeLa cells. J Biol Chem 271:31210-31214

Berger F, Brownlee C (1993) Ratio confocal imaging of free cytoplasmic calcium gradients in polarizing and polarized *Fucus* zygotes. Zygote 1:9-15

Bibikova TN, Blancaflor EB, Gilroy S (1999) Microtubules Regulate Tip Growth and Orientation in Root Hairs of *Arabidopsis thaliana*. Plant J 17:657-665

Bibikova TN, Jacob T, Dahse I, Gilroy S (1998) Localized changes in apoplastic and cytoplasmic pH are associated with root hair development in *Arabidopsis thaliana*. Development 125:2925-2934

Bibikova TN, Zhigilei A, Gilroy S (1997) Root hair growth in *Arabidopsis thaliana* is directed by calcium and an endogenous polarity. Planta 203:495-505

Blackbourn HD, Barker PJ, Huskinsson NS, Battey NH (1992) Properties and partial protein sequence of plant annexins. Plant Physiol 99:864-871

Bonnet HT, Newcomb EH (1966) Coated vesicles and other cytoplasmic components of growing root hairs of radish. Protoplasma 62:59-75

Brownlee C, Manison NFH, Anning R. (1998) Calcium, polarity and osmoregulation in *Fucus* embryos: One messenger, multiple messages. Ex Biol Online 3

Brownlee C, Pulsford AL (1988) Visualization of the cytoplasmic Ca^{2+} gradient in *Fucus serratus* zygotes: correlation with cell ultrastructure and polarity. J Cell Sci 91:249-256

Brownlee C, Wood JW (1986) A gradient of free cytosolic calcium in rhizoid cell of *Fucus serratus*. Nature 320:624-626

Bush DS (1995) Calcium regulation in plant cells and its role in signaling. Ann Rev Plant Physiol Plant Mol Biol 46:95-122

Cai G, Moscatelli A, Cresti M (1997) Cytoskeletal organization and pollen tube growth. Trends Plant Sci 2:86-91

Calvert CM, Gant SJ, Bowles DJ (1996) Tomato annexins p34 and p35 bind to F-actin and exhibit nucleotide phosphodiesterase activity inhibited by phospholipid binding. Plant Cell 8:333-342

Cardenas L, Vidali L, Domingues J, Perez H, Sanches F, Hepler PK, Quinto C (1998) Rearrangement of actin microfilaments in plant root hairs responding to *Rhizobium etli* nodulation signals. Plant Physiol 116:871-877

Carroll AD, Moyen C, Van Kesteren P, Tooke F, Battey NH, Brownlee C (1998) Ca^{2+}, Annexins, and GTP modulate exocytosis from maize root cap protoplasts. Plant Cell 10 1267-1276

Clark GB, Turnwald S, Tirlapur UK, Haas CJ, von der Mark K, Roux SJ, Scheuerlein R (1995) Polar distribution of annexin-like proteins during phytochrome-mediated initiation and growth of rhizoids in the ferns *Dryopteris* and *Anemia*. Planta 197:376-384

Cyr RJ (1994) Microtubules in plant morphogenesis: Role of the cortical array. Annu Rev Cell Biol 10:153-180

de Konick P, Schulman H (1998) Sensitivity of CaM kinase II to the frequency of Ca^{2+} oscillations. Science 279:227-230

de Ruijter NCA, Rook MB, Bisseling T, Emons AMC (1998) Lipochito-oligosaccharides re-initiate root hair tip growth in *Vicia sativa* with high calcium and spectrin-like antigen at the tip. Plant J 13:341-350

de Ruijter NCA, Bisseling T, Emons AMC (1999) Rhizobium Nod factors induce an increase in sub-apical fine bundles of actin filaments in Vicia sativa root hairs within min. Mol Plant Microbe Interactions 12:829-832

Delmer DP, Potikha TS (1997) Structures and functions of annexins in plants. Cell Mol Life Sci 53:546-553

Derksen J, Rutten T, Van Amstel T, de Win A, Doris F, Steer M (1995) Regulation of pollen tube growth. Acta Bot Neerl 44:93-119

Dolmetsch RE, Lewis RS, Goodnow CC, Healy JI (1997) differential activation of transcription factors induced by Ca^{2+} response amplitude and duration. Nature 386:855-858

Doonan JH, Cove DJ, Lloyd CW (1988) Microtubules and microfilaments in tip growth: evidence that microtubules impose polarity on protonemal growth in *Physcomitrella patens*. J Cell Sci 89:533-540.

Doris FP, Steer MW (1996) Effects of fixatives and permeabilization buffers on pollen tubes: implication for localization of actin microfilaments using phalloidin staining. Protoplasma 195:25-36

Ehrhardt DW, Wais R, Long SR (1992) Depolarization of alfalfa root hair membrane potential by *Rhizobium meliloti* Nod factors. Science 256:998-1000

Ehrhardt DW, Wais R, Long SR (1996) Calcium spiking in plant root hairs responding to Rhizobium nodulation signals. Cell 85:673-681

Emons AMC (1987) The cytoskeleton and secretory vesicles in root hairs of *Equisetum* and *Limnobium* and cytoplasmic streaming in root hairs of *Equisetum*. Ann Bot 60:625-632

Emons AMC, Derksen J (1986) Microfibrils, microtubules, and microfilaments of the trichoblast of *Equisetum hyemale*. Acta Bot Neerl 35: 311-320

Emons AMC, Wolters-Arts JA, Traas J, Derksen J (1990) The effects of colchicine on microtubules and microfibrils in root hairs. Acta Bot Neerl 39:19-27

Epel D, Dube F (1987) Intracellular pH and cell proliferation. In: Boynton AL, Leffert HM (Eds) Control of animal cell proliferation. Academic Press, New York, pp 363-393

Felle HH (1989) pH as a second messenger in plants. In: WF Boss. DJ Morre (Eds) Second messengers in plant growth and development. AR Liss, New York, pp 145-166

Felle HH, Hepler PK (1997) The cytosolic Ca^{2+} concentration gradient of *Sinapsis alba* root hairs as revealed by Ca^{2+} selective microelectrode tests and Fura-dextran ratio imaging. Plant Physiol 114:39-45

Felle HH, Konorosi E, Konorosi A, Schultze M (1995) Nod signal induced plasma membrane potential changes in alfalfa root hairs are differentially sensitive to structural modifications of lipochitooligosaccharide. Plant J 7:939-947

Felle HH, Konorosi E, Konorosi A, Schultze M (1996) Rapid alkalinization in alfalfa root hairs in response to rhizobial lipooligosaccharide signals. Plant J 10:295-301

Felle HH, Kondorosi E, Kondorosi A, Schultze M (1998) The role of ion fluxes in Nod factor signaling in *Medicago sativa*. Plant J 13:455-463

Fisher DD, Gilroy S, Cyr RJ (1996) Evidence for opposing effects of calmodulin on cortical microtubules. Plant Physiol 112:1079-1087

Fowler JE, Quatrano RS (1997) Plant cell morphogenesis: plasma membrane interactions with the cytoskeleton and cell wall. Annu Rev Cell Dev Biol 13:697-743

Franklin-Tong VE, Drobak BK, Allan AC, Watkins PAC, Trewavas AJ (1996) Growth of pollen tubes of *Papaver rhoeas* is regulated by a slow-moving calcium wave propagated by inositol 1,4,5-triphosphate. Plant Cell 8:1305-1321

Friedman HH, Spiegelstein H, Goldschmidt EE, Halevy AH (1990) Flowering response in *Pharbitis nil* to agents affecting cytoplasmic pH. Plant Physiol 94:114-119

Galway ME, Heckman JJW, Schifiefelbein JW (1997) Growth and ultrastructure of *Arabidopsis* root hairs: the rhd3 mutation alters vacuole enlargement and tip growth. Planta 201:209-218

Garril A, Jackson SL, Lew RR, Heath IB (1993) Ion channels and tip growth: stretch activated channels generate a Ca^{2+} gradient essential for hyphal growth. Eur J Cell Biol 60:358-365

Gehring CA, Irving RI, Kabbara AA, Parish RW, Boukli NM, Broughton WJ (1997) Rapid, plateau-like increases in intracellular free calcium are associated with Nod-factor induced root hair deformation. Mol Plant Microbe Interactions 10:791-802

Girbardt M (1969) Die Ultrastruktur der Apikalregion Pilzyphen. Protoplasma 67:413-441

Gruenberg J, Emans N (1993) Annexins in membrane traffic. Trends Cell Biol 3:224-227

Hagen LM (1998) The fission yeast microtubule cytoskeleton. J Cell Sci 111:1603-1612

Hardingham GE, Chawla S, Johnson C, Bading H (1997) Distinct functions of nuclear and cytoplasmic calcium in the control of gene expression. Nature 385:260-264

Hedrich R (1994) Voltage-dependent chloride channels in plant cells: identification, characterization and regulation of a guard cell anion channel. Curr Topics Membr 42:1-33

Herrmann A, Felle HH (1995) Tip growth in root hair cells of *Sinapis alba* L.: significance of internal and external Ca^{2+} and pH. New Phytol 129:523-533

Heslop-Harrison J, Heslop-Harrison Y, Cresti M, Tiezzi A, Moscatelli A (1988) Cytoskeletal elements, cell shaping and movement in the angiosperm pollen tube. J Cell Sci 91:49-60

Holdaway-Clarke TL, Feijo JA, Hackett GR, Kunkel JG, Hepler PK (1997) Pollen tube growth and the intracellular cytosolic calcium oscillate in phase while extracellular calcium influx is delayed. Plant Cell 9:1999-2010

Holzinger A, Callaham DA, Hepler PK, Meindl U (1995) Free calcium in *Micrasterias*: local gradients are not detected in growing lobes. Eur J Cell Biol 67:363-371

Homann U, Tester M (1997) Ca^{2+}-independent and Ca^{2+}-GTP-binding protein-controlled exocytosis in a plant cell. Proc Natl Acad Sci USA 94:6565-6570

Hyde GJ, Heath IB (1995) Ca^{2+}-dependent polarization of axis establishment in the tip-growing organism, *Saprolegnia ferax*, by gradients of the ionophore A23187. Eur J Cell Biol 67:356-362

Ikebuchi NW, Wiseman DM (1990) Calcium-dependent regulation of actin filament bundling by lipocortin-85. J Biol Chem 265:3392-342

Jaffe L, Nuccitelli R (1977) Electrical control of development. Annu Rev Biophys Bioeng 6:445-476

Jones DL, Gilroy S, Larsen PB, Howell SH, Kochian LV (1998) Effect of aluminum on cytoplasmic Ca^{2+} homeostasis in root hairs of *Arabidopsis thaliana* (L.). Planta 206:378-87

Jones DL, Shaff JS, Kochian LV (1995) Role of calcium and other ions in directing root hair tip growth in *Limnobium stoloniferum*. I. Inhibition of tip growth by aluminum. Planta 197:672-680

Kiermayer O (1981) Cytoplasmic basis for morphogenesis in *Micrasterias*. In: Kiermayer O (Ed.) Cytomorphogenesis in plants. Springer-Verlag, Wien, pp 147-189

Klusener B, Boheim G, Lib H, Engelberth J, Weiler EW (1995) Gadolinium-sensitive, voltage-dependent calcium release channels in the endoplasmic reticulum of a higher plant mechanoreceptor organ. EMBO J 14:2708-2714

Knight H, Trewavas AJ, Read N (1993) Confocal microscopy of living fungal hyphae microinjected with Ca^{2+}-sensitive fluorescent dyes. Mycol Res 97:1505-1515

Kohno T, Shimmen T (1987) Ca^{2+}-induced fragmentation of actin filaments in pollen tubes. Protoplasma 141:461-491

Kohno T, Shimmen T (1988) Accelerated sliding of pollen tube organelles along Characeae actin bundles regulated by Ca^{2+}. J Cell Biol 106:1539-43

Kropf DL (1997) Induction of polarity in Fucoid zygotes. Plant Cell 9:1647-1659

Kropf DL, Henry CA, Gibbon BC (1995) Measurement and manipulation of cytosolic pH in polarizing zygotes. Eur J Cell Biol 68:297-305

Kropf DL, Bisgrove SR, Hable WE (1998) Cytoskeletal control of polar growth in plant cells. Curr Opin Cell Biol 10:117-122

Kropf DL, Quatrano RS (1987) Localization of membrane-associated calcium during development of fucoid algae using chlorotetracycline. Planta 171:158-170

Kuhteiber WM, Jaffe LF (1990) Direction of extracellular calcium gradients with a calcium specific vibrating electrode. J Cell Biol 110:1565-1573

Kurkdjian AC (1995) Role of differentiation of root epidermal cells in Nod factor (from *Rhizobium meliloti*)-induced root hair depolarization of *Medicago sativa*. Plant Physiol 107:783-790

Lancelle SA, Hepler PK (1992) Ultrastructure of freeze substituted pollen tubes of *Lilium longiflorum*. Protoplasma 167:215-230

Levina NN, Lew RR, Heath IB (1994) Cytoskeletal regulation of ion channel distribution in the tip-growing organism *Saprolegnia ferax*. J Cell Sci 107:127-134

Levina NN, Lew RR, Hyde GJ, Heath IB (1995) The roles of Ca^{2+} and plasma membrane ion channels in hyphal tip growth of *Neurospora crassa*. J Cell Sci 108:3405-3417

Li YQ, Zhang HQ, Pierson ES, Huang FY, Linskens HF, Hepler PK, Cresti M (1996) Enforced growth-rate fluctuation causes pectin ring formation in the cell wall of *Lilium longiforum* pollen tubes. Planta 200:41-49

Lloyd CW, Wells B (1985) Microtubules are at the tips of root hairs and form helical patterns corresponding to inner wall fibrils. J Cell Sci 75:225-238

Love J, Brownlee C, Trewavas AJ (1997) Ca^{2+} and calmodulin dynamics during photopolarization in *Fucus serratus* zygotes. Plant Physiol 115:249-261

Malho R (1998a) Expanding tip-growth theory. Trends Plant Sci 3:40-43

Malho R (1998b) Pollen tube guidance – the long and winding road. Sex Plant Reprod 11:242-244

Malho, R (1998c) Role of 1,4,5-inositol trisphosphate-induced Ca^{2+} release in pollen tube orientation. Sex Plant Reprod 11:231-235

Malho R, Read ND, Pais MS, Trewavas AJ (1994) Role of cytosolic free calcium in the reorientation of pollen tube growth. Plant J 5:331-341

Malho R, Read ND, Trewavas AJ, Pais MS (1995) Calcium channel activity during pollen tube growth and reorientation. Plant Cell 7:1173-1184

Malho R, Trewavas AJ (1996) Localized apical increases of cytosolic free calcium control pollen tube orientation. Plant Cell 8:1935-1949

Malho R, Moutinho A, van der Luit A, Trewavas AJ (1998) Spatial characteristics of calcium signalling: The calcium wave as a basic unit in plant cell calcium signalling. Phil Trans Roy Soc London B 353:1463-1473

Masucci JD, Schiefelbein JW (1994) The rhd6 mutation of *Arabidopsis thaliana* alters root hair initiation through an auxin and ethylene associated process. Plant Physiol 106:1335-1346

Meekes HTHM (1985) Ultrastructure, differentiation and cell wall texture of trichoblasts and root hairs of *Ceratopteris thalictroides* (L) Brongn. (Parkeriaceae). Aquatic Bot 21:347-362

Messerli M, Robinson KR (1997) Tip localized Ca^{2+} pulses are coincident with peak pulsatile growth rates in pollen tubes of *Lilium longiflorum*. J Cell Sci 110:1269-1278

Miller DD, de Ruijter NCA, Emons AMC (1997a) From signal to form: aspects of the cytoskeleton plasma membrane cell wall continuum in root hair tips. J Exp Bot 48:1881-1896

Miller D, Hable W, Gottwald J, Ellard-Ivey M, Demura T, Lomax T, Carpita N (1997b) Connections: the hard wiring of the plant cell for perception, signaling, and response. Plant Cell 9:2105-2117

Miller DD, Callaham DA, Gross DJ, Hepler PK (1992) Free Ca^{2+} gradient in growing pollen tubes of *Lilium*. J Cell Sci 101:7-12

Miller DD, de Ruijter NCA, Bisseling T, Emons AMC (1999) The role of actin in root hair morphogenesis: studies with lipochito-oligosaccharide as a growth stimulator and cytochalasin as an actin perturbing drug. Plant J 17: 141-154

Mizukami M, Wada S (1983) Morphological anomalies induced by antimicrotubule agents in *Bryopsis plumosa.* Protoplasma 114:151-162

Moutinho A, Love J, Trewavas AJ, Malho R (1998a) Distribution of calmodulin protein and mRNA in growing pollen tubes. Sex Plant Reprod 11:131-139

Moutinho A, Trewavas AJ, Malho R (1998b) Relocation of a Ca^{2+}-dependent protein kinase activity during pollen tube reorientation. Plant Cell 10:1499-510

Muir SR, Sanders D (1997) Inositol-1,4,5-trisphosphate-sensitive Ca^{2+}-release across nonvacuolar membranes in cauliflower. Plant Physiol 114:1511-1521

Mylona P, Pawlowski K, Biesseling T (1995) Symbiotic nitrogen fixation. Plant Cell 7:869-885

Nuccitelli R, Jaffe LF (1976) Current pulses involving chlorine and potassium efflux relieve excess pressure in *Pelvetia* embryos. Planta 131:315-32

Obermeyer G, Weisenseel MH (1991) Calcium channel blocker and calmodulin antagonists affect the gradient of free calcium ions in growing pollen tubes. Eur J Cell Biol 56:319-327

Oppenheimer DG, Pollock MA, Vacik J, Szymanski DB, Ericson B, Feldmann K, Marks MD (1997) Essential role of a kinesin-like protein in *Arabidopsis* trichome morphogenesis. Proc Natl Acad Sci USA 94:6261-6266

Peterson RL, Farquhar ML (1996) Root hairs: specialized tubular cells extending root surfaces. Bot Rev 62:2-33

Picton JM, Steer MW (1983) Evidence for the role of Ca^{2+} ions in tip extension in pollen tubes. Protoplasma 115:1-17

Pierson ES, Miller DD, Callaham DA, Shipley AM, Rivers BA, Cresti M, Hepler PK (1994) Pollen tube growth is coupled to the extracellular calcium ion flux and the intracellular calcium gradient: effect of BAPTA-type buffers and hypertonic media. Plant Cell 6:1815-1828

Pierson ES, Miller DD, Callaham DA, vanAken J, Hackett G, Hepler PK (1996) Tip-localized calcium entry fluctuates during pollen tube growth. Dev Biol 174:160-173

Pingret JL, Journet EP, Barker DG (1998) *Rhizobium* Nod factor signaling: evidence for a G-protein-mediated transduction mechanism. Plant Cell 10:659-671

Plieth C, Sattelmacher B, Hansen U-P (1997) Ca^{2+}-H^{+}-exchange buffers in green algae. Protoplasma 198:107-124

Pollard HB, Guy HR, Arispe N, de la Fuente M, Lee G, Rojas EM, Pollard JR, Srivastava M, Zhang-Keck ZY, Merezhinskaya N (1992) Calcium channel and membrane fusion activity of synexin and other members of Annexin gene family. Biophys J 62:15-18

Pu R, Robinson R (1998) Cytoplasmic calcium gradients and calmodulin in the early development of the fucoid alga *Pelvetia compressa.* J Cell Sci 111:3197-3207

Putnam-Evans C, Harmon AC, Palevitz BA, Fechheimer M, Cormier MJ (1989) Calcium-dependent protein kinase is localized with F-actin in plant cells. Cell Motil Cytoskel 12:12-22

Rathore KS, Cork RJ, Robinson KR (1991) A cytoplasmic gradient of Ca^{2+} is correlated with the growth of lily pollen tubes. Dev Biol 148:612-619

Ridge RW (1988) Freeze substitution improves the ultrastructural preservation of legume root hairs. Bot Mag 101:427-441

Ridge RW (1992) A model of legume root hair growth and *Rhizobium* infection. Symbiosis 14:359-373

Roberts DM, Harmon A (1992) Calcium-modulated proteins targets of intracellular calcium signals in higher plants. Annu Rev Plant Physiol Plant Mol Biol 43:375-414

Robinson KR (1996) Calcium and the photopolarization of *Pelvetia* zygotes. Planta 198:378-384

Robinson KR, Jaffe LF (1973) Ion movements in a developing Fucoid egg. Dev Biol 35:349-361

Robinson KR, Jaffe LF (1975) Polarizing Fucoid eggs drive a calcium current through themselves. Science 187:70-75

Roncal T, Ugalde UO, Irastorza A (1993) Calcium-induces conidiation in *Penicillium cyclopium*: calcium triggers cytosolic alkalinization at the hyphal tip. J Bacteriol 175:879-886

Rosen WG, Gawlik SR, Dashek WV, Siegesmund KA (1964) Fine structure and cytochemistry of *Lilium* pollen tubes. Am J Bot 51:61-71

Sater AK, Alderton JM, Steinhardt (1994) An increase in intracellular pH during neural induction in *Xenopus*. Development 120:433-442

Satterlee JS, Sussman MR (1999) Unusual membrane-associated protein kinases in higher plants. J Memb Biol 164:205-213

Sato S, Ogasawara Y, Sakuragi S (1995) The relationship between growth, nucleus migration and cytoskeleton in root hairs of radish. In: F Balishka, M Ciamporova, O Gasparikova, PW Barlow (Eds) Structure and function of roots. Kluwer Academic Press, Dordrecht, pp 69-74

Schiefelbein JW, Shipley A, Rowse P (1992) Calcium influx at the tip of growing root-hair cells of *Arabidopsis thaliana*. Planta 187:455-459

Schiefelbein JW, Somerville C (1990) Genetic control of root hair development in *Arabidopsis thaliana*. Plant Cell 2: 235-243

Schmiedel C, Schnepf E (1979) Side branch formation and orientation in the caulonema of the moss, *Funaria hyrometrica*; experiments with inhibitors and with centrifugation. Protoplasma 101:47-59

Schnepf E (1986) Cellular polarity. Annu Rev Plant Physiol 37:23-47

Seiler S, Nargang FE, Steinberg G, Schiwa M (1997) Kinesin is essential for cell morphogenesis and polarized secretion in *Neurospora crassa*. EMBO J 16:3025-3034

Shaw S, Quatrano, RS (1996) Polar localization of a dihydropyridine receptor on living *Fucus* zygotes. J Cell Sci 109:335-342

Sievers A, Schnepf E (1981) Morphogenesis and polarity in tubular cells with tip growth. In: Kiermayer O (Ed) Cytomorphogenesis in Plants. Springer-Verlag, New York, pp 265-299

Smith RD, Wilson JE, Walker JC, Baskin TI (1994) Protein phosphatase inhibitors block root hair growth and alter cortical cell shape of *Arabidopsis* roots. Planta 194:516-524

Speksnijder JE, Miller AL, Weisenseel MH, Chen TH, Jaffe LF (1989) Calcium buffer injections block Fucoid egg development by facilitating calcium diffusion. Proc Natl Acad Sci USA 86:6607-6611

Staxen I, Pical C, Montgomery LT, Gray JE, Hetherington AM, McAinsh MR (1999) Abscisic acid induced oscillations in guard cell cytosolic free calcium that involve phosphoinositide-specific phospholipase C. Proc Natl Acad Sci USA 96:1779-1784

Taylor AR, Manison N, Brownlee C (1997) Regulation of channel activity underlying cell volume and polarity signals in *Fucus*. J Exp Bot 48:579-588

Taylor AR, Manison N, Fernandez C, Wood J, Brownlee C (1996) Spatial organization of calcium signalling involved in cell volume control in the *Fucus* rhizoid. Plant Cell 8:2015-2031

That TC, Rossier C-T, Barja F, Turian G, Roos UP (1988) Induction of multiple germ tubes in *Neurospora crassa* by anti-tubulin agents. Eur J Cell Biol 46:68-79

Tominaga M, Morita K, Sonobe S, Yokata E, Shimmen T (1997) Microtubules regulate the organization of actin filaments at the cortical region in root hair cells of *Hydrocharis*. Protoplasma 199:83-92

Traas JA, Braat P, Emons AM, Meekes H, Derksen J (1985) Microtubules in root hairs. J Cell Sci 76:303-320

van Amstel ANM, Derksen J (1993) The complex helical texture of the secondary cell wall of *Urtica dioica* root hairs is not controlled by microtubules: A quantitative analysis. Acta Bot Neerl 42:141-151

Ward J, Schroeder JI (1994) Calcium-activated K^+ channels and calcium-induced calcium release by slow vacuolar ion channels in guard cell vacuoles implicated in the control of stomatal closure. Plant Cell 6:669-683

White PJ (1998) Calcium channels in the plasma membrane of root cells. Annals Bot 81:173-183

Wu Q, Sandrock T, Turgeon BG, Yoder OG, Wirsel SG, Aist JR (1998) A fungal kinesin required for organelle motility, hyphal growth, and morphogenesis. Mol Biol Cell 9:89-101

Wymer CL, Bibikova TN, Gilroy S (1997) Cytoplasmic free calcium distributions during the development of root hairs of *Arabidopsis thaliana*. Plant J 12:427-39

Yamashita RA, May GS (1998) Motoring along the hyphae: molecular motors and the fungal cytoskeleton. Curr Opin Cell Biol 10:74-79

Chapter 10
pH Regulation in and by Root Hairs

Hubert H. Felle and Almut Herrmann

Botanisches Institut, Justus-Liebig-Universität, Senckenbergstr. 17, D-35390 Giessen, Germany

1 Introduction

Plants live in a proton world. Since all life is based on aqueous chemistry and because water spontaneously ionizes, protons are part of any aqueous solution and cannot be removed completely from their milieu: their free concentration must be regulated. Being both substrate and product of metabolic pathways, they have the apparent potential to play a synergistic role between cellular compartments or substrates, which otherwise share no common effector. Protons are ions which possess the ability to translocate energy across membranes, either as electrical charge, as chemical gradient or jointly as proton motive force ("pmf"). This emerges from their importance as an energy link during adenosine triphosphate (ATP) synthesis in energy-conserving membranes (thylakoids and cristae) on the one hand, and from their role as the major ion during primary active membrane transport to form a general energy potential on the other hand. A wide variety of transporters use this energy to drive the translocation of ions, required for cellular function, or of organic substrates (e.g., sugars) from which the cells may draw and regain basic energy.

As root hairs develop at the utmost periphery of a plant, they mediate between the symplast, the apoplast, and the rhizosphere by reaching out into this space through rapid tip growth and the formation of a fine aqueous layer on the rhizoplane. In this layer substances are dissolved, sorted, taken up or excluded from transport or, in case the substance represents a signal, reacted to otherwise. Since this unstirred layer, which functionally includes the cell wall, is adjacent to the plasma membrane and is only some micrometres in dimension, its ionic composition must be tightly controlled to guarantee proper transport, whereby pH plays a fundamental role.

Key words. apoplast, auxin, calcium, elongation, G-protein, ion-selective microelectrode, *Medicago sativa*, Nod-factor, pH-regulation, root hairs, second messenger, *Sinapis alba*, tip growth, *Zea mays*

2 Proton-Dependent Transport

Root hairs, through their enlarged surface, rapid growth, and specific transporters are superior tools to take up ions, nutrients, and phytohormones from the soil. Through active H^+ extrusion large parts of roots and root hairs acidify their cell walls, the adhering water film and thus the rhizosphere to bring substances to be taken up into a transportable form, to facilitate their uptake through specialized transport systems (H^+symport), and to provide the energy, necessary to drive such transport.

2.1 Active Proton Extrusion

Proton extrusion by the plasma membrane H^+ ATPase (proton pump) is one of the most prominent properties of root hairs. As can be shown by vibrating probe analysis, net proton currents of several pmol cm^{-2} s^{-1} leave the cells and make minerals available for uptake. Apart from this, the protons which are translocated across the plasma membrane, generate a membrane potential and at the same time build up a transmembrane H^+ gradient. Both quantities can be changed independently of each other and can act as such, but as a sum they amount to the "pmf", a thermodynamic entity, which drives the transport and intracellular accumulation of a wide array of substances. As shown in Fig. 1, with decreasing external pH the "pmf" gets more and more negative, until a pH is reached where pH change and depolarization just balance. It is interesting and certainly more than accidental that the maximal "pmf" is attained around pH 5, whereas the pH of the apoplast, of the adhering water film, and of the bulk are in equilibrium (see below).

2.2 Proton Symport

Apart from the regulatory role of proton symport for pH-homeostasis (see below), the transport of inorganic ions, organic substrates, and phytohormones together with protons is abundant in root hairs, and most important for the entire plant. Although it can be shown that root hairs will take up sugars and amino acids when available, anions such as nitrate, phosphate, sulphate, and also abscisic acid or auxins appear to be the main symport transportees. There is a background of a rather negative membrane potential, and anions cannot overcome this energy barrier by themselves, i.e., these anions cannot enter the root hairs through channels, and they need a driver ion, the proton. It may be of secondary interest here, but it should be mentioned briefly, that most of these anions, when added to root hairs, transiently depolarize the cells and alkalize the external space, thus boosting the idea that these substances are not simply symported with the amount of protons that neutralize their negative charges, but with at least one charge in excess. This uptake of protons together with anions has consequences for the cytoplasm in that it is rapidly acidified (Felle 1994), and hence will or may cause changes in the activities of other proton-dependent transporters and in pH-dependent metabolic interchanges.

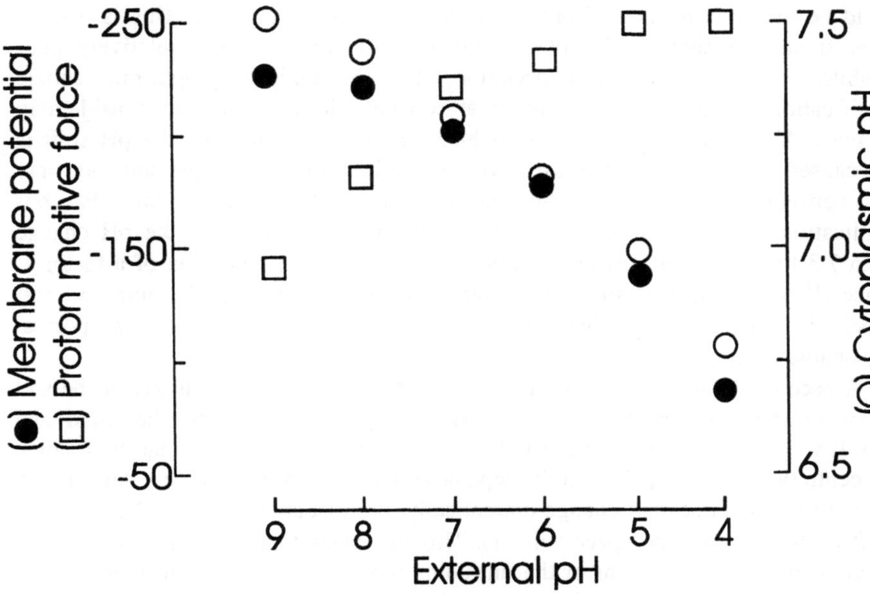

Fig. 1. Proton-motive-force (*open squares*) of *Sinapis alba* root hairs as the sum of the membrane potential (*filled circles*) and the pH-gradient across the plasma membrane, given in millivolts as calculated from the cytoplasmic pH (*open circles*) and extracellular pH, as indicated

3 Cytosolic pH Regulation

Although pH regulation of the cytosol is inherent in all cells, root hairs, according to their specific functions within the rhizosphere, should possess highly sophisticated strategies to work pH homeostasis. One of the major challenges roots and root hairs quite frequently experience is a low oxygen supply, a problem which is increasingly difficult in cases of heavy rain or flooding. Under these extreme conditions, which may last for an extended period of days, the root hairs have to maintain functionality, i.e., the primary active transporter of the plasma membrane, the H^+ ATPase, has to be kept active. In case this is not achieved, the root hairs cease to grow, and transport of ions, nutrients, and phytohormones will rapidly decline. It can be demonstrated that, as long as the oxygen supply is not too low or shut off completely, the H^+ ATPase works quite efficiently. A common observation is that the cytoplasmic pH is rapidly decreased by 0.6 units under anoxia or hypoxia (Fox et al. 1995; Felle 1996), which at first glance appears to go along with an inhibition of oxidative phosphorylation and consequently a deactivated H^+ ATPase. Since experiments indicate that deactivation of the pump and cytoplasmic acidification occur simultaneously, the cytoplasmic acidification could be the consequence of protons leaking into the cytoplasm. This is not the case. It can be shown with oligomycin or antimycin A, both relatively poor inhibitors of oxidative phosphorylation, that the typical cytoplasmic anoxic acidification of 0.6 units takes place at concentrations too low to considerably decrease ATP supply for the pump, which strongly indicates that the pH shift is not caused by the pump but must have metabolic reasons. This pH shift not only has consequences for glycolytic reactions (Fox et al. 1995), but also has implications for the pump activity (Felle 1996). With respect to the pH change from 7.3 to 6.7, this means an increase in free proton concentration of a factor 4. Since H^+ is the transport substrate of the H^+ ATPase, this gives the pump a boost which, in spite of a possible drop in [ATP], will keep substrate transport at reasonable rates.

A second challenge for root hairs are the high ionic concentrations of the soil fluids or their concentration changes (including pH) which can be rapid and massive. Electrophysiological studies on root hair cells show that high ionic concentrations or low pH strongly depolarize the plasma membrane. Amazingly, the cytosolic pH does not change dramatically, but just about 0.1 to 0.2 pH per pH shift in the extracellular space (Fig. 1). This can mean that the plasma membrane is relatively tight for protons or that intruding protons are rapidly consumed. Such consumation can be mastered by three different mechanisms: (1) Proton export, i.e., by the plasma membrane proton pump into the extracellular space, by the tonoplast proton pumps into the vacuole or by other proton transporters into various intracellular compartments. Since protons are transport substrates, a cytosolic acidification usually stimulates these transporters. Whether export of protons is a reasonable strategy to dispose of excess acid depends entirely on the question whether the exported protons are removed from the extracellular space or remain there to acidify this space. In the latter case the futile circle will result in a deactivation of the pump. (2) Cytosolic buffering: at pH 7 cytosolic proton buffering is in the range of 50 to 70 mM H^+/pH. Such values are obtained by challenging the cytosolic pH with weak acids which, due to the higher membrane solubility of the protonated over the dissociated form, enter the cell in their protonated form and release protons which acidify the cytosol according to the pK_a of the acid and its concentration added externally. (3) Metabolic proton consumption. Although we know the essentials of metabolism and the places where protons are formed or consumed by the individual chemical reactions, its contribution to pH-regulation is only partly understood. This is because we are looking at a complicated network of proton-consuming and -generating reactions, which determine the basic cytosolic pH. Thus, permanent cytosolic pH shifts for the purpose to initiate a change in growth or an adaption to hypoxia must come from metabolism, whereas pH perturbations from outside are tempered by cytoplasmic buffers and reversed by transport.

4 The Internal pH of Root Hairs: a Cytosolic Tip-to-Base Gradient?

It is well established that tip-growing cells develop a tip-to-base Ca^{2+} gradient (see Chapter 9 this volume). With pH this seems less clear. Gibbon and Kropf (1994) have demonstrated a pH-gradient in *Pelvetia* rhizoid cells, i.e., they observed a lower pH at the tip than in the remainder of the cell. Also in pollen tubes, Feijo et al. (1999) demonstrate pH differences within the tip which, however, were not simple gradients from tip to base, but an acidic tip is followed by an alkaline band behind the tip and a "normal" pH (slightly alkaline) for the rest of the cell. In root hairs tip-to-base pH gradients are not demonstrated yet. Thus, microelectrode tests carried out on *Sinapis* root hairs did not clearly show the existence of a substantial pH difference between the apex and the regions behind (Herrmann and Felle 1995), although it should be said that the sensitive tip of a pH microelectrode only measures a small volume within a cell, i.e., the demonstration of a pH gradient requires several separate measurements. On the other hand, if substantial pH-gradients in root hairs really existed, they should have been demonstrated in one way or the other with some of the available techniques by now. There are indications, however, that the pretreatment of the root hairs and their state of growth may considerably influence the pH. Vibrating probe tests revealed inwardly as well as outwardly directed H^+-fluxes, which seemed to depend on the external ionic conditions chosen (Herrmann and Felle 1995). Weisenseel et al. (1979) and Jones et al. (1995) demonstrated inwardly directed H^+ fluxes in most of the cells tested and in rare cases proton efflux. Since tip growth exists with both net H^+ influx and H^+ efflux, a direct and regulating influence of these net fluxes can be safely excluded. Although Ca^{2+} net fluxes into the tip are usually brought into connection with the intracellular $[Ca^{2+}]$ gradient in root hairs (Felle and Hepler 1997; Wymer et al. 1997), this may not be true for pH. This holds even though a number of cellular processes central to tip growth such as cytoskeletal organization, vesicle fusion, and enzyme activity are sensitive to pH or say to changes in pH. Since this pH sensitivity is usually observed as "inhibitory", a direct regulatory role of pH is not necessarily given. In any case, there is no obvious reason why the apex of a growing root hair should clearly be different from the regular well-controlled cytosolic pH of 7.2 ± 0.2, and it would be an error to infer from inwardly directed net proton fluxes a more acidic cytosol, or to conclude from such fluxes an alkaline exterior. As noted above, proton net extrusion could be attributed to a pump, but proton influx cannot, and suggestions that a reduced pump activity in this region may be causing the apparent H^+ influx would not be to the point and indeed have not been supported by data (Harold and Caldwell 1990).

Despite some reports on the existence of intracellular pH gradients in other tip growing cells (Turian 1979; Roncal et al. 1993; Gibbon and Kropf 1994; Feijo et al. 1999) their general importance for tip growth has not yet been established. A rigorous quantitative spatial analysis of cytoplasmic pH in three taxonomically diverse types of tip-growing cells (fungal hyphae, pollen tubes, and fern rhizoids) has been carried out and in none of them were pronounced pH gradients within the apical 50 μm found, indicating that pH gradients may not be crucial to tip growth in plant and fungal cells in general (Parton et al. 1997). The observation that manipulation of the cytoplasmic pH may inhibit tip growth (Herrmann and Felle 1995; Parton et al. 1997) may show that a well-regulated cytoplasmic pH is a prerequisite for proper tip growth, but would not point to the causally linked involvement of a pH gradient therein.

5 External pH Regulation

Apart from these interactive pH changes, the root hair space is a well-regulated pH-sphere. It can easily be demonstrated that the pH within that space differs considerably from the pH in the surrounding medium or in the larger rhizosphere. In that respect, pH values around 5 to 5.5 appear optimal, since the pHs of medium, root hair space, and surface (and finally apoplast) are in an equilibrium then (Fig. 2). In this context it should be mentioned that ferric chelate reductase, which is thought to be localized in root hairs, has its optimum exactly in that pH range (Moog et al. 1995). The main role in holding this equilibrium is due to the activity of the plasma membrane H^+ ATPase which constantly exports H^+ from the cytoplasm into the external spaces and as such contributes to external acidification and pH homeostasis.

Since root hairs are in close contact with their surrounding soil and the adhering water film is rather thin, low pump activity should rapidly acidify this layer to a very low pH. Experience shows that this is not the case, i.e., external pH must be regulated through proton feedback: either the proton pump must be regulated down, or H^+ reentry takes place. This problem is illustrated in Fig. 2: under conditions where external pH is above 5, surface pH is lower than the bulk pH and one finds it easy to argue that this is due to pump activity. However, in conditions where external pH is lower than 5, surface pH is not more but less acidic than the surrounding medium, indicating that the H^+ pump cannot be responsible. There are two ways to envisage H^+-import, either through H^+ channels or by H^+ symport. Although H^+ channels exist for instance in certain H^+ATPases, free H^+ channels in root hairs appear highly unlikely for energy reasons. H^+ entry on its own means dissipation of valuable metabolic energy and need not be considered seriously. It is conceivable that in specialized cells or intracellular compartments, well-regulated H^+ channels exist as an overflow valve or to produce heat, but at the plasma membrane of a transporting cell such an arrangement would largely be counterproductive and an energy waste. Therefore, H^+ co-transporters are probably the regulatory force there. In this context, especially anion symporters which translocate Cl^- or NO_3^- with transport stoichiometries of nH^+/anion may contribute considerably to the H^+ balance. The more acidic the solution becomes, the higher the inwardly directed proton motive force, as long as the voltage generated by the H^+ export thermodynamically exeeds the electrical force due to H^+ reentry (Fig. 1). In fact it can be demonstrated that in a solution with no membrane permeable anions present the pH drops to much

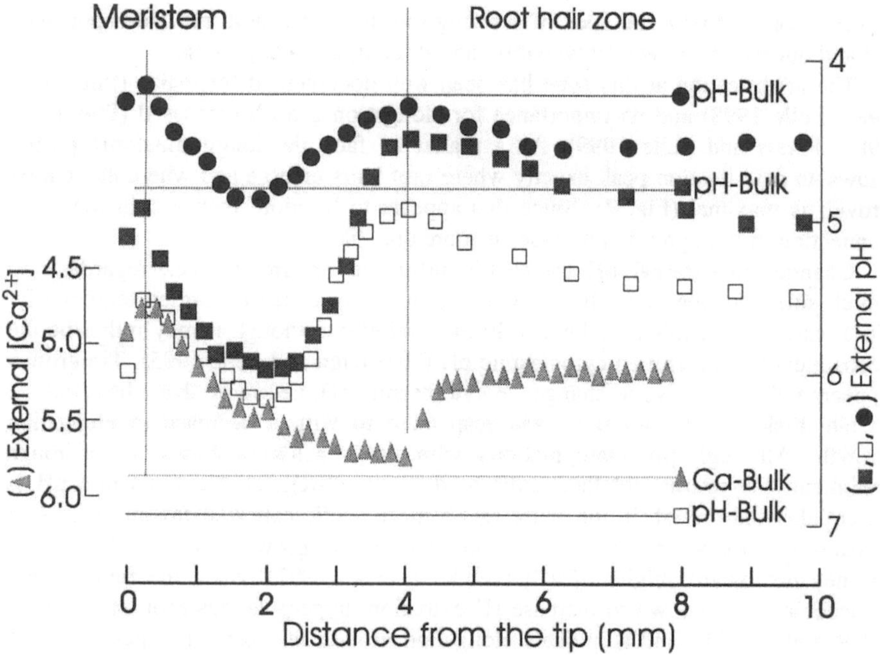

Fig. 2. Longitudinal profiles of pH, and free [Ca^{2+}] *(filled triangles)* of *Zea mays*, measured with ion-selective microelectrodes at a distance of 5 μm from the root surface at three different bulk pHs *(filled circles, filled squares, open squares)*, as indicated by the *straight lines* (Felle unpublished 1998)

lower values, indicating the role proton symporters play in apoplastic pH homeostasis.

6 A Role of External pH for Tip Growth?

Bibikova et al. (1998) demonstrated with confocal imaging that the pH of the cell wall acidifies at the root hair initiation site. This acidification is present from the first morphological indications of localized growth, but not before, and is maintained to the point where the process of root hair initiation ceases and tip growth begins. Preventing the wall acidification with pH buffers arrests the initiation process but growth resumes when the wall is returned to an acidic pH. Cytoplasmic pH was found to be elevated from 7.3 to 7.7 at the initiation site, and this elevation coincided with the acidification of the wall. Preventing a localized increase in cytosolic pH with butyrate, however, inhibited neither the wall acidification nor the initiation process. In contrast, there is no detectable gradient in pH associated with the apex of growing root hairs, but both elevated apoplastic pH and butyrate treatment irreversibly inhibit the tip growth process.

The acidification at this zone has been well documented for maize (Pilet et al. 1983; Felle 1998) and its importance for elongation growth discussed (Pilet et al. 1983; Peters and Felle 1999). As a matter of fact, the longitudinal pH profile shows an acidification peak exactly where root hairs emerge and where elongation growth is maximal (Fig. 2). Since this appears to be more than a coincidence, a connection of both growth processes is more than likely.

Changes in external pH transiently influence tip growth, i.e., regardless to which side, alkaline or acidic, growth is stopped or at least retarded temporarily. To a lesser extent this may be a truly external effect, though it may rather be the consequence of an altered cytoplasmic pH (Herrmann and Felle 1995). Tip growth appears rather sensitive to changes in cytoplasmic pH, i.e., more than about 0.3 pH to either side is not tolerated and responded to with a decrease in elongation growth. Although pH manipulations with weak acids or bases are a rough treatment, the results obtained indicate that a well-regulated cytoplasmic pH is essential. Acid growth in the sense that a more acidic cell wall favors tip growth apparently does not take place, i.e., root hairs do not grow towards acidic regions as they do towards higher [Ca^{2+}] (Bibikova et al. 1997). Also, the fungal toxin fusicoccin (FC), known to increase H^+ extrusion, hyperpolarizes root hairs (Marrè 1979; Felle 1982), and stimulates elongation growth in most cell types of higher plants, but apparently has no stimulating influence on tip growth in root hairs. Although it is possible that a well-regulated pH on both sides of the plasma membrane is an important factor of tip growth in root hairs, an active and specific role of pH in this growth type remains so far obscure.

7 Root Hairs, Antennas in the Rhizosphere

Root hairs are chemosensors that make a first encounter with the conditions of the rhizosphere they grow into (Boller 1995). In doing so, they recognize the presence of other organisms (their own kind, microorganisms or fungi) or their cell wall fractions and respond accordingly with either defense reactions or symbiotic responses. Within the wide range of such immediate responses, which include, for instance, programmed cell death, oxidative burst, production of phytoalexins, ion fluxes, etc, extracellular alkalinization is quite common.

As a number of reports state, the alkalinization response is not confined to root hairs, but involves also cells from all kinds of plant tissues. As such, Felix et al. (1993) demonstrated that lipochito-oligosaccharides, so-called Nod-factors, but also chitotetraose, their oligosaccharide backbone, induced such alkalinization in tomato suspension cultured cells, and Nürnberger et al. (1994) showed that parsley likewise reacted with alkalinization to an encounter with oligopeptides, just to give a few examples. Out of this common defense reaction, in which Nod-factors act as elicitors, root hairs of legumes have developed host-specific symbiotic partnerships with rhizobial bacteria. Thus, in root hairs of alfalfa the sulfated lipochito-oligosaccharide tetramer elicits a variety of strong reactions (depolarization, ion fluxes, alkalinization) of transient character, whereas

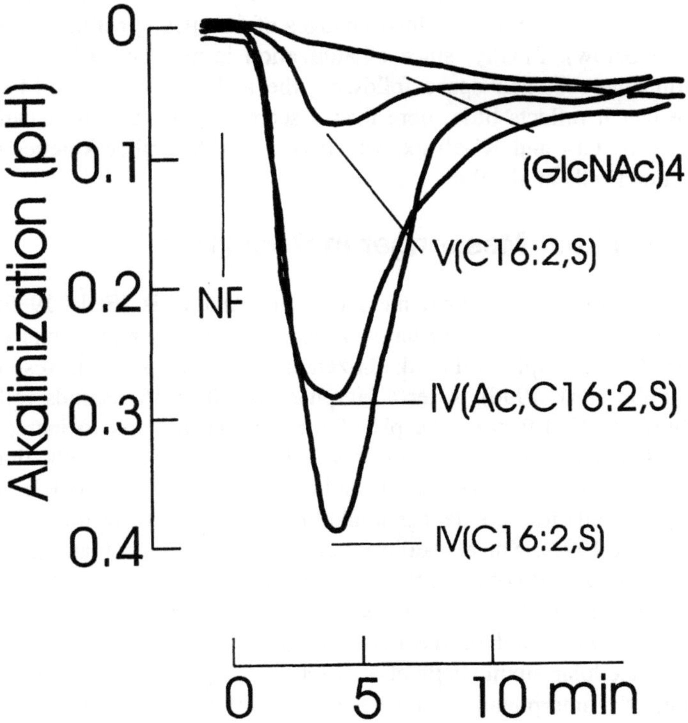

Fig. 3. Alkalinization of the root hair space of *Medicago sativa* after the addition of different Nod-factors (NF), as indicated, at 0.1 μM each, or of chitotetraose at 1 μM to the external solution. Measured with a pH-sensitive microelectrode (Felle et al. unpublished 1998)

chitotretraose causes only a minor persistent alkalinization. Typically, this alkalinization starts 10 to 15 s after the first encounter of the root hairs with the Nod-factor, peaks after about 3 min, strongly depends on the chemical decoration of the Nod-factor, and is refractory to a second addition of the Nod-factor (Fig. 3). The interpretation of this root hair space alkalinization is still a matter of debate. It is conceivable that the weak response to even high concentrations of chitotetraose is the remains of an original unspecific but stronger reaction to attacking microorganisms, which has been developed further in two stages to distinguish between friend and foe. As such, this could be an immediate way to make external conditions unfavorable for the attacking organism: a pH change could lead to a deactivation of certain ion channels or alter affinities for cell wall-destroying enzymes, all of which are reactions that buy time for the attacked organism to fight the intruder with more effective weapons. The distinction between potential attackers and a symbiotic partner could then be achieved by suppression of the defense response, or by transforming it into a transient, as demonstrated. On the other hand, the transient could also be a signal to trigger preformed reactions, in the manner we know it from intracellular changes of free [Ca^{2+}] acting as a second messenger (see below). Finally, since alkalinization is not the earliest response observed within the root hair space following the addition of the Nod-factor, it could also be that it reflects little more than a secondary reaction to a variety of ion fluxes such as Cl^-- and K^+-efflux, which occur with similar kinetics during Nod-factor action (Felle et al. 1998).

8 pH, a Cellular Messenger in Root Hairs?

pH has been proposed as a second messenger in plant cells (Felle 1989; Blatt 1992). Are there pH-signals in root hairs, which initiate or trigger signal chains? Yes, there are. One example is the alkalinization of the cytoplasm in response to Nod-factors (Felle et al. 1996). After a lag-phase of 10 to 15 seconds following external addition of Nod-factors, the pH of the cytoplasm starts to increase just before the cells depolarize. It is not clear yet which function this alkalinization actually has, however, we know that cytosolic pH controls K^+ channels (inward rectifier) in guard cells to the extent that alkalinization blocks inward K^+ currents (Grabov and Blatt 1997). Since alkalinization remains persistent as long as the Nod-factors are present, it could be that this pH change reflects a metabolic shift which goes along with an altered metabolism, required to host the rhizobia.

A second example would be the cytosolic alkalinization in root hairs in the presence of extracellular auxin (Felle et al. 1996). As argued above with the Nod-factor example, the interpretation of such a pH shift is not simple, because its origin is uncertain. Since at the same time the extracellular space also alkalizes, transport of protons is excluded, i.e., protons must have been consumed metabolically. Substances like auxin have a profound impact on tip growth. As shown in Figure 4, membrane depolarization, cytosolic alkalinization and tip growth appear in a causal relationship. The auxin signal in root hairs appears to be mediated by G proteins, which is indicated by two separate observations: (1) after injection of $GDP_{\beta}S$ into a root hair, which should block the respective signal chain, the auxin-induced alkalinization is strongly reduced (Fig. 4b). (2) MAS 7, which has been reported to potently stimulate G-proteins, acidifies the cytosol and completely blocks the response to indole acetic acid (IAA). Although the exact rationale behind these effects has not been found yet, they indicate that a cytosolic pH change resulting from an external signal may be more than just a side effect, and appears to be an integrative link within a signal transduction chain connected with tip growth.

9 Technical Aspects

Measuring the concentrations of free protons or any other ion inside living cells or their movements outside cells leading to electrical currents requires the use of highly sophisticated techniques, which include ionselective microelectrodes, fluorescent dyes, NMR-spectroscopy, and vibrating probe analysis. Since the

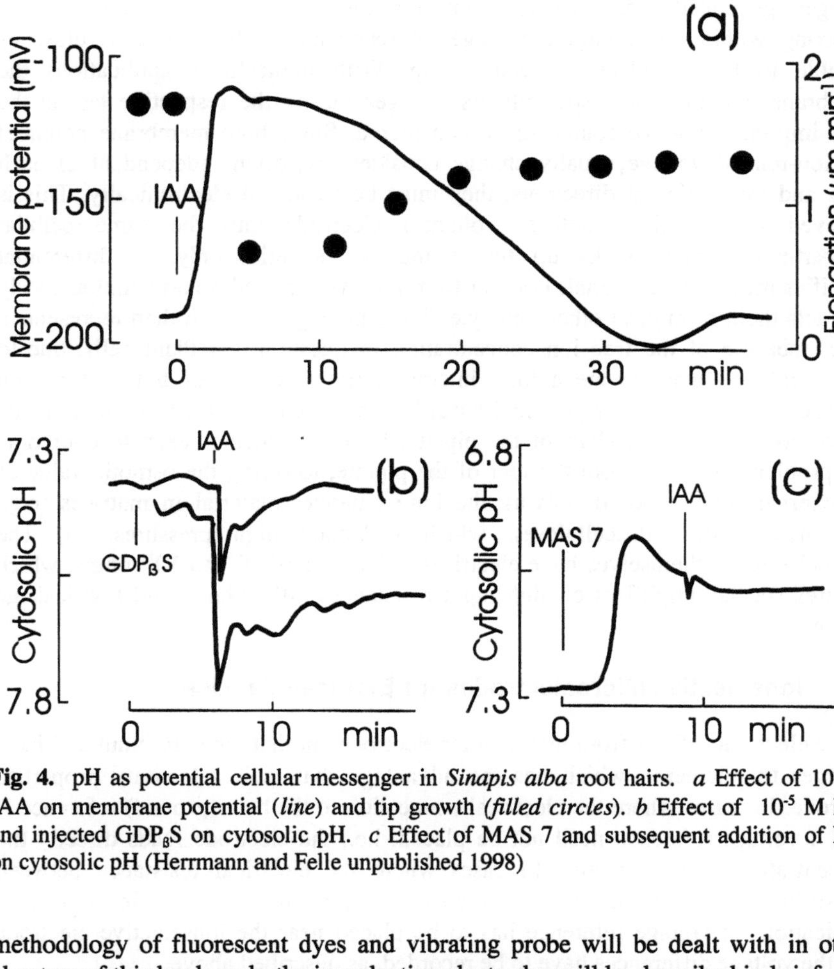

Fig. 4. pH as potential cellular messenger in *Sinapis alba* root hairs. *a* Effect of 10^{-5} M IAA on membrane potential (*line*) and tip growth (*filled circles*). *b* Effect of 10^{-5} M IAA and injected $GDP_{\beta}S$ on cytosolic pH. *c* Effect of MAS 7 and subsequent addition of IAA on cytosolic pH (Herrmann and Felle unpublished 1998)

methodology of fluorescent dyes and vibrating probe will be dealt with in other chapters of this book, only the ionselective electrodes will be described in brief.

9.1 Ionselective Microelectrodes for Intracellular Use in Plant Cells

The fabrication, and the intracellular as well as extracellular use, of these electrodes has been described in detail (Felle and Bertl 1986; Felle 1987; 1994; 1998; Felle et al. 1998). These electrodes are glass pipettes with submicrometer tips in which a so-called liquid-ion-exchanger membrane is placed. This membrane separates the solution to be tested from the reference solution in the pipette of a given concentration. The sensor within the membrane binds and releases the respective ion and thus creates characteristic voltages at the solution/membrane interfaces. These voltages are calibrated and compared with those in the test solutions (i.e., cytosol). To obtain a continuous intracellular recording three conditions have to be met: (1) as an electrode measures not only voltages generated from the ion concentration differences, but all voltages occurring within the circuit, a voltage reference has to be placed as close as possible to the ionselective electrode tip. With intracellular applications the membrane potential and especially its changes during the respective test is the most important voltage source to be considered. Since both membrane potential and ion-related voltage usually change considerably, often independent of each other and into different directions, they must be separated electronically. This is achieved by inserting another (voltage-) electrode into the same cellular compartment, which picks up the membrane potential only. A differential amplifier measures the signals coming from the two electrodes and simultaneously subtracts them from each other, thus yielding a net signal, which then represents a direct measure of the free ion concentration in question. (2) Plant cells, due to their rigid cell wall, create a high internal turgor pressure of 5 to 10 bar on average. This pressure is a potential hazard to the electrode as it tends to shift the sensor cocktail into the shaft of the pipette. Several strategies exist to overcome this problem: pressure from the rear of the pipette, lowering the osmotic value of the exernal solution, or dissolving the ion-exchanger material in matrices (e.g., polyvinylchloride, nitrocellulose), which withstand high pressures. (3) The microelectrodes themselves have electrical resistances of 10^{10} to 10^{11} ohms, which requires special amplifiers of high input resistances (10^{15} ohms) and low leakage currents.

9.2 Ionselective Microelectrodes for Extracellular Use

These electrodes differ from intracellular electrodes in that they are blunt and have a larger tip diameter, which has the advantage that their selectivity properties improve. In cases where the electrical connection to the bath (ground) reference is low, a voltage reference need not be placed, i.e., the electrode reads directly the concentration-related voltage. In cases where the electrical resistance between measuring electrode and bath reference is high, for instance in apoplastic applications, a voltage reference has to be placed near the ionselective electrode and the voltage differences have to be recorded, as described above.

References

Bibikova TN, Zhigilei A, Gilroy S (1997) Root hair growth in *Arabidopsis thaliana* is directed by calcium and an endogenous polarity. Planta 203: 495-505

Bibikova TN, Jacob T, Dahse I, Gilroy S (1998) Localized changes in apoplastic and cytoplasmic pH are associated with root hair development in *Arabidopsis thaliana*. Dev 125: 2925-2934

Blatt (1992) K^+ channels of stomatal guard cells. Characteristics of the inward rectifier and its control by pH. J Gen Physiol 99: 615-644

Boller T (1995) Chemoperception of microbial signals in plant cells. Annu Rev Plant Mol Biol 46: 189-214

Feijo JA, Hackett GR, Kunkel JG, Hepler PK (1999) Growing pollen tubes posess a constitutive alkaline band in the clear zone and a growth-dependent acidic tip. J Cell Biol 144: 4483-496

Felix G, Regenass M, Boller T (1993) Specific perception of subnanomolar concentrations of chitin fragments by tomato cells: induction of extracellular alkalinization, changes in protein phosphorylation, and establishment of a refractory state. Plant J 4: 307-316

Felle HH (1982) Effects of fusicoccin upon membrane potential, resistance and current-voltage characteristics in root hairs of *Sinapis alba*. Plant Sci Lett 25: 219-225

Felle HH (1987) Proton transport and pH control in *Sinapis alba* root hairs: a study carried out with double-barrelled pH microelectrodes. J Exp Bot 38: 340-354

Felle HH (1989) pH as a second messenger in plants. In: Boss W, Morré DJ (Eds) Second Messengers in Plant Growth and Development. Alan Liss, New York, pp. 145-166

Felle HH (1994) The H^+/Cl^- symporter in root hair cells of *Sinapis alba*. An electrophysiological study using ion-selective microelectrodes. Plant Physiol 106: 1131-1136

Felle HH (1996) Control of cytoplasmic pH under anoxic conditions and its implication for plasma membrane transport in *Medicago sativa* root hairs. J Exp Bot 47: 967-973

Felle HH (1998) The apoplastic pH of *Zea mays* root cortex as measured with pH-sensitive microelectrodes. J Exp Bot 49: 987-995

Felle H, Bertl A (1986) The fabrication of H^+-selective liquid-membrane microelectrodes for use in plant cells. J Exp Bot 37: 1416-1428

Felle HH, Hepler PK (1997) The cytosolic Ca^{2+}- concentration gradient of *Sinapis alba* root hairs as revealed by Ca^{2+}- selective microelectrode tests and fura-dextran ratio imaging. Plant Physiol 114: 39-45

Felle HH, Kondorosi E, Kondorosi A, Schultze M (1996) Rapid alkalinization in alfalfa root hairs in response to rhizobial lipochito-oligosaccharide signals. Plant J 10: 295-301

Felle HH, Kondorosi E, Kondorosi A, Schultze M (1998) The role of ion fluxes in Nod-factor signaling in *Medicago sativa*. Plant J 13: 455-463

Fox GG, McCallan NR, Ratcliffe RG (1995) Manipulating cytoplasmic pH under anoxia: a critical test of the role of pH in the switch from aerobic to anaerobic metabolism. Planta 195: 323-330

Gibbon BC Kropf DL (1994) Cytosolic pH gradients associated with tip growth. Science 263: 1419-1421

Grabov A, Blatt MR (1997) Parallel control of the inward-rectifier K^+ channel by cytosolic free Ca^{2+} and pH in *Vicia* guard cells. Planta 201: 84-95

Harold FM, Caldwell JH (1990) Tips and currents: electrobiology of apical growth. In: Heath IB (ed.) Tip Growth in Plant and Fungal Cells. Academic Press, London, pp. 59-89

Herrmann A, Felle HH (1995) Tip growth in root hair cells of *Sinapis alba* L.: significance of internal and external Ca^{2+} and pH. New Phytol 129: 523-533

Jones DL, Shaff JE, Kochian LV (1995) Role of calcium and other ions in directing root hair tip growth in *Limnobium stoloniiferum*. Planta 197: 672-690

Marre, E (1979) Fusicoccin: A tool in plant physiology. Annu Rev Plant Physiol 30: 273-288

Moog PR, van der Kooij TA, Brüggemann W, Schiefelbein JW, Kuiper PJC (1995) Responses to iron deficiency in *Aradopsis thaliana*: The turbo iron reductase does not depend on the formation of root hairs and transfer cells. Planta 195: 505-513

Nürnberger T, Nennstiel D, Jabs T, Sacks WR, Hahlbrock K, Scheel D (1994) High affinity binding of a fungal oligopeptide elicitor to parsley plasma membranes triggers multiple defense responses. Cell 78: 449-460

Parton RN, Fischer S, Malho R, Papasouliotis O, Jelitto TC, Leonard T, Read ND (1997) Pronounced cytoplasmic pH gradients are not required for tip growth in plant and fungal cells. J Cell Sci 110:1187-1198

Peters and Felle (1999) The correlation of profiles of surface pH and relative elemental growth in maize roots. Plant Physiol. 121: 905-912

Pilet PE, Versel JM, Mayor G (1983) Growth distribution and surface pH patterns along maize roots. Planta 158: 398-402

Roncal T, Ugalde UO, Irastorza A (1993) Calcium-induced conidiation in *Penicillium cyclopium*: calcium triggers cytosolic alkalinization at the hyphal tip. J Bacteriol 175: 879-887

Turian G (1979) Cytochemical gradients and mitochondrial exclusion in the apices of vegetative hyphae. Experientia 35: 1164-1166

Weisenseel MH, Dorn A, Jaffe LF (1979) Natural H^+ currents traverse growing roots and root hairs of barley (*Hordeum vulgare* L.). Plant Physiol 64: 512-518

Wymer CL, Bibikova TN, Gilroy S (1997) Cytoplasmic free calcium distribution during the development of root hairs of *Arabidopsis thaliana*. Plant J 12: 427-439

Chapter 11
Evidence for the Uptake of Non-Essential Ions and Essential Nutrient Ions by Root Hairs and Their Effect on Root Hair Development

R. Larry Peterson and Kevin J. Stevens

Department of Botany, University of Guelph, Guelph, Ontario N1G 2W1, Canada

1 Introduction

Root hairs are tubular extensions of epidermal cells found on roots of most vascular plant species (Cormack 1962; Hofer 1991; Peterson and Farquhar 1996). There has been considerable debate as to the role of root hairs in nutrient uptake and observations of the normal growth of shoots of two of three root hair mutant lines of *Zea mays* L. (maize) suggest that they may not always play a significant role (Wen and Schnable 1994). Root hairs when present, however, extend the absorbing surface of roots and therefore presumably affect the uptake of nutrients in the vicinity of the root cylinder (Jungk 1991). Clarkson (1991) argues that if the density of root hairs is great enough, the depletion zones of adjacent hairs will overlap thus most available ions in the soil nutrient solution between adjacent root hairs will be available for uptake. The distance away from the depletion zone surrounding the root cylinder to which root hairs elongate extends the region of the rhizosphere from which nutrients can be absorbed (Nye 1966; and Fig. 1). There has been some interest, therefore, in using genotypes of agricultural species with long root hairs to enhance nutrient uptake. This characteristic, as well as the ability of root hairs to grow into small soil pores and into soil particles, could positively affect nutrient uptake. This would be of particular importance in terms of ions such as phosphate that are bound to soil fractions and are essentially immobile (Jungk 1991).

The effects of various essential nutrient ions and non-essential ions on root hair initiation and elongation could be significant. Soils containing high levels of certain nutrients or polluting ions may impede root hair initiation and/or elongation leading to reduced capacity for ion uptake. It is well documented that there is a patchiness to the distribution of nutrients in soils and that roots respond

Key words. calcium, elongation, initiation, iron, nitrogen, phosphorus, uptake to this by altering branching and root hair development (see reviews by Lynch 1995; Peterson 1992; Peterson and Farquhar 1996).

The absorptive surface area of roots also may be extended through their association with soil-borne fungi. Roots have evolved with mutualistic symbiotic fungi to form various types of mycorrhizas (Peterson and Farquhar 1994; Smith and Read 1997). The extraradical mycelium that develops can extend the absorbing surface of roots significantly (Fig. 1) and the narrow diameter of hyphae ensures that more of the soil volume can be accessed. In plant species capable of forming arbuscular mycorrhizal associations, there appears to be an inverse relationship between the degree of root branching and root hair production with dependency on mycorrhiza formation (Baylis 1975). When uptake of nutrients into roots is considered for plants in natural settings, the role of mycorrhizas must be taken into account.

This chapter reviews the literature concerning the acquisition of nutrients by root hairs and the effects of essential nutrient ions and non-essential ions on root hairs. The very early literature on these topics will not be reviewed and readers are directed to the reviews by Cormack (1949, 1962) for the pertinent literature.

2 Uptake of Essential Nutrient Ions by Root Hairs

The body of literature on uptake of essential nutrient ions by roots is extensive and beyond the scope of this chapter; only papers dealing specifically with uptake by root hairs will be reviewed. There are however relatively few essential ions that have been investigated in terms of their uptake by root hairs, and as such this section will be limited to a discussion of phosphorus, calcium, nitrogen, potassium and iron uptake by root hairs.

2.1 Phosphorus

The importance of root hairs in nutrient uptake, including phosphorus (P), has been suggested for a considerable length of time (Nye 1966), yet until recently there has been only indirect evidence for the uptake of P by root hairs. Early studies using autoradiography to map P depletion around root systems, found a correlation between P uptake and root hairs in *Allium cepa* var. Mustang (Bhat and Nye 1974b; Brewster et al. 1976), *Brassica napus* (Bhat and Nye 1973; Bhat and Nye 1974b; Bhat et al. 1976; Brewster et al. 1976; Misra et al. 1988), *Gossypium hirsutum* cv. Sicot (Misra et al. 1988) and wheat (Lewis and Quirk 1967). These studies clearly show that zones of depletion are greater surrounding root systems with root hairs, or in root systems with long root hairs as opposed to short root hairs, but are not able to unequivocally show a causal relationship between root hairs or root hair length and P uptake.

Attempts to model P uptake by root systems often underestimated the realized P uptake when the root systems were considered "smooth cylinders" (Ernani et al. 1994; Itoh and Barber 1983a,b). However, when root hair characteristics were taken into account and these parameters included in the models, there was much closer agreement between predicted and actual values (Itoh and Barber 1983 a,b;

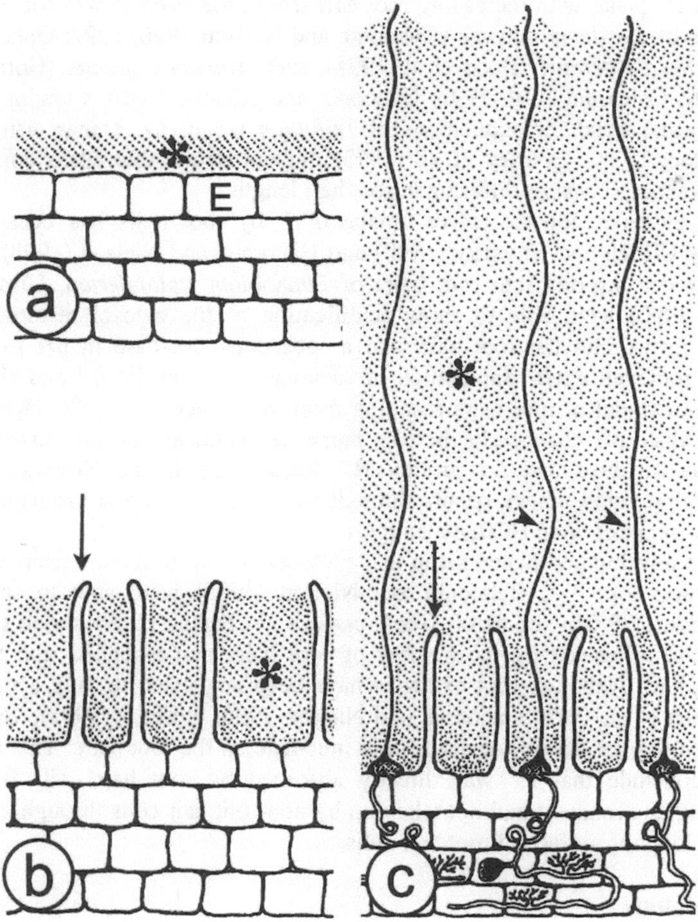

Fig. 1. Schematic representations of the interface between roots and the soil. *a* Root without root hairs. Epidermal cells (*e*) interact directly with the rhizosphere and there is a narrow depletion zone (*asterisk*) surrounding the root cylinder. *b* Root with root hairs (*arrow*). The depletion zone (*asterisk*) extends approximately to the length of root hairs. *c* Root with root hairs (*arrow*) and extraradical fungal hyphae (*arrowheads*) of an arbuscular mycorrhizal association. The depletion zone (*asterisk*) extends a considerable distance from the root cylinder

Föhse et al. 1991). Calculations in the latter study suggest that in soils with low available P, root hair uptake may account for up to 90% of total P accumulation (Föhse et al. 1991). In the model proposed by Itoh and Barber (1983a), P uptake was influenced little by root hairs shorter than 0.05 mm, doubled with root hairs 0.3 mm and increased 4 fold with root hairs 0.6 mm in length. Further evidence suggesting the importance of root hair length in P uptake has been gained by studies intent on improving crop yields or land reclamation. A trend towards an increase in P uptake with increasing root hair length has been shown for cultivars or varieties of *Hordeum vulgare* (Gahoonia and Nielsen 1996, 1997; Gahoonia et al. 1997), *Secale cereale* (Baon et al. 1994), and *Triticum aestivum* (Horst et al. 1993, 1996; Gahoonia et al. 1997; Gahoonia and Nielsen 1996). Caradus (1981) found a similar trend for populations of *Trifolium repens* cv. Tamar differing in root hair length, but Gourley et al. (1993) did not detect this when comparing three cultivars of *T. repens* differing in root hair length.

Strong evidence for the active uptake of P by root hairs has been shown recently by Ullrich and Novacky (1990) and Gahoonia and Nielsen (1998). Using microelectrodes inserted into root hairs of *Limnobium stoloniferum*, Ullrich and Novacky (1990) were able to show acidification of the cytosol resulting from phosphate influx into the root hair, and a subsequent increase in pH following phosphate removal. After examination of vacuolar and cytosolic pH and electrical membrane potential of root hairs resulting from the uptake of K^+, Cl^-, $H_2PO_4^-$ and NO_3^-, and since pH changes in the root hairs are dependent on the fluxes of the strong ions involved and not strictly H^+ fluxes, Ullrich and Novacky (1990) concluded that anion uptake occurs through an H^+/A^- cotransport mechanism and not a K^+/A^- symport or an $HC0_3^-/A^-$ antiport.

Using a novel experimental setup that excluded all but root hairs from a source of the radioisotope P^{32}, Gahoonia and Nielsen (1998) were able to show the accumulation of P^{32} in shoots of *Secale cereale* cv. Petkus II. Phosphorus uptake was directly influenced by the number of root hairs that extended into the P^{32}-containing soil. Although root exudates have been suggested to play a role in P uptake (Bhat et al. 1976; Gahoonia and Nielsen 1996; Hoffland 1992), Gahoonia and Nielsen did not notice a water film or mucilage on the root hairs. This allowed them to conclude that P^{32} was directly absorbed by root hair cells from the surrounding environment and not taken up by non-root hair cells through capillary transport along the surface of root hair cells.

2.2 Calcium

Calcium is essential for many aspects of plant structure and growth (White 1998) including tip growth of root hairs (Hermann and Felle 1995). Since Ca^{2+} must be obtained from the soil solution, root hairs may be involved in their uptake and transport to underlying cells in the root. Using a Ca^{2+}-specific vibrating microelectrode, Schiefelbein et al. (1992) located a Ca^{2+} gradient at the tip of growing *Arabidopsis thaliana* root hairs and the direction of the gradient indicated an influx of Ca^{2+} into the tip region. Use of a Ca^{2+} channel blocker in the nutrient solution stopped tip growth and concomitantly a Ca^{2+} gradient was not detected. These results indicate that Ca^{2+} channels located in the plasma membrane at the tip of root hairs are involved in the uptake of this ion. White (1998) has recently reviewed the literature supporting this view.

Root hairs of soybean accumulate approximately eight times the amount of Ca^{2+} compared to the rest of the root tissues analyzed together (Werner et al. 1985). This is indirect evidence for root hairs being involved in Ca^{2+} uptake from the soil solution. *Rhizobium japonicum*, a symbiotic bacterium involved in nitrogen fixation, may be able to utilize this store of Ca^{2+} after invasion of root hairs for their subsequent growth (Werner et al. 1985). There is considerable interest in determining cytosolic levels of Ca^{2+} in plant cells in general because of the many regulatory roles of this cation in intracellular processes. Calcium gradients in root hairs are covered in detail in Chapter 9, this volume.

2.3 Nitrogen

Root hairs of *Oryza sativa* L. (rice) and *Zea mays* L. (maize) preferentially take up and accumulate the fluorescent pH indicator probe, BCECF (2',7'-bis-(2-carboxyethyl)-5[and –6]-carboxyfluorescein) in the cytosol (Kosegarten et al. 1997). This feature was used to monitor shifts in cytosolic pH by fluorescence ratio changes when roots were immersed in ammonia solutions at various external pH. Treatment with 2 mM ammonia at an external pH of 7.0 shifted the cytosolic pH in root hairs of both species to be more alkaline, with the response being somewhat faster in rice than maize. At an external pH of 5.0 there was no change. At the higher pH, the NH_4^+ /NH_3 equilibrium is shifted in favor of NH_3 which is easily transported across the plasma membrane.

The large root hairs of *Limnobium stoloniferum* (G.F. Mey) Griseb. have been used extensively to study uptake of ions, including NO_3 (Ullrich and Novacky 1990). Using microelectrodes inserted into single root hairs, these authors found that NO_3 provided in a bathing solution containing 1 mM $Ca(NO_3)_2$ caused a depolymerization of the plasma membrane; there was an increase in cytosolic pH accompanying NO_3 uptake and an enhanced uptake of K^+.

Root hairs of *Lycopersicon esculentum* L. cv. Moneymaker (tomato) developed on aeroponically grown plants, were used to study uptake systems for nitrogen (Lauter et al. 1996). Using the *Arabidopsis* clones AtNrt1 and AtNrt2 as heterologous probes, these authors isolated three genes from a root hair-specific tomato cDNA library. The three were expressed specifically in root tissues, two primarily in root hairs. Two genes (LeNrt 1-1, LeNrt 1-2) appear to be nitrate transporters, while the third (LeAmt 1)is likely an ammonium transporter. The expression of these genes in root hairs suggests that they are involved in uptake of nitrogen from the soil solution.

2.4 Potassium

Potassium is a macronutrient for plant growth and plays an essential role in a number of metabolic processes (Gassmann and Schroeder 1994). These authors used protoplasts derived from root hairs of wheat (*Triticum aestivum* L. cv Scout) combined with patch-clamping to show that inward-rectifying K^+ channels occur in the plasma membrane and these provide a mechanism for low affinity K^+ uptake. *Limnobium stoloniferum* root hairs were used by Grabav and Böttger (1994) to investigate K^+ channels in the plasma membrane, particularly in reference to the effect of alterations in inwardly and outwardly rectifying currents on K^+ transport across membranes. It is clear that the tip region of root hairs is subject to membrane changes induced by substances in the bathing medium and that these changes alter ion uptake (see also Ullrich and Novacky 1990).

2.5 Iron

A number of grapevine rootstocks showing various degrees of chlorosis resistance were tested for iron uptake using a solution containing radioactive FeEDTA (prepared by adding EDTA to $^{59}FeCl_3$) in relation to root diameter and root hair density (Bavaresco et al. 1991). The rootstock showing the most chlorosis resistance had the highest Fe uptake and correlated with this had the highest number of root hairs.

Iron accumulates in root hairs of soybeans to a level that is ten times that found in other root tissues whereas in wheat root hairs the accumulation is only three times that of other root tissues (Werner et al. 1985). This accumulation of iron in soybean root hairs might be advantageous to the growth of *Rhizobium japonicum*, a symbiotic nitrogen fixing bacterium known to have a high requirement for this ion (Werner et al. 1985).

3 Effects of Essential Nutrient Ions on Root Hair Development

The effects of various essential and other ions on root hair initiation and elongation have often been recorded in a general way in experiments involving the growth of plants in hydroponic nutrient solutions and soil. In many of these experiments, it is difficult to determine the effects of specific ions on root hairs separately from abiotic factors such as pH and aeration. Nevertheless, there is a body of literature that shows that root hairs are influenced by various constituents in the substrate in which roots grow. These effects may be important either directly or indirectly in nutrient uptake by roots.

3.1 Phosphorus

Hetrick (1991) states that of all the possible methods for altering root system morphology to improve nutrient acquisition, changes in root hair morphology are the least metabolically costly. Root hair characteristics should, therefore, show a level of plasticity in response to changes in nutrient availability. In terms of phosphorus (P) nutrition, however, few studies have critically examined the relationship between P availability and root hair development. Additionally, comparisons between studies are hampered by the various methods of expressing P availability and differences in experimental design (i.e., solution vs. soil culture). Although 3 separate studies have included an assessment of the effects of P availability on root hair density in *Brassica napus*, the results are inconclusive. Bole (1973) found that root hair density of *Brassica napus* was directly related to P availability; at high P (23 ppm extractable P in soil amended with 45 mg P/ pot) roots averaged 5 root hairs mm^{-1}, while root hairs were absent at the lowest P levels (5 ppm extractable P without amendment). In contrast, Brewster et. al

(1976) noted an abundance of root hairs on *B. napus* roots growing at low P levels (10^{-6} and 10^{-5} M) and an absence of root hairs or underdeveloped root hairs at the higher P levels (10^{-4} and 10^{-3} M). Bhat and Nye (1974a) state that P availability had no effect on root hair density, but did affect root hair distribution; *B. napus* plants grown at low P levels (2.4 x 10^{-6} mol/ml) possessed an area directly behind the root tip lacking root hairs while root hair distribution extended to closer to the tip of roots exposed to higher levels of P availability (15.1 x 10^{-6} mol/ml). Root hairs, when present in the low P treatments were, however, 4-5 times longer than those in the high P levels.

A trend towards increasing root hair density and length with decreasing P availability (1000-2 μM P) has been shown for *Lycopersicon esculentum* cv Eurocross B, *Spinacea oleraceae* cv Medania and *Brassica oleracea* cv Quinta (Föhse and Jungk 1983). Although root hair length decreased with increasing P for each species, root hairs were present at the highest P level for all three of the test species.

The time required for a response to changing P availability has not been fully explored. However, in a recent study of *Arabidopsis thaliana*, Bates and Lynch (1996) found a significant increase in root hair length 8 days following sowing in low P compared to high P treatments (1 mmol m^{-3} vs 1000 mmol m^{-3}). Sixteen days after sowing, plants in the low P treatment had root hairs approximately 3 times longer than those in the high P treatment. It should be noted that sampling did not occur prior to the 8th day so the actual time of divergence in root hair characteristics is unknown. The increase in root hair length was suggested to result from an increase in the rate and duration of root hair growth. For *A. thaliana*, a 30% greater root hair density in low P plants has been attributed to a higher proportion of epidermal cells which developed into root hairs, rather than a proportional increase in the number of epidermal cells and root hair cells (Bates and Lynch 1996).

Some debate exists regarding whether the root hair response is dependent on local P availability of individual roots, or the overall P status of the entire plant. Using a combination of well and transfer studies Bates and Lynch (1996) concluded that for *Arabidopsis thaliana*, local P availability is the contributing factor in root hair response. Plants grown under high P conditions produced roots with short root hairs, but when transferred to a low P environment, produced shorter root hairs. The opposite was true for plants grown in low P solution and transferred to nutrient solutions with a higher P concentration. Placing individual root tips into wells with high or low P concentrations resulted in the section of root in the well producing short and long root hairs respectively. Roots outside the well also produced root hairs that were long or short if in low or high P solutions respectively. In contrast, Föhse and Jungk (1983) suggest that root hair formation is more dependent on overall P concentration of the plant rather than local dependency. Their study involved a split root experiment in which varying proportions of a root system were placed in high P (1000 μM) or low P (2 μM) nutrient solutions. Root systems lacked root hairs if 75-100% of the root system was exposed to the high P treatment, but when 50% of the roots were exposed to low P, root hairs were produced on roots in both high and low P chambers. A dependency on local P concentrations is, however, also suggested from their data (Fig. 4 in Föhse and Jungk 1983).

The mechanisms responsible for the phosphorus response in root hairs, though not extensively studied, appears to involve auxins (Bates and Lynch 1996). Indole-3-acetic acid (IAA) applied to plants in their high P treatment resulted in the formation of long root hairs, while the auxin inhibitor 2-(p-chlorophenoxy)-2-methylpropionic acid (CMPA) caused a reduction in root hair length in plants grown under low P conditions

3.2 Calcium

Early work focused on the effects of Ca^{2+} on two aspects of root hair development: plasticity of that portion of the epidermal cell wall that is involved in the outgrowth of a root hair bulge and tip growth of elongating hairs (see Cormack 1949, 1962 for a summary of this work). At this time there was debate as to the presence of calcium in root hair cell walls; this was resolved by autoradiographic evidence (Cormack et al. 1963; Tanaka and Woods 1972). More recently, Jaunin et al. (1991) were able to show, using X-ray microanalysis, that trichoblasts (root hair forming epidermal cells) of *Zea mays* L. had higher levels of calcium than adjacent atrichoblasts (non-root hair forming epidermal cells) and that most of the calcium was localized in the cell wall.

There is considerable evidence that Ca^{2+} affects root hair initiation and length (Ewens and Leigh 1985). In *Avena sativa* L. (oats), nutrient solutions deficient in calcium lead to a decrease in the length and number of root hairs (Tanaka and Woods 1972). Unlike many situations in which strontium can replace calcium in physiological processes, this is not the case for root hair formation in *A. sativa* (Tanaka and Woods 1973). Plants of *Zea mays* respond to either $CaCl_2$ or $CaSO_4$ in nutrient solution by an increase in root number and an increase in the number of roots producing hairs; a pretreatment with EGTA, known to bind Ca^{2+} ions, decreased the number of roots producing hairs compared to the control (Jaunin and Hofer 1988). This same study showed that there was an increase in the percentage of epidermal cells forming root hairs when roots were treated with 10 mol m^{-3} $CaCl_2$ compared to a buffered control. In *Arabidopsis thaliana* (L.) Heynh., root hair length is dependent on Ca^{2+} concentration with maximum elongation occurring at concentrations between 0.3 - 3.0 mM (Schiefelbein et al. 1992).

3.3 Nitrogen

Nitrogen availability is one of the major factors controlling plant growth and consequently an enormous literature exists concerning forms of nitrogen that can be absorbed by plant roots and the mechanisms involved in uptake. Less research has involved the direct effects of nitrogen compounds on root hair induction and growth.

Nitrate at concentrations greater than 10 mmol m^{-3} reduced the length of *Brassica napus* (rape) root hairs (Föhse and Jungk 1983). In experiments to determine which ions in nutrient solutions have the most impact on root hair growth in *Triticum aestivum* (wheat), Ewens and Leigh (1985) found that nitrate can have significant effects on root hair length, but is only one of several factors affecting this. Of several grass species studied to determine the effects of either nitrate or ammonium on root hair development, length of hairs was affected only for *Holcus lanata* L.; there was a significant increase in length with decreasing nitrate concentration (Robinson and Rorison 1987). However, when both root hair density and root hair length were considered together (expressed as specific root hair length) then *Deschampsia flexuosa* (L.) Trin. and *Poa annua* L. had the highest value and this was in nutrient solutions containing ammonium as the nitrogen source. These values increased with decreasing nitrogen concentration. Of particular interest in this work is the observation that differences in root hair length and density do not usually translate into an effect on relative plant growth rate (Robinson and Rorison 1987).

Nitrite is usually present in soils at very low concentrations, but soils with low oxygen content, caused by either water-logging or compaction, may accumulate significant levels of NO_2^- (Zsoldos et al. 1993). Root hairs in *Triticum aestivum* L. cv. GK Öthalom (wheat) are sensitive to NO_2^- (Zsoldos et al. 1993). If this effect proves to be common to other plant species then this may be important when plants are exposed to water-logging or other stresses that cause low oxygen conditions.

3.4 Iron

Characteristic morphological and structural alterations in roots have been documented in a number of plant species when grown under iron deficiency conditions and some of these features have been correlated with physiological changes in these roots (Briat et al. 1995). Root hair induction triggered by iron deficiency has been recorded for several dicotyledonous angiosperms (Römheld and Marschner 1981; Landsberg 1982; Bell et al. 1988; Rosenfield et al. 1991), as has the modification of root hair structure by the elaboration of wall ingrowths lined by plasma membrane and the concomitant increase in the number of mitochondria and other cellular organelles (Landsberg 1982, 1986, 1989). These features, typical of transfer cells, also occur in epidermal cells not forming root hairs when roots are exposed to nutrient solutions deficient in iron (Römheld and Kramer 1983). Chaney et al. (1992), concluded, however, that root hair induction observed in *Lycopersicon esculentum* L. (tomato) was related to chlorosis rather than being part of the Fe-stress response. The physiological changes that have been reported to be correlated with the increase in root hair development are typical of those plant species classified as Strategy I plants in terms of response to iron deficiency (Briat et al. 1995). These plants, dicotyledonous and non-grass monocotyledonous species, show inducible ferric reductase and enhanced ATPase activity (Briat et al. 1995). These changes in roots of these species lead to an increased reductase activity in the rhizosphere which in turn makes iron available via the reduction of Fe^{3+} compounds to Fe^{2+}, the form in which iron can be absorbed readily by roots (Kochian 1991). An histochemical procedure for the induced ferric reductase activity was demonstrated on root hairs of tomato plants grown in an iron deficient nutrient solution (Bell et al. 1988).

Similarly, the changes reported to be correlated with observed structural changes in transfer cell-like root hairs are enhanced H^+ excretion (Römheld and Marschner 1981; Römheld and Kramer 1983; Landsberg 1986), excretion of various reducing compounds (Römheld and Marschner 1981; Bell et al. 1988; Rosenfield et al. 1991) and the resultant increase in uptake of iron. More recently, however, it has been shown that a mutant of *Arabidopsis thaliana* lacking the ability to initiate root hairs, also fails to form transfer cells under iron deficient conditions (Moog et al. 1995). This mutant still showed induction of ferric chelate reductase when roots were exposed to iron deficient nutrient solutions (Moog et al. 1995). These authors conclude, therefore, that physiological and morphological changes induced by iron deficiency are controlled by separate mechanisms.

3.5 Zinc

There is a paucity of information on the effects of zinc on root hairs. Two cultivars, one tolerant and one sensitive, of the grass *Festuca rubra* L. responded differently to zinc treatment in terms of distance back from the meristem at which root hairs are initiated (Powell et al. 1988). Root hairs in the sensitive cultivar were initiated closer to the meristem-root cap junction than in the tolerant cultivar at the same zinc concentration and this was correlated with a concomitant earlier differentiation of the first xylem tracheary element in the sensitive cultivar. The authors suggest that the sensitive cultivar may respond to zinc by decreasing the number of times derivative cells divide before elongating and therefore root hair initiation appears closer to the meristem.

3.6 Magnesium

Magnesium does not have any effect on root hair development in oats as shown by deleting magnesium from the nutrient solution (Tanaka and Woods 1973).

4 Uptake of Non-Essential Ions and Their Effects on Root Hair Development

4.1 Aluminum

There is a considerable volume of literature on the effects of aluminum on plant productivity because of the prevalence of agricultural and other soils with low pH and the concomitant release of Al^{3+} into the soil solution. The exact mechanism of aluminum toxicity is debatable but it is known that the addition of aluminum to nutrient solutions generally causes an inhibition in root hair elongation (Care 1995; Jones et al. 1995, 1998) and in some cases higher concentrations of aluminum lead to inhibition of root hair initiation (Care 1995). Genotypes of *Trifolium repens* L. with either the short or long root hair phenotype reacted equally to aluminum (Care 1995). Root hairs can recover their normal growth rate if aluminum is removed from the nutrient solution (Jones et al. 1995), indicating that over the short term at least, Al^{3+} at fairly low concentrations is not toxic. Supporting this view is the observation that aluminum at a concentration of 50 μM had no effect on cytoplasmic streaming in root hairs of *Triticum aestivum* (wheat) for periods of time up to 465 minutes (Sattelmacher et al. 1993). Al^{3+} does block K^+ channels in protoplasts derived from root hairs, but whether this is directly correlated with toxicity symptoms in intact seedlings is not clear (Gassmann and Schroeder 1994). Likewise the relationship between aluminum toxicity and the observed changes in Ca^{2+} concentration in root hairs does not appear to be strictly correlated (Jones et al. 1995, 1998). Low concentrations of aluminum are detrimental to nodulation in *Glycine max* (L.) Merr. (soybean) and this has been attributed to its effect on root hair development (Brady et al. 1990; Hecht-Buchholz et al. 1990). An interaction between calcium concentration and aluminum concentration in terms of root hair development has been shown also for soybean (Brady et al. 1993). The reduction in root hair formation at a total activity of monomeric Al species of 2 μM in a solution with 500 μM Ca could be overcome by increasing the Ca concentration to 2000 μM.

Attempts have been made to select plants that show aluminum tolerance and in species such as *Trifolium repens* L. root hair length has been one of the traits of interest (Care 1995; Wheeler 1995). Lines of this species with either long or short root hairs have been selected but from experiments to date it does not appear that either of these traits confers tolerance to Al^{3+} (Care 1995; Wheeler 1995).

In E107, a pleiotropic mutant of *Pisum sativum* cv. 'Sparkle', high concentrations of aluminum are accumulated in roots and shoots and correlated with this, walls of root hairs and mucilage on the root surface stain for aluminum indicating its location in these regions of the root (Guinel and LaRue 1993).

4.2 Lead

Lead applied in nutrient solutions reduces the frequency of root hairs in *Raphanus sativa* L. (Lane and Martin 1980) and in *Fagus* seedlings (Breckle 1991). Accumulation of lead has been demonstrated in root hairs of *Zea mays* L. seedlings exposed to this heavy metal by a histochemical procedure (Sieghardt 1984).

5 Salinity

Although salinity is becoming a major problem in many agricultural soils because of irrigation (Epstein et al. 1980) there has been little work on the effects of NaCl on root hairs of crop or other species. In *Medicago sativa* L. (lucerne), root hair number was reduced with concentrations of NaCl between 70 - 100 mol m^{-3} and hairs that formed were short and swollen (Lakshmi-Kumari et al. 1974). A study of the effects of sodium chloride on the nodulation process in *Vicia faba* L. showed that root hair number was reduced and the proportion of hairs containing infection threads also decreased (Zahran and Sprent 1986). If these effects occur with other legume species, and if they ultimately lead to a reduction in nitrogen fixation, then saline soils could lead to lower yields in a number of crops. In *Nicotiana tabacum* L. (tobacco), root hairs formed in an aerated hydroponic system to which NaCl was added to bring the concentration up to either 100 or 200 mol m^{-3} had various morphologies including bladder-like projections from root-hair forming sites (Tyerman et al. 1989). The functional significance of these structures was not determined although one possibility suggested by the authors is that these swollen cells may accumulate NaCl and thus reduce the transport of Na and Cl ions into the rest of the root. Short, bladder-like root hairs with wall ingrowths and cytological features typical of transfer cells developed in the halophyte species, *Atriplex hastata* L., grown in nutrient solutions to which NaCl was added (Kramer et al. 1978). These authors suggest that since these cytological modifications were only found in plants exposed to NaCl, these structural changes are an adaptation to salinity. They further suggest that the development of transfer cells might be related to the selective uptake of K^+ rather than Na^+ by roots of this species when grown under saline conditions.

Calcium is important in maintaining membranes and therefore in ion transport (Cramer et al. 1985). An early response of *Gossypium hirsutum* L. (cotton) root hairs to treatment with NaCl is the replacement of membrane-associated Ca^{++} by Na^+, as determined by using chlorotetracyline as a fluorescent probe for membrane-bound Ca^{++} (Cramer et al. 1985). This leads to leakiness of the plasma membrane and an efflux of cytosolic K^+. The loss of K^+ is detrimental to cells in that this cation is essential for maintaining turgor and is a cofactor for many enzymes.

6 Conclusions

Root hairs continue to be studied in terms of nutrient relationships of plants and there is increasing evidence, using modern methods employing microelectrodes and ratio imaging with fluorescent probes, that certain essential nutrient ions can be taken up directly into root hairs. With this in mind, there is some effort to choose plant genotypes with more and longer root hairs to enhance nutrient uptake. It should be stressed however, that in natural environments most plants associate with mycorrhiza fungi, and since these fungi have the potential to significantly alter nutrient uptake dynamics, this should be taken into consideration when evaluating the role of root hairs in ion uptake. This has not been the case in many studies since experimental systems often involve seedlings growing in nutrient culture under somewhat sterile conditions.

Many factors, including non-essential ions and essential nutrient ions, affect the initiation and elongation of root hairs. Ions such such as lead and aluminum which are generally detrimental to root hair development, may have substantial consequences in terms of nutrient uptake dynamics in areas where there is significant ion accumulation resulting from anthropogenic or other biological or abiotic activities. There is considerable evidence that the concentration of many essential nutrient ions affects root hair initiation and elongation although it is less clear if this is a major factor in natural soils; few studies have examined the combined effects of different ions and ion concentrations on root hair characteristics. Additionally, the patchy nature of essential nutrient ions in soils also may affect root hair distribution along root systems, which in turn may influence nutrient uptake.

Studies involving root hair mutants, particularly *Arabidopsis*, have proven and should continue to prove valuable in determining the factors involved in hair development and the roles root hairs play in nutrient acquisition.

ACKNOWLEDGMENTS

We thank the Natural Sciences and Engineering Research Council of Canada for financial support, Lewis Melville for creating Fig. 1 and Heather Zurbrigg for formatting the manuscript.

References

Baon JB, Smith SE, and Alston AM (1994) Growth response and phosphorus uptake of rye with long and short root hairs: Interactions with mycorrhizal infection. Plant Soil 167:247-254

Bates TR, Lynch JP (1996) Stimulation of root hair elongation in *Arabidopsis thaliana* by low phosphorus availability. Plant Cell Environ 19:529-538

Bavaresco L, Fregoni M, Fraschini P (1991) Investigations on iron uptake and reduction by excised roots of different grapevine rootstocks and a *V. vinifera* cultivar. Plant Soil 130:109-113

Baylis GT (1975) The magnolioid mycorrhiza and mycotrophy in root systems derived from it. In: Sanders FE, Mosse B, Tinker PB (Eds) Endomycorrhizas. Academic Press, London, pp 373-389

Bell PF, Chaney RL, Angle JS (1988) Staining localization of ferric reduction in roots. J Plant Nutr 11:1237-1252

Bhat KKS, Nye PH (1973) Diffusion of phosphate to plant roots in soil. 1. Quantitative autoradiography of the depletion zone. Plant Soil 38:161-175

Bhat KKS, Nye PH (1974a) Diffusion of phosphate to plant roots in soil. 2. Uptake along the roots at different times and the effect of different levels of phosphorus. Plant Soil 41:365-382

Bhat KKS, Nye PH (1974b) Diffusion of phosphate to plant roots in soil. 3. Depletion around onion roots without root hairs. Plant Soil 41:383-394

Bhat KKS, Nye PH, Baldwin JP (1976) Diffusion of phosphate to plant roots in soil. 4. The concentration distance profile in rhizosphere of roots with root hairs in a low-P soil. Plant Soil 44:63-72

Bole JB (1973) Influence of root hairs in supplying soil phosphorus to wheat. Can J Soil Sci 53:169-175

Brady DJ, Edwards DG, Asher CJ, Blamey FPC (1993) Calcium amelioration of Aluminium toxicity effects on root hair development in soybean (*Glycine max* (L.) Merr. New Phytol 123:531-538

Brady DJ, Hecht-Buchholz C, Asher CJ, Edwards DG (1990) Effects of low activities of Aluminium on soybean (*Glycine max*). I. Early growth and nodulation. In: van Beusichem ML (Ed) Plant nutrition - physiology and applications. Kluwer Academic, Dordrecht, pp 329-334

Breckle S-W (1991) Growth under stress-heavy metals. In: Waisel Y, Eshel A, Kafkafi U (Eds) Plant roots the hidden half. Marcel Dekker, New York, pp 351-373

Brewster JL, Bhat KKS, Nye PH (1976) The possibility of predicting solute uptake and plant growth response from independently measured soil and plant characteristics. 5. The growth and phosphorus uptake of rape in soil at a range of phosphorus concentrations and a comparison of results with the predictions of a simulation model. Plant Soil 44:295-328

Briat J-F, Fobis-Loisy I, Grignon N, Lobréaux S, Pascal N, Savino G, Thoiron S, von Wirén N, van Wuytswinkel O (1995) Cellular and molecular aspects of iron metabolism in plants. Biol Cell 84:69-81

Caradus JR (1981) Effect of root hair length on white clover growth over a range of soil phosphorus levels. N Z J Agric Res 24:353-358

Care DA (1995) The effect of aluminium concentration on root hairs in white clover (*Trifolium repens* L.). Plant Soil 171:159-162

Chaney RL, Chen Y, Green CE, Holden MJ, Bell PF, Luster DG, Angle JS (1992) Root hairs on chlorotic tomatoes are an effect of chlorosis rather than part of the adaptive Fe-stress-response. J Plant Nutr 15:1857-1875

Clarkson DT (1991) Root structure and sites of ion uptake. In: Waisel Y, Eshel A, Kafkafi U (Eds) Plant roots the hidden half. Marcel Dekker, New York, pp 417-453

Cormack RGH (1949) The development of root hairs in angiosperms. Bot Rev 15:583-612

Cormack RGH (1962) The development of root hairs in angiosperms II. Bot Rev 28:446-464

Cormack RGH, Lemay P, Maclachlan GA (1963) Calcium in the root-hair wall. J Exp Bot 14:311-315

Cramer GR, Läuchli A, Polito VS (1985) Displacement of Ca^{2+} by Na^{+} from the plasmalemma of root cells. Plant Physiol 79:207-211

Epstein E, Norlyn JD, Rush DW, Kingsburg RW, Kelley DB, Cunningham GA, Wrona AF (1980) Saline culture of crops: a genetic approach. Science 210:399-404

Ernani PR, Santos JCP, Kaminski J, Rheinheimer DS (1994) Prediction of phosphorus uptake by a mechanistic model in a low phosphorus highly weathered soil as affected by mycorrhizae inoculation. J Plant Nutr 17:1067-1078

Ewens M, Leigh RA (1985) The effect of nutrient solution composition on the length of root hairs of wheat (*Triticum aestivum* L.). J Exp Bot 36:713-724

Föhse D, Claassen N, Jungk A (1991) Phosphorus efficiency of plants. 2. Significance of root radius, root hairs and cation-anion balance for phosphorus influx in seven plant species. Plant Soil 132:261-272

Föhse D, Jungk A (1983) Influence of phosphate and nitrate supply on root hair formation of rape, spinach and tomato plants. Plant Soil 74:359-368

Gahoonia TS, Care D, Nielsen NE (1997) Root hairs and phosphorus acquisition of wheat and barley cultivars. Plant Soil 191:181-188

Gahoonia TS, Nielsen NE (1996) Variation in acquisition of soil phosphorus among wheat and barley genotypes. Plant Soil 178:223-230

Gahoonia TS, Nielsen NE (1997) Variation in root hairs of barley cultivars doubled soil phosphorus uptake. Euphytica 98:177-182

Gahoonia TS, Nielsen NE (1998) Direct evidence on participation of root hairs in phosphorus (P^{32}) uptake from soil. Plant Soil 198:147-152

Gassmann W, Schroeder JI (1994) Inward-rectifying K^{+} channels in root hairs of wheat. A mechanism for aluminum-sensitive low-affinity K^{+} uptake and membrane potential control. Plant Physiol 105:1399-1408

Gourley CJP, Allan DL, Russelle MP (1993) Differences in response to available phosphorus among white clover cultivars. Agron J 85:296-301

Grabov A, Böttger M (1994) Are redox reactions involved in regulation of K^+ channels in the plasma membrane of *Limnobium stoloniferum* root hairs? Plant Physiol 105:927-935

Guinel FC, LaRue TA (1993) Excessive aluminium accumulation in the pea mutant E107 (brz). Plant Soil 157:75-82

Hecht-Buchholz C, Brady DJ, Asher CJ, Edwards DG (1990) Effects of low activities of aluminium on soybean (*Glycine max*). II Root cell structure and root hair development. In: Van Beusichem ML (Ed) Plant nutrition-physiology and applications. Kluwer Academic, Dordrecht, pp 335-343

Hermann A, Felle HH (1995) Tip growth in root hair cells of *Sinapis alba* L.: significance of internal and external Ca^{2+} and pH. New Phytol 129:523-533

Hetrick BAD (1991) Mycorrhizas and root architecture. Experientia 47:355-362

Hofer R-M (1991) Root hairs. In: Waisel Y, Eshel A, Kafkafi U (Eds) Plant roots the hidden half. Marcel Dekker, New York, pp 129-148

Hoffland E (1992) Quantitative evaluation of the role of organic acid exudation in the mobilization of rock phosphate by rape. Plant Soil 140:279-289

Horst WJ, Abdou M, Wiesler F (1993) Genotypic differences in phosphorus efficiency of wheat. Plant Soil 155/156:293-296

Horst WJ, Abdou M, Wiesler F (1996) Differences between wheat cultivars in acquisition and utilization of phosphorus. Zeitschrift fur Pflanzenernährnahrung und Bodenkunde 159:155-161

Itoh S, Barber SA (1983a) A numerical solution of whole plant nutrient uptake for soil-root systems with root hairs. Plant Soil 70:403-413

Itoh S, Barber SA (1983b) Phosphorus uptake by six plant-species as related to root hairs. Agron J 75:457-461

Jaunin F, Hofer R-M (1988) Calcium and rhizodermal differentiation in primary maize roots. J Exp Bot 39:587-593

Jaunin F, Rouelle-Rossier VB, Hofer R-M (1991) Cytochemical and X-ray microanalytical studies of calcium in the extending zone, particularly in the rhizodermis of *Zea mays* roots. Physiol Plant 81:15-23

Jones DL, Gilroy S, Larsen PB, Howell SH, Kochian LV (1998) Effect of aluminum on cytoplasmic Ca^{2+} homeostasis in root hairs of *Arabidopsis thaliana* (L.) Planta 206:378-387

Jones DL, Shaff JE, Kochian LV (1995) Role of calcium and other ions in directing root hair tip growth in *Limnobium stoloniferum* I. Inhibition of tip growth by aluminum. Planta 197:672-680

Jungk AO (1991) Dynamics of nutrient movement at the soil-root interface. In: Waisel Y, Eshel A, Kafkafi U (Eds) Plant roots the hidden half. Marcel Dekker, New York, pp 455-481

Kochian LV (1991) Mechanisms of micronutrient uptake and translocation in plants. In: Mortvedt JJ (Ed) Micronutrients in Agriculture Ed. 2. Soil Science Society of America, Madison, WI, pp 229-296

Kosegarten H, Grolig F, Wieneke J, Wilson G, Hoffmann B (1997) Differential ammonia-elicted changes of cytosolic pH in root hair cells of rice and maize as monitored by 2',7' -bis-(2carboxyethyl)-5 (and -6)-Carboxyfluorescein-fluorescence ratio. Plant Physiol 113:451-461

Kramer D, Anderson WP, Preston J (1978) Transfer cells in the root epidermis of *Atriplex hastata* L. as a response to salinity: a comparative cytological and X-ray microprobe investigation. Aust J Pl Physiol 5:739-747

Lakshmi-Kumari M, Singh CS, Subba-Rao NS (1974) Root hair infection and nodulation in lucerne (*Medicago sativa*) as influenced by salinity and alkalinity. Plant Soil 40:261-268

Landsberg E-C (1982) Transfer cell formation in the root epidermis: a prerequisite for Fe-efficiency. J Plant Nutr 5:415-432

Landsberg E-C (1986) Function of rhizodermal transfer cells in the Fe stress response mechanism of *Capsicum annuum* L. Plant Physiol 82:511-517

Landsberg E-C (1989) Proton efflux and transfer cell formation as a response to Fe deficiency of soybean in nutrient solution culture. Plant Soil 114:53-61

Lane SD, Martin ES (1980) Further observations on the distribution of lead in juvenile roots of *Raphanus sativus*. Z Pflanzenphysiol 97:145-152

Lauter F-R, Ninnemann O, Bucher M, Riesmeier JW, Frommer WB (1996) Preferential expression of an ammonium transporter and of two putative nitrate transporters in root hairs of tomato. Proc Natl Acad Sci USA 93:8139-8144

Lewis DG, Quirk JP (1967) Phosphate diffusion in soil and uptake by plants. 3. P^{31}-movement and uptake by plants as indicated by P^{32}-autoradiography. Plant Soil 27:445-453

Lynch J (1995) Root architechture and plant productivity. Plant Physiol. 109:7-13

Misra RK, Alston AM, Dexter AR (1988) Role of root hairs in phosphorus depletion from a macrostructured soil. Plant Soil 107:11-18

Moog PR, van der Kooij TAW, Brüggemann W, Schiefelbein JW, Kuiper PJC (1995) Responses to iron deficiency in *Arabidopsis thaliana*: the turbo iron reductase does not depend on the formation of root hairs and transfer cells. Planta 195:505-513

Nye PH (1966) The effect of nutrient intensity and buffering power of a soil, and the absorbing power, size and root hairs of a root, on nutrient absorption by diffusion. Plant Soil 25:81-105

Peterson RL (1992) Adaptations of root structure in relation to biotic and abiotic factors. Can J Bot 70:661-675

Peterson RL, Farquar ML (1994) Mycorrhizas-Integrated development between roots and fungi. Mycologia 86:311-326

Peterson RL, Farquhar ML (1996) Root hairs: specialized tubular cells extending root surfaces. Bot Rev 62:1-40

Powell MJ, Davies MS, Francis D (1988) Effects of zinc on meristem size and proximity of root hairs and xylem elements to the root tip in a zinc-tolerant and a non-tolerant cultivar of *Festuca rubra* L. Ann Bot 61:723-726

Robinson D, Rorison IH (1987) Root hairs and plant growth at low nitrogen availabilities. New Phytol 107:681-693

Römheld V, Kramer D (1983) Relationship between proton efflux and rhizodermal transfer cells induced by iron deficiency. Z Pflanzenphysiol 113:73-83

Römheld V, Marschner H (1981) Iron deficiency stress induced morphological and physiological changes in root tips of sunflower. Physiol Plant 53:354-360

Rosenfield C-L, Reed DW, Kent MW (1991) Dependency of iron reduction on development of a unique root morphology in *Ficus benjamina* L. Plant Physiol 95:1120-1124

Sattelmacher B, Heinecke I, and Muhling KH (1993) Influence of minerals on cytoplasmic streaming in root hairs of intact wheat seedlings (*Triticum aestivum* L). Plant Soil. 156: 107-110

Schiefelbein JW, Shipley A, Rowse P (1992) Calcium influx at the tip of growing root-hair cells of *Arabidopsis thaliana*. Planta 187:455-459

Sieghardt H (1984) Eine anatomische-histochemische Studie zur Bleiverteilung in Primärwurzeln von *Pisum sativum* L. Mikroscopie 41:125-131

Smith SE, Read DJ (1997) Mycorrhizal symbiosis. Academic Press, San Diego

Tanaka Y, Woods FW (1972) Root and root hair growth in relation to supply and internal mobility of calcium. Bot Gaz 133:29-34

Tanaka Y, Woods FW (1973) Root and root hair growth of oats: replaceability of calcium. Can J Bot 51:1655-1659

Tyerman SD, Oats P, Gibbs J, Dracup M, Greenway H (1989) Turgor-volume regulation and cellular water relations of *Nicotiana tabacum* roots grown in high salinities. Aust J Pl Physiol 16:517-531

Ullrich CI, Novacky AJ (1990) Extra- and intracellular pH and membrane potential changes induced by K^+, Cl^-, $H_2PO_4^-$, and NO_3^- uptake and fusicoccin in root hairs of *Limnobium stoloniferum*. Plant Physiol 94:1561-1567

Wen T-J, Schnable PS (1994) Analyses of mutants of three genes that influence root hair development in *Zea mays* (Gramineae) suggest that root hairs are dispensable. Am J Bot 81:833-842

Werner D, Kuhlmann K-P, Gloystein F, Richter F-W (1985) Calcium, iron and cobalt accumulation in root hairs of soybean (*Glycine max*). Z Naturforsch 40:912-913

Wheeler DM (1995) Effect of root hair length on aluminum tolerance in white clover. J Plant Nutr 18:955-958

White PJ (1998) Calcium channels in the plasma membrane of root cells. Ann Bot 81:173-183

Zahran HH, Sprent JI (1986) Effects of sodium chloride and polyethylene glycol on root-hair infection and nodulation of *Vicia faba* L. plants by *Rhizobium leguminosarum*. Planta 167:303-309

Zsoldos F, Haunold E, Vashegyi A, and Herger P. 1993. Nitrite in the root-zone and its effects on ion uptake and growth of wheat seedlings; physiologia plantarum. Physiol Plant 89:26-631

Chapter 12
Specification of Root Hair Cells

John Schiefelbein

Department of Biology, 830 North University Avenue, University of Michigan, Ann Arbor, MI 48109, USA

1 Introduction

The initial step in the formation of a root hair is the specification of a newly-formed epidermal cell to differentiate as a root hair cell. This represents an example of a central problem in developmental biology; namely, how do particular cell types acquire their identity? During root development in most species, two cell types arise in the epidermis: root-hair cells (which derive from trichoblasts) and non-hair cells (which derive from atrichoblasts). The simplicity, accessibility, and conspicuous cell types of the root epidermis has led to its wide use as a model to understand the problem of cell specification in plants. This chapter will review the information garnered from these studies over the past century, with emphasis on the molecular genetic findings from recent research on the specification of root-hair cells in *Arabidopsis thaliana*.

2 Mechanisms Controlling Root Hair Cell Specification

Roots develop in a largely continuous fashion from controlled cell proliferation, differentiation, and morphogenesis at the root apex. In the epidermis, cells typically form organized columns (or files) along the length of the root, which result from repeated transverse divisions in the meristematic region (Esau 1965; Cutter 1978). As their fate is specified and they age, epidermal cells become progressively more differentiated. Trichoblasts may often be distinguished from atrichoblasts long before the physical outgrowth of the root hair because of differences in cell size, nucleic acid content, protein content, and enzymatic activities in the immature cells.

The examination of root epidermis development in numerous flowering plants has shown that the distribution of root-hair cells and non-hair cells varies in different species. This has led to the identification of three basic mechanisms that

Key words. *Arabidopsis*, atrichoblast, cell fate, embryogenesis, epidermis, hypocotyl, trichoblast, trichome operate to control root-hair cell specification. To some extent, the use of these mechanisms reflects evolutionary relationships among plants, and each will be discussed below.

2.1 Random Mechanism

In some plants, there is no apparent pattern to the distribution of root hair and non-hair cells, and any newly-formed epidermal cell appears to be capable of producing a root hair (Leavitt 1904). This type of specification mechanism generates a random distribution of hairs along the root, and it is characteristic of most dicotyledonous plants (Leavitt 1904; Row and Reeder 1957; Clarke et al. 1979). Little is known about the molecular basis for this type of root hair specification, although the proportion of cells that produce root hairs is rather constant under a given set of conditions, which implies that environmental factors play an important role in this mechanism. The effect of environment factors and hormones on hair specification is discussed later in this chapter.

2.2 Asymmetric Cell Division Mechanism

In a second group of species, which includes many monocotyledons, root epidermal cell specification is linked to an asymmetric cell division. The smaller of the two daughter cells from the division is densely cytoplasmic and differentiates into a root hair cell, whereas the larger one is less densely cytoplasmic and generates one or more mature non-hair cells (Sinnott and Bloch 1939; Avers 1963; Cutter and Feldman 1970). These observations suggest that this specification mechanism relies on the establishment of cellular polarity in the mother cell to generate an unequal distribution of cytoplasm and the asymmetric mitosis.

Interestingly, the arrangement of the asymmetric division may vary in different species. For example, in *Phleum pratense*, cytoplasm becomes concentrated at the distal end of the cell (the end nearest to the root apex) and subsequently the cell divides to form a small distal cell (which ultimately produces a root hair) and a large proximal cell (Avers 1963; Ivanov and Filippenko 1976). The arrangement of the asymmetric division is reversed in epidermal cells of *Hydrocharis morsus-ranae*, which produces trichoblasts as the proximal daughter cell and atrichoblasts as the distal daughter (Cutter and Feldman 1970).

Another notable aspect of this specification mechanism is that when the asymmetric cell divisions are inhibited by gamma-irradiation, all epidermal cells are able to form root hairs (Ivanov and Filippenko 1976). This suggests that the asymmetric mitosis is actually required for the production of the atrichoblasts (non-hair cells), not the trichoblasts. Although the cellular features of this type of cell specification mechanism have been described, nothing is currently known about the underlying molecular events.

2.3 Cell Position-Dependent Mechanism

In a third group of plants, which includes *Arabidopsis* and other members of the Brassicaceae, a distinct position-dependent pattern of epidermal cell types is generated (Cormack 1935; Bunning 1951; Dolan et al. 1994; Galway et al. 1994). In this case, trichoblasts form in the intercellular space between underlying cortical cells (epidermal cells located outside an anticlinal cortical cell wall), whereas atrichoblasts develop over a single cortical cell (epidermal cells located outside a periclinal cortical cell wall). The simple correlation between cell position and cell type differentiation in the epidermis of Brassicaceae species implies that cell-cell communication events are critical for the establishment of cell identity.

To help define the putative cell-cell communication event(s), surgical experiments have been performed with roots of *Sinapis* to partially separate the epidermis from the underlying tissues (Bunning 1951; Barlow 1984). All epidermal cells that differentiated from the separated tissue were found to produce root hairs, suggesting that these cells had been released from the influence of a negative regulator of root hair development. It follows that the normal distribution of epidermal cell types may be explained by the action of a root-hair inhibitor produced by cortical cells which preferentially affects the epidermal cells lying directly over them (Bunning 1951; Barlow 1984).

Due to relatively recent work with the small crucifer *Arabidopsis*, a great deal is currently known about the cellular and molecular basis of this type of root hair cell specification mechanism. The next section of this chapter will focus entirely on this topic.

3 Molecular Genetic Control of *Arabidopsis* Root Hair Cell Specification

3.1 Development of the Root Epidermis of *Arabidopsis*

The *Arabidopsis* primary root possesses a set of basic tissues in a radial array that is striking in its simplicity and regularity. From outside to inside, there are single-cell-thick rings of epidermis, cortex, endodermis, and pericycle tissues surrounding the central stele, with a constant number of eight cells per ring for the cortex and endodermis layers (Dolan et al. 1993). The origin of these cells can be traced back to specific sets of cells in the meristematic region, termed initials, which are formed and organized during embryonic development (Dolan et al. 1993; Scheres et al. 1994; Baum and Rost 1996). Together, the entire cluster of initials and the quiescent center cells at the center of the cluster make up the "promeristem," which is believed to represent the minimal group of root meristem cells.

The *Arabidopsis* root epidermis is generated from a set of 16 initials, termed epidermal/lateral root cap initials because they also give rise to the cells of the lateral root cap (Dolan et al. 1993; Scheres et al. 1994). The immediate daughter cells produced from these initials undergo secondary transverse divisions in the meristematic region of the root, and these divisions (typically 5 or 6 rounds per daughter cell) serve to generate additional cells within the same file (Baum and Rost 1996; Berger et al. 1998b). Furthermore, anticlinal longitudinal divisions occasionally occur and result in an increase in the number of epidermal cell files; this activity causes the observed number of epidermal cell files to vary from 18 to 22 (Galway et al. 1994; Baum and Rost 1996; Berger et al. 1998b).

Like other species of the Brassicaceae, *Arabidopsis* only develops root hairs on epidermal cells that are located outside anticlinal cortical cell walls. Because the *Arabidopsis* primary root always possesses eight files of cortical cells, there are eight root hair cell files and approximately 10 to 14 non-hair cell files (Dolan et al. 1994; Galway et al. 1994). Prior to hair formation, the differentiating trichoblasts can be distinguished from the atrichoblasts by their greater rate of cell division (Berger et al. 1998b), reduced cell length (Dolan et al. 1994; Masucci et al. 1996), enhanced cytoplasmic density (Dolan et al. 1994; Galway et al. 1994), reduced rate of vacuolation (Galway et al. 1994), cell surface ornamentation (Dolan et al. 1994), and cell wall epitopes (Freshour et al. 1996). It is likely that some of these cellular differences are linked; for example, the enhanced cell division rate in trichoblasts could lead to their greater cytoplasmic density and reduced vacuolation rate.

3.2 *Arabidopsis* Genes Controlling Root Hair Cell Specification

Several mutations have been identified in *Arabidopsis* that disrupt the normal production of root epidermal cell types. Two of these patterning mutants, *transparent testa glabra* (*ttg*) and *glabra2* (*gl2*) possess root hairs on essentially every root epidermal cell, which implies that the normal role of the *TTG* and *GL2* genes is either to promote non-hair cell differentiation or to repress root hair cell differentiation (Galway et al. 1994; DiCristina et al. 1996; Masucci et al. 1996). These mutations do differ in their specific effects on non-hair cell differentiation; the *ttg* mutations alter all aspects of non-hair differentiation (including the cell division rate, cytoplasmic density, and vacuolation rate) whereas the *gl2* mutations only affect the final cell morphology and do not affect the earlier cellular phenotypes (Galway et al. 1994; Masucci et al. 1996; Berger et al. 1998b). Thus, the TTG may be considered an earlier acting and wider-ranging component required for position-dependent non-hair cell differentiation.

The *TTG* gene has recently been cloned and found to encode a small protein with WD40 repeats (A. Walker, personal communication). Although the protein sequence does not provide a clear mechanistic understanding of TTG's role, it is known that *ttg* mutations can be functionally complemented by expression of the maize R cDNA (under the control of the strong cauliflower mosaic virus 35S promoter) in *Arabidopsis* (Lloyd et al. 1992; Galway et al. 1994). These data indicate that TTG is likely to activate an R homolog (a protein related to basic helix-loop-helix transcriptional activators; Ludwig et al. 1989) in *Arabidopsis* to specify the non-hair cell fate.

The *GL2* gene encodes a homeodomain transcription factor protein (Rerie et al. 1994), and it is preferentially expressed in the differentiating non-hair epidermal cells within the meristematic and elongation regions of the root

(Masucci et al. 1996). *GL2* gene expression has been shown to be influenced by the TTG and the maize R protein, with *ttg* mutations causing a reduction in *GL2* promoter activity and *35S::R* expression causing ectopic *GL2* promoter activity (Hung et al. 1998). Interestingly, despite the reduction in *GL2* promoter activity by *ttg* mutations, the appropriate cell position-dependent pattern of *GL2* expression is maintained, which implies that TTG is not required to specify the positional information for *GL2* expression (Hung et al. 1998). Taken together, the emerging picture is that TTG acts at an early stage in epidermis development to activate an R-like transcription factor which in turn positively regulates the expression of *GL2* (and perhaps other as yet unidentified genes) in a cell position-dependent manner to specify the non-hair cell type.

A third *Arabidopsis* gene affecting root epidermis cell specification has been identified and is designated *CAPRICE* (*CPC*) (Wada et al. 1997). The *cpc* mutant plants possess a small number of root hair cells, implying that CPC is a positive regulator of the root hair cell fate. Interestingly, the *gl2* mutation is epistatic to *cpc*, which suggests that CPC acts in the TTG/GL2 pathway as a negative regulator of GL2. A possible explanation for CPC's negative action is provided by the nature of its gene product; *CPC* encodes a small protein with a Myb-like DNA binding domain but without a typical transcriptional activation domain (Wada et al. 1997). Thus, CPC may inhibit *GL2* transcription by binding to its promoter and blocking its activation.

These molecular genetic findings lead to a simple view of the control of root epidermal cell fate in *Arabidopsis*. In this model, the relative activity of two competing transcription factors is critical for determining the fate of the epidermal cells. The activity of one of these transcription factors (dependent on the TTG and R-homolog proteins) is proposed to promote the expression of *GL2* (and probably other genes) within the immature epidermal cells located outside single cortical cells and to lead to non-hair cells. The other activity (involving the CPC protein) is proposed to inhibit *GL2* expression in the immature epidermal cells located between two cortical cells and to lead to the development of root-hair cells.

Future work will be necessary to test this simple model and to expand our understanding of the molecular control of root hair cell specification. In particular, it will be necessary to identify the putative R homolog from *Arabidopsis* and test the notion that it regulates *GL2*, and it will be important to examine the accumulation pattern of the *TTG* and *CPC* RNA/protein. Furthermore, the molecular basis of the signals that promote these transcription factor activities will need to be defined.

3.3 Nature of the Positional Information

3.3.1 Timing of Positional Cues

In addition to the molecular components of the epidermal cell specification pathway, other key issues that have not yet been resolved include the timing and directionality of the positional signaling that directs epidermal cell fate in *Arabidopsis*. One of the important questions is: at what point during epidermis development is the positional information presented to the cells to cause them to begin to adopt a particular fate? In this regard, it has long been known that trichoblasts and atrichoblasts exhibit cytological differences within the meristematic region of the root, which indicates that positional information must be provided at an early stage of root development (Cormack 1962; Cutter 1978; Dolan et al. 1994; Galway et al. 1994). This view has been supported by the recent use of two reporter gene fusions, a *GL2::GUS* fusion construct (Masucci et al. 1996) and an enhancer-trap *GFP* construct (line J2301; Berger et al. 1998c), each of which express their reporter gene in a position-dependent manner (within epidermal cells located outside a periclinal cortical cell wall) within the meristematic region of the root. Careful examination using these sensitive reporters reveals position-dependent gene expression within, or just one cell beyond, the epidermal/lateral root cap initials, which implies that positional information may be provided (and cell fates begin to be defined) within these initial cells (Masucci et al. 1996; Berger et al. 1998a).

The presence of differential gene expression within the initial cells of the root meristem led to the possibility that positional information may be initially provided to the epidermal cells during embryogenesis, at the time the basic root structure and meristem initials are formed (Scheres et al. 1994). Indeed, the analysis of the J2301 enhancer-trap *GFP* (Berger et al. 1998a) and the *GL2::GUS* (Y. Lin and J. Schiefelbein, unpublished, 1998) reporters show that the epidermal cell specification pattern becomes established during embryonic root development in *Arabidopsis*. For each of these reporters, expression is first detected at the heart stage, when the epidermal tissue becomes distinct from the ground tissue, and expression is later detected in a position-dependent pattern that mirrors the post-embryonic pattern (Berger et al. 1998a; Y. Lin and J. Schiefelbein, unpublished, 1998). Thus, it appears that the positional information is provided during embryonic root development to establish the proper pattern of gene activities that ultimately leads to appropriate post-embryonic cell type differentiation.

An interesting question arises from the identification of positional information during embryogenesis: is the positional information only provided to the developing epidermis during embryonic root development, or is the positional information provided both embryonically and post-embryonically? The former case is a possibility because embryonic positional information may establish two specific types of epidermal initials in the root meristem: ones that divide to produce only root-hair cells and ones that generate only non-hair cells. Because of the regular transverse divisions that occur to generate the epidermal cell files, the outcome of this process would be the same as for the alternate possibility; namely, the production of a position-dependent pattern of root-hair cell files and non-hair cell files in the mature *Arabidopsis* root.

To determine whether positional information is, in fact, provided to epidermal cells post-embryonically, two sorts of experiments were conducted. In one, a detailed analysis of peculiar epidermal cell clones was performed (Berger et al. 1998a). The clones examined were ones derived from rare post-embryonic longitudinal divisions of epidermal cells, which causes the two resulting daughter cells to occupy different positions relative to the underlying cortical cells. In these clones, the cells were observed to express marker genes and to exhibit characteristics that are appropriate for cells in their new position. In a second set of experiments, specific differentiating epidermal cells were subjected to laser ablation such that neighboring epidermal cells were able to invade the available space (Berger et al. 1998a). Regardless of the original state of the ablated cell or invading cell (trichoblast or atrichoblast), the ultimate characteristics of the invading cell were nearly always determined by its new location rather than its old. Therefore, in each of these sets of experiments, cells had effectively undergone a post-embryonic change in their position and, in response, exhibited a change in their developmental fate. This supports the view that positional information is provided post-embryonically, not only embryonically, to ensure appropriate cell specification in the *Arabidopsis* root epidermis.

3.3.2 Directionality of Positional Cues

In addition to the timing of the positional information, an important issue relates to the directionality of the putative signals. For example, the position-dependent pattern of epidermal cell types could arise because cortical cells signal to one or both of the epidermal cells, because the epidermal cells signal to one another, and/or because underlying tissue provides a diffusible signal to one or both of the epidermal cells. As described earlier in this chapter, the classic surgical experiments of Bunning (1951) appears to support the view that underlying tissues provide a signal to induce the non-hair cell fate.

Recently, the laser ablation of specific cells was employed to gain further insight into the directionality of the positional signals that define the epidermal cell types in *Arabidopsis* (Berger et al. 1998a). In one set of experiments, plants harboring the J2301 enhancer-trap *GFP* reporter were subjected to ablations in which immature epidermal cells were isolated from their neighbors within the same file or in adjacent files. In nearly every case, the isolated cells, which had lost contact with their epidermal neighbors, maintained the same reporter gene expression and differentiated according to their original position (Berger et al. 1998a). This result suggests that communication between living epidermal cells is not required to maintain the cell's appropriate choice of cell fate.

In a second set of cell ablations in the J2301 line, specific cortical cells were ablated such that the overlying immature epidermal cell(s) were isolated. Regardless of the original state of the isolated epidermal cell (trichoblast or atrichoblast), the ablation of the underlying cortical cell(s) had no effect on their future GFP expression or morphogenesis (Berger et al. 1998a). Thus, signaling between living cortical and epidermal cells is not required for the maintenance of proper cell-type differentiation.

It is important to note that although no evidence concerning cell communication was obtained in these ablation experiments, they do not rule out the possibility that such communication takes place. For example, it may be that the immature epidermal cells do not require positional information to maintain their developmental fate, only to change their developmental fate. It would therefore be very interesting (though technically challenging) to conduct similar ablation experiments during the time in which the positional information is being established in the embryo or during the time in which rare longitudinal divisions occur that change the relative position of the epidermal cells.

3.4 Genetic Overlap in the Control of Root and Shoot Epidermal Cell Specification in *Arabidopsis*

3.4.1 Overlap Between the Root and Hypocotyl Epidermis

During the course of work on the root epidermis, an unexpected relationship became apparent between cell specification in the root and the above-ground tissues of the *Arabidopsis* plant. The most striking similarities were observed in the epidermis of the hypocotyl. Although hypocotyl epidermal cells do not produce root hairs, there are two distinct files of epidermal cells in the *Arabidopsis* hypocotyl that arise in a position-dependent manner (Wei et al., 1994; Gendreau et al. 1997; Hung et al. 1998; Berger et al. 1998c). One type of hypocotyl cell file preferentially includes stomatal cells and are present outside an anticlinal cortical cell wall, whereas the other type of hypocotyl cell file possesses non-stomatal cells and are located outside a periclinal cortical cell wall. This means that cells of the hypocotyl epidermis, like the root epidermis, undergo position-dependent cell differentiation to generate a common pattern of cell types throughout the apical-basal axis of the *Arabidopsis* seedling.

The similarity in cell specification in the root and hypocotyl epidermis is also apparent when one considers the molecular components employed. Both the *ttg* and *gl2* mutations significantly alter the patterning of the hypocotyl cell types, causing a greater proportion of ectopic stomata (stomata located outside a periclinal cell wall) (Hung et al. 1998; Berger et al. 1998c). Furthermore, both the *GL2::GUS* and the J2301 enhancer-trap *GFP* reporter genes are preferentially expressed in epidermal cells located outside the periclinal cortical cell wall of the root and hypocotyl (Hung et al. 1998; Berger et al. 1998c). This parallel pattern of gene activity indicates that the *TTG/GL2* pathway is employed in both organs of the seedling to ensure that cells located outside a periclinal cortical cell wall differentiate into non-hair cells in the root and non-stomatal cells in the hypocotyl epidermis.

The similar pattern of specialized and non-specialized epidermal cells in both the root and hypocotyl is the result of common developmental activities during embryogenesis. This is not surprising, given the common origin of the root and hypocotyl epidermis from a basal set of protodermal cells at the globular stage of the *Arabidopsis* embryo (Scheres et al. 1994). Analysis of the *GL2::GUS* and J2301 enhancer-trap *GFP* reporter genes demonstrates that both the future hypocotyl and root express these markers in a similar fashion (Berger et al. 1998a; Y. Lin and J. Schiefelbein, unpublished data). Therefore, the specification of cell fate in the epidermis of both organs is controlled by a common position-dependent mechanism that is established during embryogenesis.

3.4.2 Overlap Between the Root and Leaf Epidermis

In addition to affecting the hypocotyl epidermis, the *TTG* and *GL2* genes are known to also affect trichome formation in the shoot epidermis of *Arabidopsis* (Marks 1994; Galway et al. 1994; Masucci et al. 1996; Larkin et al. 1997). This evidence for overlap between the root and leaf epidermis was unexpected because the patterning of cell types in these two tissues appears to be quite different; the root epidermis mechanism appears to rely on the positional relationship between epidermal cells and underlying cortical cells whereas the leaf epidermis mechanism relies on sensing trichome density. A further interesting aspect of this relationship is that TTG and GL2 control epidermal hair formation in opposite ways in the root and leaf. They are required for the formation of hairless cells in the root and hair-bearing (trichome) cells in the leaf. Although the meaning of this root/leaf overlap is unclear, it may be that TTG and GL2 represent general epidermis transcriptional regulators that have been recruited to participate in very different cell-type specification mechanisms during the evolution of epidermis development in angiosperms.

4 Environmental and Hormonal Effects on Root Hair Cell Specification

4.1 Effect of Environmental Factors

The production of root hair or non-hair cells can be dramatically modified by environmental factors. One of the most important factors is the humidity level. A maximum number of root hairs typically form on roots grown in humid air; whereas roots subjected to drought conditions or a flooded environment generally possess fewer hairs (Uphof 1962). Because trichoblasts and atrichoblasts are still typically formed under these conditions, the humidity effect is likely to act after the specification mechanism has dictated the basic trichoblast/atrichoblast pattern.

Additional environmental factors may influence root hair formation, including light (Cutter 1978), oxygen concentration (Cormack 1962), carbon dioxide concentration (Cormack 1962), calcium (Cormack 1962; Tanaka and Woods 1972), and mycorrhizal associations (Mexal et al. 1979; Peterson 1992). The mechanism responsible for the epidermal cell response to these environmental factors is unclear at the present time.

4.2 Effect of Plant Hormones

A great deal of evidence supports the role of plant hormones, particularly auxin and ethylene, in root hair development (Cormack 1949, 1962; Abeles 1973; Barbieri and Galli 1993; see also Chapter 7 this volume). In *Arabidopsis*, the results from numerous pharmacological and genetic experiments indicate that these hormones can promote root hair cell differentiation. For example, aminoethoxyvinylglycine (AVG, an ethylene biosynthesis inhibitor) or Ag^+ (an inhibitor of ethylene perception) blocks root hair formation (Masucci and

Schiefelbein 1994; Tanimoto et al. 1995) and 1-amino-cyclopropane-1-carboxylic acid (ACC) induces some ectopic root hairs in *Arabidopsis* (Tanimoto et al. 1995). Further, mutations affecting the *CONSTITUTIVE TRIPLE RESPONSE* (*CTR1*) locus, which encodes a Raf-like protein kinase proposed to negatively regulate the ethylene signal transduction pathway (Kieber et al. 1993) causes root hairs to form on epidermal cells which are normally hairless (Dolan et al. 1994). Also, the hairless root phenotype of the *dwarf (dwf;* auxin-resistant) and *auxin resistant2 (axr2*; auxin, ethylene, and abscisic acid resistant) mutants implicate auxin in root hair formation (Mizra et al. 1984; Wilson et al. 1990). Finally, the hairless *rhd6* mutant phenotype can be suppressed by the inclusion of ACC (an ethylene precursor) or indole-3-acetic acid (IAA, an auxin) in the growth media (Masucci and Schiefelbein 1994).

Given the involvement of hormones in root epidermal cell differentiation, an attractive possibility was that they may act as a diffusible signal(s) responsible for the position-dependent cell specification. According to this notion, TTG and GL2 would represent downstream molecular components that are regulated by the hormone(s) in a cell position-dependent manner. To test this possibility, the relationship between the *TTG/GL2* pathway (negative regulators of root hair cell differentiation) and the ethylene/auxin-related pathway (positive regulators of root hair cell differentiation) were analyzed (Masucci and Schiefelbein 1996). In contrast to the above model, the results from epistasis tests and *GL2* promoter-reporter gene analyses show that the ethylene/auxin pathway does not regulate the *TTG/GL2* pathway. In addition, studies of the developmental timing of the hormone effects indicate that the ethylene and auxin pathways promote root hair outgrowth after epidermal cell-type characteristics have developed (Masucci and Schiefelbein 1996). Taken together, the results suggest that the *TTG/GL2/CPC* pathway acts upstream of, or independently from, the ethylene/auxin pathway to define the pattern of cell types in the root epidermis. One of the implications of this proposal is that the newly-formed epidermal cells in the *Arabidopsis* root may initially be defined as a trichoblast or atrichoblast (due to the action of the *TTG/GL2/CPC* pathway) although the final pattern of epidermal cell types may be determined by the action of hormones (perhaps governed by or intertwined with the effect of environmental factors).

5 Concluding Remarks

A general outcome from studies of root epidermis cell specification is an appreciation for the importance of the specification of the non-hair cells. Because the needs of the plant can apparently be satisfied by having only a portion of the root epidermal cells produce root hairs, plant species have evolved mechanisms to produce root-hair cells and non-hair cells in appropriate proportions. Interestingly, rather than defining their hair/non-hair distribution by using mechanisms that regulate root-hair cell production, plants appear to use mechanisms that control the production of non-hair cells. For example, the asymmetric division mechanism employed by many monocotyledons is apparently used to generate non-hair cells, because in the absence of this asymmetric division, only root-hair cells are produced (Ivanov and Filippenko 1976). As another example, the cell position-dependent mechanism employs a genetic pathway (involving the TTG/GL2 proteins of *Arabidopsis*) to specifically regulate non-hair cell formation; mutations blocking this pathway lead to plants composed solely of root-hair cells (Galway et al. 1994; Masucci et al. 1996). Thus, it is likely that the particular distribution of root-hair and non-hair cells in different plant species is due to the evolution of various mechanisms that serve to promote non-hair cell specification.

Another general feature highlighted by these studies is the inherent plasticity or flexibility in the developmental fate of root epidermal cells. There are now numerous examples whereby cells that embark on one differentiation program may be caused to switch to the alternate program upon a change in their position or in response to external factors. This includes the effect of laser ablation (Berger et al. 1998a) or ACC treatment (Tanimoto et al. 1995; Masucci and Schiefelbein 1996) on root epidermal cell differentiation. It is tempting to speculate that plasticity of this sort may be important to ensure proper root-hair cell production in the event of extreme environmental conditions or rare abnormal cell divisions.

ACKNOWLEDGMENTS

We thank our colleagues in the root development community, including Fred Berger, Liam Dolan, Yan Lin, Alan Lloyd, David Marks, Myeong Min-Lee, Ben Scheres, Takuji Wada, and Amanda Walker for their helpful discussions and for providing manuscripts prior to publication.

References

Abeles FB (1973) Ethylene in plant biology. Academic Press, New York and London

Avers CJ (1963) Fine structure of *Phleum* root meristem cells. II. Mitotic asymmetry and cellular differentiation. Am J Bot 50:140-148

Barbieri P, Galli E (1993) Effect on wheat root development of inoculation with an *Azospirillum brasilense* mutant with an altered indole-3-acetic acid production. Res Microbiol 144:69-75

Barlow PW (1984) Positional controls in root development. In: Barlow PW, Carr DJ (Eds) Positional controls in plant development. Cambridge University Press, Cambridge, pp 281-318

Baum SF, Rost TL (1996) Root apical organization in *Arabidopsis thaliana*. 1. Root cap and protoderm. Protoplasma 192:178-188

Berger F, Haseloff J, Schiefelbein J, Dolan L (1998a) Positional information in root epidermis is defined during embryogenesis and acts in domains with strict boundaries. Current Biology 8:421-430

Berger F, Hung C-Y, Dolan L, Schiefelbein J (1998b) Control of cell division in the root epidermis of *Arabidopsis thaliana*. Dev Biol 194:235-245

Berger F, Linstead P, Dolan L, Haseloff J (1998c) Stomata patterning on the hypocotyl of *Arabidopsis thaliana* is controlled by genes involved in the control of root epidermis patterning. Dev Biol 194:226-234

Bunning E (1951) Uber die Differenzierungsvorgange in der Cruciferenwurzel. Planta 39:126-153

Clarke KJ, McCully ME, Miki NK (1979) A developmental study of the epidermis of young roots of *Zea mays*. Protoplasma 98:283-309

Cormack RGH (1935) The development of root hairs by *Elodea canadensis*. New Phytol 34:19-25

Cormack RGH (1949) The development of root hairs in angiosperms. Bot Rev 15:583-612

Cormack RGH (1962) Development of root hairs in angiosperms. II. Bot Rev 28:446-464

Cutter EG (1978) The epidermis. In: Plant anatomy. Clowes & Sons, London, pp 94-106

Cutter EG, Feldman LJ (1970) Trichoblasts in *Hydrocharis*, I. Origin, differentiation, dimensions and growth. Am J Bot 57:190-201

DiCristina MD, Sessa G, Dolan L, Linstead P, Baima S, Ruberti I, Morelli G (1996) The *Arabidopsis* Athb-10 (GLABRA2) is an HD-Zip protein required for regulation of root hair development. Plant J 10:393-402

Dolan L, Janmaat K, Willemsen V, Linstead P, Poethig S, Roberts K, Scheres B (1993) Cellular organisation of the *Arabidopsis thaliana* root. Development 119:71-84

Dolan L, Duckett C, Grierson C, Linstead P, Schneider K, Lawson E, Dean C, Poethig S, Roberts K (1994) Clonal relations and patterning in the root epidermis of *Arabidopsis*. Development 120:2465-2474

Esau K (1965) Plant anatomy. John Wiley and Sons, New York

Freshour G, Clay RP, Fuller MS, Albersheim P, Darvill AG, Hahn MG (1996) Developmental and tissue-specific structural alterations of the cell-wall polysaccharides of *Arabidopsis thaliana* roots. Plant Physiol 110:1413-1429

Galway ME, Masucci JD, Lloyd AM, Walbot V, Davis RW, Schiefelbein JW (1994) The *TTG* gene is required to specify epidermal cell fate and cell patterning in the *Arabidopsis* root. Dev Biol 166:740-754

Gendreau E, Traas J, Desnos T, Grandjean O, Caboche M, Hofte H (1997) Cellular basis of hypocotyl growth in *Arabidopsis thaliana*. Plant Physiol 114:295-305

Hung CY, Lin Y, Zhang M, Pollock S, Marks MD, Schiefelbein J (1998) A common position-dependent mechanism controls cell-type patterning and GLABRA2 regulation in the root and hypocotyl epidermis of *Arabidopsis*. Plant Physiol 117:73-84

Ivanov VB, Filippenko VN (1976) Role of asymmetric mitosis in differentiation of rhizodermis cells of festucoid grasses. Dokl Akad Nauk SSSR 230:712-715

Kieber JJ, Rothenberg M, Roman G, Feldman KA, Ecker JR (1993) *CTR1*, a negative regulator of the ethylene response pathway in *Arabidopsis*, encodes a member of the Raf family of protein kinases. Cell 72:427-441

Larkin JC, Marks MD, Nadeau J, Sack F (1997) Epidermal cell fate and patterning in leaves. Plant Cell 9: 1109-1120

Leavitt RG (1904) Trichomes of the root in vascular cryptograms and angiosperms. Proc Boston Soc Nat Hist 31:273-313

Lloyd AM, Walbot V, Davis RW (1992) *Arabidopsis* and *Nicotiana* anthocyanin production activated by maize regulators *R* and *C1*. Science 258:1773-1775

Ludwig SR, Habera LF, Dellaporta SL, Wessler SR (1989) *Lc*, a member of the maize *R* gene family responsible for tissue-specific anthocyanin production, encodes a protein similar to transcriptional activators and contains the myc-homology region. Proc Natl Acad Sci USA 86:7092-7096

Marks MD (1994) The making of a plant hair. Curr Biol 4:621-623

Masucci JD, Schiefelbein JW (1994) The *rhd6* mutation of *Arabidopsis thaliana* alters root-hair initiation through an auxin- and ethylene-associated process. Plant Physiol. 106:1335-1346

Masucci JD, Schiefelbein JW (1996) Hormones act downstream of *TTG* and *GL2* to promote root hair outgrowth during epidermis development in the *Arabidopsis* root. Plant Cell 8:1505-1517

Masucci JD, Rerie WG, Foreman DR, Zhang M, Galway ME, Marks MD, Schiefelbein JW (1996) The homeobox gene *GLABRA2* is required for position-dependent cell differentiation in the root epidermis of *Arabidopsis thaliana*. Development 122:1253-1260

Mexal JG, Reid CPP, Burke EJ (1979) Scanning electron microscopy of lodgepole pine roots. Bot Gaz 140:318-323

Mizra JI, Olsen GM, Iversen T-H, Maher EP (1984) The growth and gravitropic responses of wild-type and auxin resistant mutants of *Arabidopsis thaliana*. Physiol Plantarum 60:516-522

Peterson RL (1992) Adaptations of root structure in relation to biotic and abiotic factors. Can J Bot 70:661-675

Row HC, Reeder JR (1957) Root-hair development as evidence of relationships among genera of gramineae. Am J Bot 44:596-601

Rerie WG, Feldmann KA, Marks MD (1994) The *GLABRA2* gene encodes a homeodomain protein required for normal trichome development in *Arabidopsis*. Genes Dev 8:1388-1399

Scheres B, Wolkenfelt H, Willemsen V, Terlouw M, Lawson E, Dean C, Weisbeek P (1994) Embryonic origin of the *Arabidopsis* primary root and root meristem initials. Development 120:2475-2487

Sinnot EW, Bloch R (1939) Cell polarity and the differentiation of root hairs. Proc Natl Acad Sci USA 25:248-252

Tanaka Y, Woods FW (1972) Root and root hair growth in relation to supply and internal mobility of calcium. Bot Gaz 133:29-34

Tanimoto M, Roberts K, Dolan L (1995) Ethylene is a positive regulator of root-hair development in *Arabidopsis thaliana*. Plant J 8:943-948

Uphof JCTH (1962) Plant hairs. In: Handbuch der Pflanzenanatomie. 2nd edition, Vol IV, Part 5. Borntraeger, Berlin, pp 1-206

Wada T, Tachibana T, Shimura Y, Okada K (1997) Epidermal cell differentiation in *Arabidopsis* determined by a Myb homolog, CPC. Science 277:1113-1116

Wei N, Kwok SF, von Arnim AG, Lee A, McNellis TW, Piekos B, Deng X-W (1994) *Arabidopsis COP8, COP10*, and *COP11* genes are involved in repression of photomorphogenic development in darkness. Plant Cell 6:629-643

Wilson A, Pickett FB, Turner JC, Estelle M (1990) A dominant mutation in *Arabidopsis* confers resistance to auxin, ethylene, and abscisic acid. Mol Gen Genet 222:377-383

Chapter 13
Genetics of Root Hair Development

Alison Cavell and Claire S. Grierson

IACR-Long Ashton Research Station, Department of Agricultural Sciences, University of Bristol, Long Ashton, Bristol BS41 9AF, U.K.

1 Introduction

1.1 Root Hair Development

Root hairs are formed by a dramatic change in cell shape. This process can be divided into a number of stages: firstly cell fate is determined, i.e., whether an epidermal cell will be a hair cell (trichoblast) or a non-hair cell (atrichoblast) (although in some species every epidermal cell is a hair cell) and the site of root hair initiation is selected. The cell wall loosens at this initiation site to form a bulge. Root hair elongation then proceeds by tip growth, a specialized growth mechanism shared by other tubular shaped cells such as pollen tubes and fungal hyphae, whereby new cell wall is deposited in a highly polar manner at the tip of the growing tube. The speed of growth and the final length of the root hair is dynamic and strongly affected by the environment. For example, root hair elongation is stimulated by low phosphorus availability (Bates and Lynch 1996).

The physiology of root hair formation is well characterized but little is known about how these cellular processes are controlled and interact with each other and the environment to form a root hair of correct shape and appropriate length.

1.2 The Use of Genetics to Study Root Hair Development

One way to address the question of how root hair development is controlled is to identify and characterize genes that influence this process. Genetics is the study of genes through their variation and is a powerful tool with which novel genes can be identified and complex developmental pathways elucidated. For example, the developmental pathway of trichome (leaf hair) morphogenesis in *Arabidopsis* has been described using a genetic approach (Hülskamp et al. 1994) and a gene involved in trichome morphogenesis, *ZWICHEL* (Oppenheimer et al. 1997) is the first plant kinesin to have a defined function. In addition to this, some genes,

Key words. *Arabidopsis*, cell, elongation, gene, genetics, morphogenesis, mutant, polarity, root hair, shape, tip growth which have been identified because of their role in cell morphogenesis, have been shown to have a wider impact than could have first been imagined. The *Drosophila FRIZZLED* gene, for example, was identified because of its effect on the polarity of hairs and bristles over the insect's body. Subsequently *FRIZZLED* genes have been identified in diverse organisms such as rat, zebrafish, chicken, sea urchin and *Caenorhabditis elegans* (Wang et al. 1996) and it has been proposed that a *FRIZZLED* signal transduction pathway exists in these organisms to control polarity.

Root hairs are amenable to genetic analysis because many mutations in root hair genes do not affect fertility and very subtle changes in root hair shape can be easily seen at the seedling stage, facilitating large screening programmes.

The use of genetics to identify genes depends on variation between different forms (alleles) of a gene. If different alleles of a gene have different effects on the root hair phenotype of the plant, then that gene can be identified as playing a role in root hair development. Allelic variation can arise naturally or can be induced experimentally through mutagenesis programs. Naturally occurring variants in root hair phenotype have been found in clover and tomato. In clover, it has been demonstrated that root hair length can be increased or decreased by selection and subsequent breeding (Caradus 1979). In tomato, a screen for strains with increased phosphate uptake capacity identified a particularly efficient strain that had root hairs that were more abundant and longer than the other strains investigated. This phenotype was called "cottony root" and inheritance studies showed that it was caused by a single recessive gene named *CRT* (Hochmuth et al. 1985).

These examples of natural variation have shown that root hair development is at least partly under genetic control. However, the frequency with which variants naturally occur is low and most success in the identification of genes involved in root hair development has come from random mutagenesis programmes. For example, three genes that influence root hair development in Maize have been identified in lines mutagenised with the *Mutator* transposon (Wen and Schnable 1994). These mutants have root hairs that are 5-10% (*rth1* and *rth3*) and 20-25% (*rth2*) of the wild type length.

1.3 *Arabidopsis* as a Model for Root Hair Development

The majority of root hair development genes have been identified in the model plant *Arabidopsis thaliana*. This is due to its fast regeneration time, small size, large seed set and excellent genome resources that make gene identification and isolation easier than in less well characterized plants. *Arabidopsis* root hairs are visible only 3-4 days after germination. The small size of seedlings means that they can be sown at high density on agar plates and large screening programmes can be carried out on mutagenized populations. In addition to this, *Arabidopsis* root hairs form in a predictable pattern, with files of trichoblasts which overlie a cleft in the underlying cortical cells (Dolan et al. 1994). Trichoblasts (hair cells) and atrichoblasts (non-hair cells) are distinguishable at an early stage of development because trichoblasts are more cytoplasmically dense than atrichoblasts and do not have the surface deposits that are visible on atrichoblasts

(Dolan et al. 1994). This can allow genes that affect the total number of root hairs to be classified as to whether they affect early differentiation of trichoblasts and atrichoblasts or whether they may be involved in the later stages of root hair development. In this chapter we aim to discuss those genes that are involved in root hair initiation or tip-growth. Genes that affect patterning and cell fate are dealt with in Chapter 12, this volume. However, cell fate and initiation are closely linked and some genes may affect both processes and may therefore be included in both chapters.

2 *Arabidopsis* Root Hair Development

2.1 Genes Involved

Twenty-four *Arabidopsis* genes that are involved in root hair development are listed in Table 1. This table includes genes that were identified because of the root hair phenotype of the mutant allele and those identified through other screens and have subsequently been shown to have a root hair phenotype. Mutants in all stages of root hair development have been identified, that is in the patterning of the root, initiation, bulge formation, tip growth and the final length of the root hair (Table 1 and Fig. 1). These genes map all over the genome (Fig. 2).

The first stage in root hair formation is initiation. This process depends on correct patterning in the root, that is the differentiation between trichoblasts (hair cells) and atrichoblasts (non-hair cells) and the correct positioning of an initiation site at the basal end of each trichoblast (in *Arabidopsis*). Presently, it is difficult to separate patterning and initiation. Trichoblasts and atrichoblasts can be distinguished by staining (Dolan et al. 1994) which can give clues about patterning in the mutant, but the precise molecular differences between trichoblasts and atrichoblasts are yet to be determined and root hair initiation may occur early on and may be integral to this differentiation.

Thirteen of the twenty four genes listed in Table 1 affect patterning and/or initiation (Fig. 1). The mutants *ctr1, erh1, pom1, rhd6 and axr2* show a wild type trichoblast/atrichoblast pattern as revealed by differential staining (Table 1), but have abnormal root hair initiation phenotypes. Other genes in this category affect the differentiation between trichoblasts and atrichoblasts. For example the mutants *erh3, rhl1, rhl2* and *rhl3* have a trichoblast/atrichoblast pattern unlike wild type and have either more or less root hairs than wild type (Table 1). A third set of genes in this category affect the position of the root hair on the trichoblast. The *rhd6, axr2* and *etr1* mutants have a basal shift in the initiation site whereas the *eto1* mutant has an apical shift (Table 1).

Table 1. *Arabidopsis* root hair genes

Gene	Mutant root hair phenotype	References to root hairs
CTR1	More root hairs than WT due to increase in ectopic root hairs (on atrichoblasts). Treatment of wild type with ethylene phenocopies *ctr1* mutation. Differential cytoplasmic staining between trichoblasts and atrichoblasts as WT.	Kieber et al. (1993), Dolan et al. (1994), Tanimoto et al. (1995), Massucci and Schiefelbein (1996)
ERH1	56% of hairs are ectopic. Some trichoblasts do not have dense cytoplasm.	Schneider et al. (1997)
POM1 /ERH2	35% of hairs are ectopic. Differential cytoplasmic staining as WT.	Hauser et al. (1995), Schneider et al. (1997)
ERH3	32% of hairs are ectopic. Root anatomy and trichoblast/atrichoblast pattern unlike WT.	Schneider et al. (1997)
RHL1	Few hairs, bald collet. Some hairs around emergence of lateral roots. Differential cytoplasm staining between trichoblasts and atrichoblasts less pronounced than WT. No bulge.	Schneider et al. (1997)
RHL2	As rhl1	Schneider et al. (1997)
RHL3	As rhl1	Schneider et al. (1997)
RHD6	Reduced number of root hairs (15-17% WT), basal shift in site of hair emergence from cell, multiple root hairs on some cells. Phenotype can be rescued by auxin or an ethylene precursor. Differential cytoplasmic staining as WT.	Masucci and Schiefelbein (1994), Masucci and Schiefelbein (1996)
AXR2	Few or no root hairs depending on conditions. Root hairs when grown in very high auxin or under low phosphorous conditions. Hairs are WT but sometimes multiple and site of hair emergence shifted basally. Differential cytoplasmic staining as WT.	Wilson et al. (1990), Massucci and Schiefelbein (1994), Massucci and Schiefelbein (1996), Bates and Lynch (1996)
AXR3	Only produces root hairs under high auxin conditions, otherwise bald.	Leyser et al. (1996)
ETR1	Basal shift in site of hair emergence. WT length distribution and no. of hairs, short hairs, only 0.5mm occasionally fail to initiate. Ethylene resistant. WT gene encodes ethylene receptor.	Massucci and Schiefelbein (1994), Massucci and Schiefelbein (1996), Pitts et al. (1998)
ETO1	Longer hairs than WT. Apical shift in site of hair emergence, normal length distribution and number of hairs, ethylene overproducer.	Pitts et al. (1998), Massucci and Schiefelbein (1996)

Table 1. *continued. Arabidopsis* root hair genes

AXR1	Deficient in elongation, little effect on initiation. Shorter hairs, more variable in length, some twin hairs. Reduced number of root hairs or WT numbers depending on conditions.	Pitts et al. (1998), Cernac et al. (1997), Masucci and Schiefelbein (1996)
RHD1	Large bulge. Root hairs are similar length to wild type but with a bulbous base.	Schiefelbein and Somerville(1990)
TIP1	Large bulge. Short, wide often branched root hairs, defective pollen tubes.	Schiefelbein et al. (1993), Ryan et al. (1998)
RHD2	Very short root hairs, which initiate to form a bulge but do not proceed to tip-growth. Calcium distribution after initiation and bulge formation is not localized as in wild type.	Schiefelbein and Somerville (1990), Wymer et al. (1997)
COW1	Short, fat, often branched root hairs. Slow elongation rate.	Grierson et al. (1997)
RHD4	Short, occasionally branched root hairs. Varying diameter (bulges and constrictions). Increased cell wall thickness at tips of hairs.	Schiefelbein and Somerville (1990), Galway et al. (1997b),
RHD3	Short, wavy and occasionally branched root hairs. Asymmetric tip-growth, abnormally distributed vesicles at the hair tip. Reduced vacuole size and increased proportion of cytoplasm in the root hair.	Schiefelbein and Somerville (1990), Galway et al. (1997a), Wang et al. (1997)
WAVY	Short, wide, curled root hairs.	Marks and Esch (1992)
SAR1	Suppresses axr1 Root hairs WT or thicker with bulge at end, sometimes ectopic. Also suppresses some aspects of aux1-7.	Pitts et al. (1998), Cernac et al. (1997)
AUX1	Some branched hairs, not as long as WT on unsupplemented medium.	Pitts et al. (1998), Cernac et al. (1997)
EIN2	Short root hairs, same number and distribution as WT.	Massucci and Schiefelbein (1994), Pitts et al. (1998), Massucci and Schiefelbein (1996)
PHYB /HY3	Root hairs longer than WT when grown in the light, same as WT when grown in the dark.	Reed et al. (1993)

The second stage of root hair formation is loosening of the cell wall at the initiation site to form a bulge. Two mutants have been identified that are affected in bulge formation; *rhd1* and *tip1* both have larger bulges than wild type (Table 1 and Fig. 1).

Tip growth commences after bulge formation to form the straight tubular structure of the root hair. Seven mutants have been identified as being affected in tip growth: *tip1, rhd2, cow1, rhd4, rhd3, wavy* and *sar1* (Table1). Root hairs on these mutants either fail to establish tip growth (*rhd2*) or produce hairs of

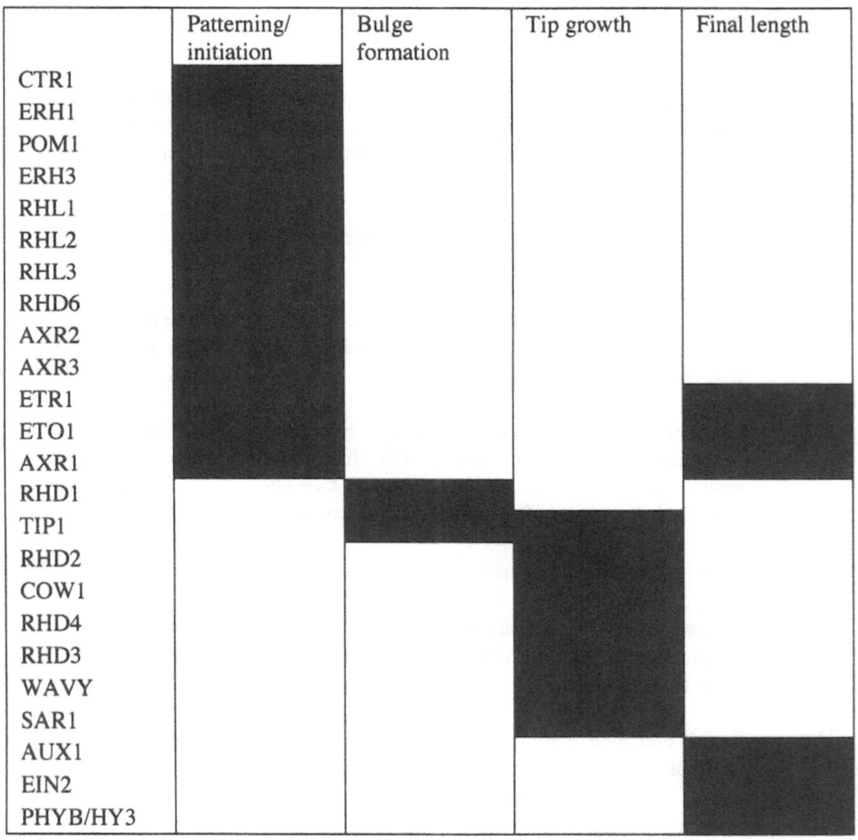

Fig. 1. Stages in root hair development in which twenty-four *Arabidopsis* genes are thought to act. *Arabidopsis* genes which have a role in root hair development are listed to the *left*. Stages in root hair development are shown along the top. Black boxes indicate where a gene is known to affect a particular stage in root hair development. Boxes are left blank if a gene is thought not to affect that stage in root hair development or if not enough information is available. A distinction is made between genes affecting tip growth whose mutants have abnormally shaped root hairs and genes affecting final length, where the mutants have root hairs of wild type shape that are abnormally long or short

abnormal shape, i.e., short, branched, fat, wavy, bulbous etc. (*tip1, cow1, rhd4, rhd3, wavy, sar1*).

The final set of mutants affect the length of the root hair only, that is the hairs are wild type shape, so tip growth is not severely affected, but the hairs are either longer (*phyB, eto1*) or shorter than wild type (*etr1, axr1, aux1, ein2*). It is interesting to note that three of these mutants are also affected in patterning and/or initiation (*etr1, eto1, axr1*) (Table 1 and Fig. 1).

2.2 Root Hair Development Pathway

Characterisation of the mutant phenotype can suggest when the wild type gene is acting, for example, the *rhd2* mutant produces a bulge but is deficient in elongation (Schiefelbein and Somerville 1990) so it may be concluded that the wild type *RHD2* gene is necessary for tip growth but not for bulge formation. This information can be used to create a pathway of gene action, i.e., genes that only affect bulge formation probably act before those that only affect tip growth.

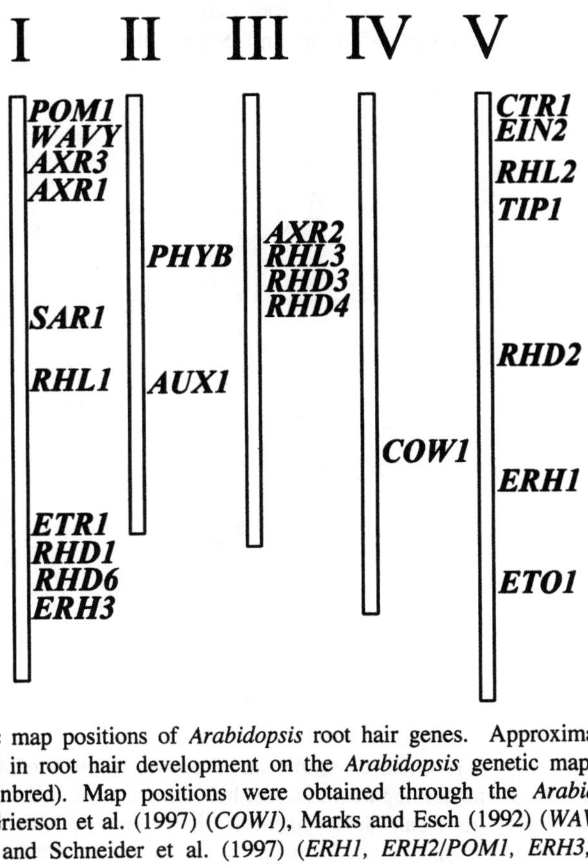

Fig. 2. Genetic map positions of *Arabidopsis* root hair genes. Approximate positions of genes involved in root hair development on the *Arabidopsis* genetic map (Classical and Recombinant Inbred). Map positions were obtained through the *Arabidopsis thaliana* database and Grierson et al. (1997) (*COW1*), Marks and Esch (1992) (*WAVY*), Ryan et al. (1998) (*TIP1)* and Schneider et al. (1997) (*ERH1, ERH2/POM1, ERH3, RHL1, RHL2, RHL3*)

However, within the pathway of root hair development, genes may interact with each other in a complex manner and may have roles in developmental stages not evident from the mutant phenotype. Genetic pathways can be better resolved by double mutant analysis. This involves crossing mutants in two different genes and selecting plants in subsequent generations that are homozygous for mutations in both genes. The double mutant then carries mutations in two genes and allows the effects of both mutations to be seen in the same plant. The phenotype of the double mutant can be used to assess the interaction between the two genes. A common type of interaction is epistasis. Here the phenotype of the double mutant looks exactly like one of the single mutant parents and it is usually interpreted as meaning that the genes act in sequence. For example, *cow1-1* is epistatic to *rhd4* because the phenotype of the double mutant is exactly like the *cow1-1* single mutant and, because the *cow1-1* single mutant root hair is the shorter, this is interpreted as meaning that *COW1* acts before *RHD4* (Grierson et al. 1997). If the double mutant phenotype is an addition of the two single-mutant phenotypes the interaction between the two genes is termed additive and suggests that the genes act in separate pathways, for example *rhd1* and *rhd3* have an additive interaction because the *rhd1*/*rhd3* double mutant has root hairs with bulbous bases (*rhd1* like) and *rhd3*-like wavy hairs (Schiefelbein and Somerville 1990). Based on both

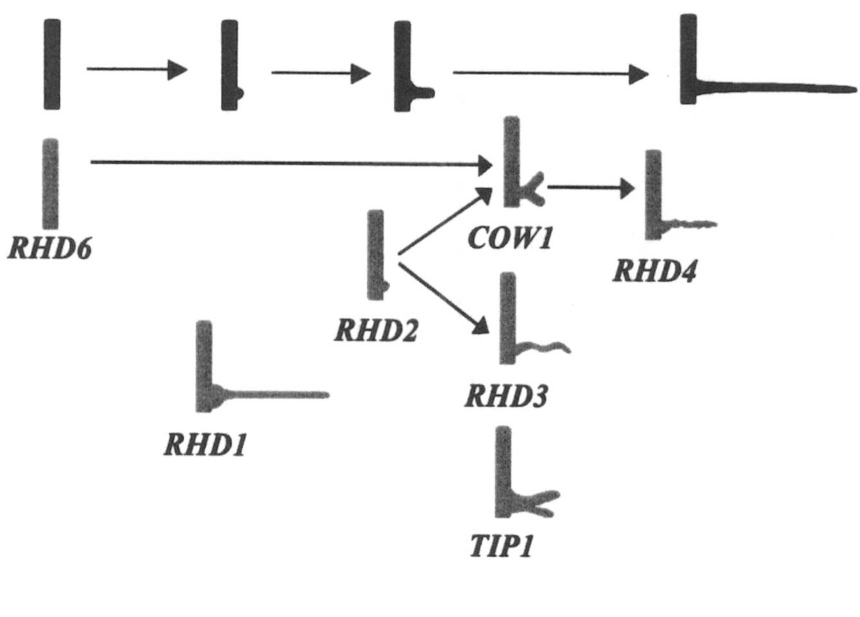

Fig. 3. Genetic pathway of root hair development. Proposed pathway of *Arabidopsis* root hair development. Wild type development of a root hair from a single cell is shown in *black*. The genetic pathway of root hair development is shown below, with genes and their corresponding mutant phenotype (*in gray*) arranged in the order they are thought to act with arrows indicating epistasis (Schiefelbein and Somerville 1990; Grierson et al. 1997)

single and double mutant analysis, two genetic pathways of root hair development have been published (Schiefelbein and Somerville 1990; Grierson et al. 1997), which together include seven genes that were identified because of their strong mutant root hair phenotype (*RHD1, RHD2, RHD3, RHD4, RHD6, TIP1* and *COW1*). Figure 3 shows this pathway and the corresponding mutant phenotypes.

The large number of auxin or ethylene resistant, sensitive or overproducing mutants listed in Table 1 reflects the fact that auxin and ethylene play an important role in the regulation of root hair development (Masucci and Schiefelbein 1994, 1996; Tanimoto et al. 1995; Cernac et al. 1997; Pitts et al. 1998). In addition to this, phosphate concentration has a strong effect on *Arabidopsis* root hair length and it is thought that auxin is involved in this response (Bates and Lynch 1996).

However, the role of auxin, ethylene and phosphate in the pathway of root hair development shown in Figure 3 is as yet unknown. It is likely that the pathway is controlled by a web of genes that interact with each other and with the environment. This might be resolved by a combination of molecular, functional and double mutant analysis in the context of a defined environment and by testing the responses of single and double mutants to auxin, ethylene and phosphate.

3 Conclusions

Genetics has proved to be useful in the identification of genes that are involved in root hair development. Twenty four *Arabidopsis* genes have been identified as being essential for normal root hair development to be completed. This is probably only a subset of the total number of genes involved in this process, many of which either have not been identified yet (mutant screens have not reached saturation) or can not be identified through mutation because of functional redundancy in the genome or because the mutant alleles are lethal. These genes may be identified using complementary approaches such as promoter or gene trapping.

Molecular characterization of root hair development genes has begun. *EIN2* has recently been cloned (Alonso et al. 1999) and has been found to be a transducer of both stress and ethylene response in *Arabidopsis* which may be the link between several different hormone signalling pathways. The *RHD3* gene has also been cloned and encodes a novel 89-kD polypeptide with putative GTP-binding domains which is expressed in all major plant organs and has a role in cell growth throughout the plant (Wang et al. 1997). This highlights the usefulness of a detailed examination of a developmental process such as root hair development for the identification of the function of novel genes. This is important now that the *Arabidopsis* genome sequence is soon to be completed as many sequenced genes are novel and have no known function.

A genetic approach has allowed a pathway of root hair development to be proposed. This may highlight which genes play a more important role in root hair development and may therefore warrant more detailed molecular characterisation.

Understanding the genetics of root hair development will allow the identification of molecular mechanisms that control the cellular processes of root hair initiation and tip growth.

References

Alonso JM, Hirayama T, Roman G, Nourizadeh S, Ecker J (1999) *EIN2*, a bifunctional transducer of ethylene and stress responses in *Arabidopsis*. Science 284:2148-2152

Bates TR, Lynch JP (1996) Stimulation of root hair elongation in *Arabidopsis thaliana* by low phosphorus availability. Plant, Cell and Environment 19:529-538

Caradus JR (1979) Selection for root hair length in white clover, *Trifolium repens*. Euphytica 28:489-494

Cernac A, Lincoln C, Lammer D, Estelle M. (1997) The *SAR1* gene of *Arabidopsis* acts downstream of the *AXR1* gene in auxin response. Development 124:1583-1591

Dolan L, Duckett CM, Grierson C, Linstead P, Schneider K, Lawson E, Dean C, Poethig S, Roberts K (1994) Clonal relationships and cell patterning in the root epidermis of *Arabidopsis*. Development 120: 2465-2474

Galway ME, Heckman JW Jr, Schiefelbein JW (1997a) Growth and ultrastructure of *Arabidopsis* root hairs: the *rhd3* mutation alters vacuole enlargement and tip growth. Planta 201:209-218

Galway ME, Lane D, Schiefelbein JW (1997b) Cell wall thickness and cell diameter vary in root hairs of the *rhd4* mutant of *Arabidopsis thaliana*. Plant Physiology 113:1624

Grierson CS, Roberts K, Feldmann KA, Dolan L (1997) The *COW1* locus of *Arabidopsis thaliana* acts after *RHD2* and in parallel with *RHD3* and *TIP1*, to determine the shape, rate of elongation, and number of root hairs produced from each site of hair formation. Plant Physiology 115:981-990

Hauser MT, Morikami A, Benfey PN (1995) Conditional root expansion mutants of *Arabidopsis*. Development 121:1237-1252

Hochmuth GJ, Gabelman WH, Gerloff GC (1985) A gene affecting tomato root morphology. Hortscience 20:1099-1101

Hülskamp M, Miséra S, Jürgens G (1994) Genetic disection of trichome cell development in *Arabidopsis*. Cell 76:555-566

Kieber JJ, Rothenberg M, Roman G, Feldmann KA, Ecker JR (1993) *CTR1*, a negative regulator of the ethylene response pathway in *Arabidopsis*, encodes a member of the raf family of protein-kinases. Cell 72:427-441

Leyser HM, Pickett FB, Dharmasiri S, Estelle M (1996) Mutations in the AXR3 gene of *Arabidopsis* result in altered auxin response including ectopic expression from the SAUR-AC1 promoter. Plant Journal 10:403-413

Marks DM, Esch JJ (1992) Trichome formation in *Arabidopsis* as a genetic model system for studying cell expansion. Current Topics in Plant Biochemistry and Physiology. 11:131-142

Masucci JD, Schiefelbein JW (1994) The *rhd6* mutation of *Arabidopsis thaliana* alters root-hair initiation through an auxin- and ethylene-associated process. Plant Physiology 106:1335-1346

Masucci JD, Schiefelbein JW (1996) Hormones act downstream of *TTG* and *GL2* to promote root hair outgrowth during epidermis development in the *Arabidopsis* root. The Plant Cell 8:1505-1517

Oppenheimer DG, Pollock MA, Vacik J, Szymanski DB, EricsonB, Feldmann K, Marks MD (1997) Essential role of a kinesin-like protein in *Arabidopsis* trichome morphogenesis. PNAS 94:6261-6266

Pitts JR, Cernac A and Estelle M (1998) Auxin and ethylene promote root hair elongation in *Arabidopsis*. The Plant Journal 16:553-560

Reed JW, Nagpal P, Poole DS, Furuya M, Chory J (1993) Mutations in the gene for the red far-red light receptor phytochrome-b alter cell elongation and physiological-responses throughout *Arabidopsis* development. Plant Cell 5:147-157

Ryan E, Grierson CS, Cavell A, Steer M, Dolan L (1998) *TIP1* is required for both tip growth and non tip growth in *Arabidopsis*. New Phytologist 138:49-58

Schiefelbein JW, Somerville C (1990) Genetic control of root hair development in *Arabidopsis thaliana*. The Plant Cell 2:235-243

Schiefelbein J, Galway M, Masucci J, Ford S (1993) Pollen tube and root-hair tip growth is disrupted in a mutant of *Arabidopsis thaliana*. Plant Physiology 103:979-985

Schneider K, Wells B, Dolan L, Roberts K (1997) Structural and genetic analysis of epidermal cell differentiation in *Arabidopsis* primary roots. Development 124: 1789-1798

Tanimoto M, Roberts K, Dolan L (1995) Ethylene is a positive regulator of root hair development in *Arabidopsis thaliana*. Plant Journal 8:943-948

Wang Y, Macke JP, Abella BS, Andreasson K, Worley P, Gilbert DJ, Copeland NG, Jenkins NA, Nathans J (1996) A large family of putative transmembrane receptors homologous to the product of the *Drosophila* tissue polarity gene *frizzled*. The Journal of Biological Chemistry 271:4468-4476

Wang H, Lockwood SK, Hoeltzel MF, Schiefelbein JW (1997) The *ROOT HAIR DEFECTIVE3* gene encodes an evolutionarily conserved protein with GTP-binding motifs and is required for regulated cell enlargement in *Arabidopsis*. Genes and Development 11:799-811

Wen T-J, Schnable PS (1994) Analyses of mutants of three genes that influence root hair development in *Zea mays* (Graminiae) suggest that root hairs are dispensible. American Journal of Botany 81:833-842

Wilson AK, Pickett FB, Turner JC, Estelle M (1990) A dominant mutation in *Arabidopsis* confers resistance to auxin, ethylene and abscisic acid. Mol Gen Genet 222:377-383

Wymer CL, Bibikova TN, Gilroy S (1997) Cytoplasmic free calcium distributions during the development of root hairs of *Arabidopsis thaliana*. Plant Journal 12:427-439

Chapter 14
Infection of Root Hairs by Rhizobia: Infection Thread Development with Emphasis on the Microtubular Cytoskeleton

Antonius C.J. Timmers

Laboratoire de Biologie Moléculaire des Relations Plantes-Microorganismes, Chemin de Borde-Rouge, BP27, 31326 Castanet-Tolosan, France

1 Introduction

The symbiotic interaction between soil bacteria of the genera *Rhizobium, Azorhizobium* or *Bradyrhizobium* (generally referred to as rhizobia) and plants of the Leguminosae family result in the formation of nodules, new organs in which the bacteria reduce nitrogen into ammonia which can subsequently be utilized by the plant (for reviews on nodulation see Mylona et al. 1995; Dénarié et al. 1996; Long 1996; Hirsch and LaRue 1997). In a large number of interactions bacterial infection takes place through the development of an infection thread in root hairs of the host plant (reviewed by Kijne 1992). In this chapter I will first briefly describe the growth and structure of root hairs and recapitulate the relevant aspects of the infection process before going into detail of the changes in the organization of the microtubular cytoskeleton in root hairs during infection with rhizobia.

2 Growth and Structure of Root Hairs

During the early study of root hairs it became clear that the tubular basal part of the hair does not grow and that growth is restricted to the tip (Cormack 1962; Belford and Preston 1961). Root hairs typically grow away from the root at a fixed angle that is species-specific (Row and Reeder 1957). Although the presence of a cytosolic Ca^{2+} gradient is necessary to sustain growth, at present, the responsible mechanism for the observed predefined orientation of growth is unknown (Bibikova et al. 1997). Root hair elongation occurs rapidly but periods of fast growth are interrupted by periods in which growth is retarded. For example, on filter paper in growth pouches root hairs of alfalfa (*Medicago sativa*) elongate

---

Key words. cell wall, cytoskeleton, GFP, infection, infection thread, microtubules, Nod factor, nodulation, *Rhizobium*, root hair curling, root hair deformation with a maximum speed of about 0.48 mm/h and reach an average size of 2.76 mm in the 10 h of their growth period (Wood and Newcomb 1989). After the first 4 h, growth slows down for about a 2-h period to 0.06 mm/h. Why elongation is interrupted temporarily is unclear but it is associated with dramatic movements of the nucleus. During periods of growth the nucleus follows the tip at a rather fixed distance but during growth retardation the nucleus migrates several times from the base of the root hair to the tip and back. However, in *Vicia sativa* root hairs the occurrence of nuclear migration has not been observed and the nucleus to tip distance was reported to be fixed at 45 μm (Lloyd et al. 1987). But in this study an analysis of the whole sequence of root hair development was not made, and nuclear migration might have been overlooked.

The developing root hair is a tip-growing cell enclosed by a cell wall, which becomes rigid just behind the root hair tip. Its protoplast has a typical polarized organization with a vesicle-rich tip functioning in cell enlargement by deposition of plasma membrane and cell wall components (Ridge 1995; Miller et al. 1997). Cell wall synthesis is restricted to a small zone at the root hair tip. For example, in *Sinapis alba* no cellulose deposition is found at regions beyond 120 μm behind the root hair tip (Belford and Preston 1961). The wall consists of two distinct layers of which the outer originates at the extreme tip and covers the whole root hair while the inner layer originates at a short distance, just below the tip (Cormack 1962). The outer layer shows the morphological characteristics of a wall defined as being a primary cell wall, and the inner wall as being a secondary cell wall (sometimes referred to as α layer and β layer respectively, Belford and Preston 1961). In the outer layer the cellulose microfibrils are oriented in various directions and embedded in a gelatinous matrix consisting of pectins, hemicellulose and proteins (Belford and Preston 1961). In contrast, axially or helicoidally oriented microfibrils are present in the inner layer (Traas et al. 1985). The cellulose microfibrils provide the main structural component of the wall while pectins serve as cementing materials whose chemical nature influences the plasticity and elasticity of growing walls (Cormack 1962). A thin cuticle-like layer of fatty material covering the external surface of living root hairs forms the interface with the rhizosphere.

The differences in the chemical and physical nature of the root hair cell wall are associated with differences in cell wall plasticity and enzyme digestibility along the root hair. The cell wall at the growing tip is softer and more easily degraded by cell wall degrading enzymes than the wall at the tubular non-growing part of the root hair. Thus, cellulase treatment of root hairs leads to a pronounced degradation of the wall at the tip region and the wall of young emerging root hairs but has little effect on the wall below the tip of root hairs at the end of their growth period. The action of pectolyase is required for complete degradation (Cocking 1985).

The root hair cytoskeleton is built of two main structural components, microtubules and microfilaments. Cortical microtubules in root hairs run from the base to the tip and are either axial or helical depending on the plant species (Traas et al. 1985). Cortical microtubules are parallel to the most recently deposited microfibrils adjacent to primary cell walls, (Wymer and Lloyd 1996) but microtubules and microfibrils do not always co-align adjacent to secondary cell walls (Emons et al. 1992). This organization suggests that cortical microtubules probably do not direct cellulose microfibrils (Seagull 1983; Emons et al. 1992) but are rather involved in determining the width of the growing root hair (Miller et al. 1997). Microtubules are not confined to the cell periphery. An endoplasmic microtubular system, consisting of thick bundles that branch within the apical dome, connects the nucleus to the tip of the root hair (Lloyd et al. 1987). A detailed understanding of the role of the cytoskeleton during initiation and growth of root hairs is currently severely hampered by the lack of knowledge of the properties of plant cytoskeleton binding proteins. Moreover, the microtubular system co-operates with the actin cytoskeleton (Tominaga et al. 1997; Hasezawa et al. 1998) but how it does this is still unclear.

Growth regulators influence the appearance of root hairs (Ridge 1995) and some induce changes comparable to the ones observed after inoculation with rhizobia (see 3.1). In clover kinetin and ABA (abscisic acid) cause root hair branching and deformations such as swelling of the tip or base, twisting, curling and bending. In *Arabidopsis thaliana* ABA elicits the "water-stress response" producing short and bulbous root hairs (Schnall and Quatrano 1992). Root hairs of *Trifolium pratense* deform in the presence of IAA (indole acetic acid) or NAA (naphthalene acetic acid) (Dart 1974) and external application of IAA at physiological concentrations transiently inhibits growth in *Sinapis alba* root hairs (Felle and Hepler 1997). Ethylene seems to be strongly involved in the regulation of root hair initiation as in *Arabidopsis* excess of this growth regulator induces the development of root hairs from atrichoblasts, root epidermal cells that normally do not form root hairs (Dolan et al. 1994).

3 The Infection Process

3.1 Root Hairs and Infection by Rhizobia

Rhizobia use different mechanisms to penetrate their host (Sprent and Faria 1988). The most peculiar and certainly the most complex one is the penetration of the root hair just below the root hair tip in a previously formed curl, the so-called shepherd's crook. In general, root hairs that become infected are those which are at a very early stage of growth at the time of inoculation. In soybean inoculated with *Bradyrhizobium japonicum* infections that lead to nodulation seem to be initiated in epidermal cells that have finished most of their elongation but have not yet reached the stage of root hair emergence (Bhuvaneswari et al. 1980; Calvert et al. 1984). In alfalfa inoculated with *Rhizobium meliloti* the most infectible root hairs are those that are up to 2 h old at the time of inoculation (Wood and Newcomb 1989). In general, the morphology of root hairs that are at the end of their growth phase at the time of inoculation becomes highly altered resulting in elongate growth, spiralling or formation of bulbous-shaped root hairs. However, mature root hairs can also become infected (G. Truchet, 1999, personal communication). Then, first a branch is formed at a place that depends on the position of the nucleus at the time of inoculation. If the nucleus lies between the base and the midpoint of the hair, a side branch opposite the position of the nucleus is formed. If the nucleus lies near the tip, the root hair later forms a tip branch. Root hair branching may serve to initiate a new growing and infectible region in fully differentiated root hairs (Bhuvaneswari et al. 1981). Thus, about 10 h after inoculation a typical zonation becomes apparent in alfalfa roots. From the root tip a non-haired zone is followed by a zone of growing root hairs similar in appearance to this zone of an uninoculated root. In the mature root hair zone, a group of tip-curled or shepherd's crook hairs is followed by a region of branched hairs and finally by a region containing elongate, spiralled, or intertwined hairs, as well as those with bulbous or spatulate shaped tips. Surprisingly, a certain number of root hairs in the responding root zone always appear to be unaffected by the presence of the rhizobia. This is especially remarkable in the zone where inoculation results in tip curling. In the zone where branches are formed a much greater proportion of the root hairs are affected and in the following zone that displays a variety of gross deformations unaffected root hairs are extremely rare.

The infectibility of a given root hair is influenced not only by its developmental stage but also by its position relative to the protoxylem poles. Rhizobia are preferentially attracted by root hairs in the receptive stage located in front of protoxylem poles, the place were also most of the nodules develop (Libbenga and Harkes 1973). Plant lectins may play a role in this process (Diaz et al. 1986).

Many of the responses of root hairs to inoculation can also be seen in reaction to treatment with Nod factors, rhizobial lipochitooligosaccharides consisting of a backbone of N-acetylglucosamine residues decorated on the two terminal residues (Dénarié et al. 1996). Nod factors are signal molecules involved in most of the early developmental responses during nodulation. However, root hairs usually considered the most susceptible to Nod factors are those terminating growth (Heidstra et al. 1994). In general these root hairs are not the ones that become infected upon inoculation with rhizobia. Therefore, the relationship between infection initiation and progression with responses to Nod factors such as root hair membrane depolarization, actin filament depolymerization and Ca^{2+} spiking, studied in these types of root hairs cannot be simply extrapolated. To verify the relation between Nod factor responses and infection these studies will have to be repeated with trichoblasts at a stage just before root hair emergence or with root hairs not older than 2h.

The changes in root hair appearance described above suggest that rhizobia influence the development of root hairs in a manner that facilitates infection (Sprent 1989). A number of observations in *Trifolium repens* suggest that Nod factors may play an important role in this process (Dazzo et al. 1996). Firstly, Nod factors promote the differentiation of root epidermal cells into root hairs. Secondly, Nod factors prolong the period of root hair elongation and delay maturation of root hairs without affecting the elongation rate per se. The window of infectibility might thereby be extended. Thirdly, Nod factors seem to disrupt the normal process of wall crystallization and maturation in elongating root hairs. This would affect the mechanical strength of the root hair wall and makes it more prone to successful attack by wall-degrading enzymes. Strikingly, clover root hairs containing multiple infection threads occur in roots treated with Nod factors before inoculation. A phenomenon which is rarely observed without Nod factor pre-treatment. Nod factors induce the production of polygalacturonase by the treated plant (Ljunggren and Fahraeus 1961) which may result in partial depolymerization of the cell wall pectin. This preparation of plant cell walls to penetration is not limited to the walls of root hairs but also occurs locally in root cortical cells at sites where infection threads will later traverse (van Spronsen et al. 1994).

3.2 Root Hair Curling

The first stage in the development of an infection thread is the formation of the shepherd's crook, a typical curling of the root hair tip by more than 360° which only forms in the presence of living homologous rhizobia (Kijne 1992). During its formation rhizobia are enclosed and progressively isolated from the rhizosphere. Within the curl an entry point for the rhizobia into the root hair is established and the infection thread is initiated. Although its importance in infection has been recognized for a long time the mechanism of curling is still a puzzle. However, it seems clear that rhizobia induce curling by redirection of tip growth (van Batenburg et al. 1986) and it involves differential and unequal root hair wall synthesis (Hubbell 1981). In the models proposed so far the place of attachment and cell wall modifying action of rhizobia is essential. At this place wall synthesis is then locally and temporarily disrupted and rapid wall growth continues in adjacent areas of the root hair tip. However, there is no indication that rhizobia specifically bind to one site of the root hair just below the root hair tip. Around the root hair tip rhizobia are embedded in a matrix and most of them are not in direct physical contact with the root hair. In this situation it might be that rhizobia have not more than only an inducing role in the process of root hair curling. If so, curling may be a direct result of the interference with the normal arrangement of the root hair cytoskeleton by compounds (signals) released by rhizobia which are able to affect the biology of the hair cell before physical contact between the two symbionts (Ridge 1992).

The formation of the shepherd's crook is not essential for infection. Rhizobia also penetrate root hairs at places where two root hairs are in close contact or at places where root hairs and epidermal cells are in contact. Very rarely infections are observed in straight root hairs (Callaham and Torrey 1981). One of the proposed roles of the curl is therefore to act as a protective mechanism against leakage of cytoplasmic material from the root hair cells during bacterial penetration (Higashi and Abe 1980). Curling might help to maintain a high concentration of cell wall modifying components. These could be hormones produced by rhizobia that may assist in stretching the root hair wall (Dart 1974) or enzymes either of plant or bacterial origin (Kijne 1992). Alternatively, as entry into the root hair is restricted by access to a hole in the curl, curling might be considered as a way of isolating a specific bacterial species from the rhizosphere and preventing unwanted species from entering the root hair.

3.3 Infection

Even under high inoculum dose, infection remains a rare event and when it occurs it does not automatically lead to nodule formation (Dart 1974). For example, in a study with alfalfa inoculated with *R. meliloti* only 52 infected root hairs were observed in a sample of 10 seedlings with an estimated total number of root hairs of 80,000. Only 27 infections resulted in the development of a nodule (Wood and Newcomb 1989).

Various models for the mechanism of entry of rhizobia into the root hair were proposed from the few ultrastructural studies of the infection process done between 1963 (Sahlman and Fahraeus 1963) and 1985 (Newcomb et al. 1979; Ridge and Rolfe 1985). The models differ essentially in the importance which is given to the role of hydrolysing enzymes from totally absent in the earlier models such as the invagination theory of Nutman (see Napoli and Hubbell 1975) and the implication of complete cell wall degradation in the later ones (Callaham and Torrey 1981). At the light microscopical level the first sign of cell infection is the development of a hyaline spot in the root hair curl as a result of local changes in the composition of the cell wall (Dart 1974). Enzymes involved in the process of wall degradation can be of both rhizobial and plant origin. Under certain conditions rhizobia are able to produce a range of wall-digesting enzymes such as pectinase, hemicellulase and cellulase (Hubbell 1981). Upon inoculation of *T. repens, T. ambiguum, Medicago sativa* and *Melilotus alba* with their homologous symbiont the plants are induced to produce polygalacturonase (Ljunggren and Fahraeus 1961) although its role in root hair invasion has been questioned (Lillich and Elkan 1968). However, ultrastructural observations on the rhizobial penetration of roots show that the cell wall-modifying action of at least the plant host is required. In those cases where the entry point is formed between an epidermal cell and a root hair it is only the root hair wall that is degraded and not the epidermal cell wall (Turgeon and Bauer 1985). In a curled root hair penetration always occurs in the same place in the region where two parts of the root hair touch one another. However, when the rhizobia are captured between two root hairs often both are infected (Kijne 1992). Even though rhizobia cannot by themselves digest a secondary cell wall (Sprent and Faria 1988) the original hair wall consisting of a primary and secondary wall is entirely degraded at the infection point (Callaham and Torrey 1981; Ridge and Rolfe 1985). Whether this cell wall-modifying action of the plant is activated by rhizobia remains to be proven, although observations indicate to a possible role of Nod factors in this respect (see above).

3.4 Infection Thread Structure and Formation

Among all the structural changes induced upon infection of root hairs by rhizobia perhaps the most intriguing is the initiation and development of the infection thread through which the bacteria enter the plant interior. The infection thread is a tip elongating structure in which invading bacteria are embedded in a matrix mainly polysaccharidic in composition surrounded by a cell wall of plant origin.

Threads vary in thickness with host species but generally the finer the root hair the finer the thread (Dart 1974). The successful infection thread is most often straight and has a certain rigidity as occasional rupturing of the root hair from its epidermal base leaves the infection thread intact and unchanged in its form (Dart 1974). Unlike root hairs the infection thread wall does not completely cover the thread along its length leaving the tip unwalled (Dart 1974; Newcomb 1981). Cellulose deposition starts just behind the tip, like the inner cell wall in root hairs. In *Pisum sativum*, immunocytological labelling indicated that the polysaccharide composition of infection thread walls in nodules closely resembles that of the surrounding cell walls (Rae et al. 1992). Most of the infection thread wall polysaccharides are common components of the cell walls of root hairs (Mort and Grover 1988). However, cell walls of infection threads are more resistant to enzymatic digestion than other cell walls in the root cortex (Higashi et al. 1987) and it seems logical that at least during the initial phase of infection and during cell wall passage in the root cortex the infection thread wall must differ from the host wall in order to prevent its digestion by bacterial enzymes involved in these processes (Sprent 1989). Moreover, ultrastructural studies show that, at least in clover infection, thread walls are continuous with the inner secondary wall of the root hair (Callaham and Torrey 1981). Unfortunately, no publication is available in which the organization of microfibrils in the cell wall of infection threads is described. In spite of this, the above mentioned observations and considerations lead to the hypothesis that during infection thread wall synthesis in the root hair the cell machinery involved in the production of secondary walls is recruited without earlier production of a primary wall.

Our knowledge of the infection thread wall is limited, but even less is known about the composition of its matrix, although the latter seems to be of both bacterial and plant origin (Sprent 1989). In pea nodules immunocytochemical analysis showed that the thread matrix has components in common with the intercellular space (VandenBosch et al. 1989). A better knowledge of the composition of the matrix may help in understanding the mechanism of growth of the infection thread. At the tip of the thread the host plasma membrane is in direct contact with the matrix and usually less so or not in contact with the invading rhizobia. Apparently, the root hair ingests the growing matrix rather than the bacteria (Kijne 1992). As a result of the close connection between infection thread tip and the invaded plant cell signal exchange is facilitated in this region and might involve the action of auxins and cytokinins synthesized by rhizobia (Libbenga and Bogers 1974) or Nod factors (Timmers et al. 1998; see below).

The formation and proliferation of the infection thread is self-regulated by the plant host but rhizobia are required for its development. Growth requires co-ordinated secretion of matrix material, plant plasma membrane and cell wall components (Rae et al. 1992). Moreover, bacterial proliferation that only takes place at the infection thread tip (Gage et al. 1996) must keep pace with the growth speed of the infection thread. Infection thread synthesis likely involves functional modification by the rhizobia of the plant endomembrane system for plasma membrane and cell wall synthesis (Robertson and Lyttleton 1982). The similar relationship between the root hair tip and the nucleus during root hair growth resembles that seen between the tip of the thread and the nucleus of the root hair during infection thread initiation and progression (Nutman 1959). During curling the nucleus migrates towards the future infection site before new cell wall synthesis begins. This migration has been thought of as being a repair response to the localized degradation of the plant cell wall and therefore the production of new cell wall material in the form of the infection thread has been regarded as a defence response (Djordjevic et al. 1987). In line with this, infection threads are structurally considered similar to plant-synthesized wound plugs and papillae formed in response to fungal invasion (Aist 1976).

Rhizobial cell surface polysaccharides possibly function as signal molecules that induce host responses for infection thread formation (Halverson and Stacey 1986). These include lipopolysaccharides (LPS) from the outer membrane, capsular polysaccharides (CPS) at the cell surface and acidic exopolysaccharides (EPS) in the growth medium (Kijne 1992). Mutations in *exo* genes involved in the production of bacterial exopolysaccharides often result in defects in initiation or maintenance of infection leading to infection thread abortion within the root hair or the epidermal cell body.

Rhizobium Nod factors clearly are important in infection thread initiation and progression. Rhizobia mutated in *nod* genes resulting in the production of modified Nod factors either exhibit a reduced capacity or a complete inability to elicit infection thread formation in their host alfalfa. Of all the plant responses, bacterial entry has the most stringent requirements for Nod factor structure while plant developmental responses such as root hair tip growth and cortical cell activation are less stringent (Ardourel et al. 1994). Infection thread development has also recently been shown to require Nod factors by treating clover roots with Nod factors in combination with inoculation by *Rhizobium leguminosarum* bv. *trifolii* (Dazzo et al. 1996). Nod factor pre-treatment significantly increased the number of infected root hairs and frequently these infected root hairs contained multiple infection threads. Apparently, Nod factors somehow modify the root hair wall and thereby promote their infectibility.

No plant genes involved in the initiation and elongation of the infection thread have been characterized at present although plant mutants defective in the process are found in nodulation mutagenesis screens in legumes such as *Melilotus alba* (Utrup et al. 1993), *Pisum sativum* (Sagan et al. 1994) and *Medicago truncatula* (Bénaben et al. 1995) and plant genes whose expression is closely associated with the infection process have been described (Mylona et al. 1995).

4 The Microtubular Cytoskeleton and Infection Thread Ontogeny

Undoubtedly, the plant cytoskeleton plays a prominent role in the initiation and the growth of the infection thread (see above). In ultrastructural studies, microtubules have been shown to be abundant at the site of infection thread initiation and along the thread's body (Goodchild and Bergersen 1966; Ridge and Rolfe 1985; Turgeon and Bauer 1985). A detailed analysis at the ultrastructural level has been made by Bakhuizen (1988) but unfortunately no clear picture of the three dimensional structure of the cytoskeleton around the infection thread has emerged from this study. Since microtubules aggregate at sites where localized wall deposition is occurring (Ridge 1990) the initiation of an infection thread might be brought about by restructuring of the cytoskeleton that targets cytoplasmic vesicles, containing wall and membrane components, to the growth point (Robertson and Lyttleton 1982). From the scarce observations available and the organization of the cytoskeleton in root hairs Ridge (1992) developed a model for the structure and the role of the cytoskeleton during root hair infection. Recently we studied the organization of the microtubular cytoskeleton during the early stages of the interaction between *R. meliloti* and *Medicago sativa* and *M. truncatula* by immunofluorescence and observation by laser confocal microscopy (Timmers et al. 1999). From this study a clear picture is now available of the changes in the organization of the microtubular cytoskeleton.

The first changes observed after inoculation of either *M. sativa* or *M. truncatula* are seen in root hairs which are located above an activated root cortex area where cells have just divided to form the nodule primordium. Root hairs elsewhere maintain their normal microtubular organization. The helical organization of the microtubular cytoskeleton is progressively replaced by an increasing concentration of microtubules around the root hair nucleus and a network of cortical and endoplasmic microtubules parallel to the axis of the root hair. Then, during migration of the nucleus to the tip of the root hair, the microtubular cytoskeleton gradually concentrates in the region between the nucleus and the root hair tip. In root hairs at the end of their growth period, at the time of inoculation, the microtubular network extends to a point next to the nucleus, where a new growth point is initiated and from which a root hair branch with a dense network of microtubules at its tip similar to the one described above is formed. During subsequent curling, microtubules organize into arrays which progressively converge to the centre of the curl, the future infection site. However, they also remain attached to the root hair tip at this moment (Fig. 1a). Just before or at the time of infection a dense network of microtubules extends into the curled root hair and the microtubular cytoskeleton is then completely recruited for the formation of the infection thread and is no longer in contact with the root hair tip (Fig. 1b). A dense network of microtubules, which is connected to the nucleus, surrounds the tip of the infection thread during its passage through the root hair. The infection thread body is covered with closely attached longitudinal arrays of microtubules (Fig. 1c). At the point where it crosses the underlying cortical cell the tip associated network enlarges and remains present after the infection thread has continued its way into the root cortex.

What can we learn about the initiation and growth of the infection thread from the changes in the organization of the microtubular cytoskeleton? During curling microtubules apparently temporarily connect the nucleus to both the root hair tip and the future infection site within the curl. Connections to the tip provide the necessary structure to continue the growth of the root hair during the formation of the curl. Computer simulations of root hair curling show that root hair growth is essential for root hair curling (van Batenburg et al. 1986). Microtubules directed to

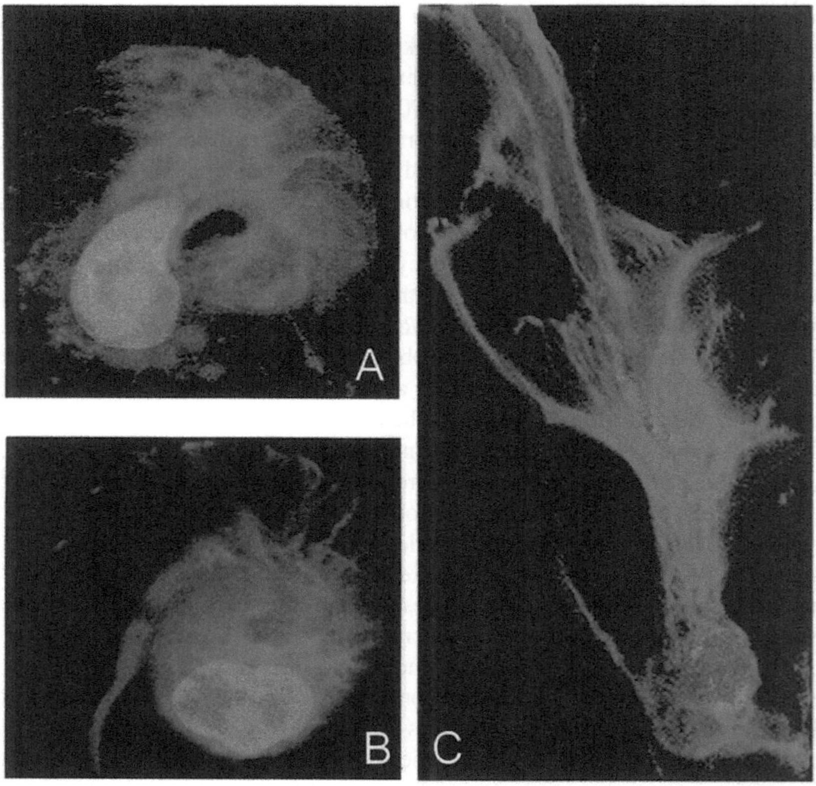

Fig. 1. The microtubular cytoskeleton visualised by immunocytochemistry during (*a*) root hair curling, (*b*) root hair infection and (*c*) infection thread progression. Microtubules are shown in green, nuclei in red and the infection thread in blue

the infection site might support the mechanism involved in local partial digestion of the cell wall and thereby prepare for the complete digestion of the cell wall by the enclosed rhizobia. However, it should be recalled that the infection point is formed below the root hair tip and most likely consists of a rigid secondary wall indigestible by rhizobia (see 3.3). The nucleus is disconnected from the root hair tip at the moment of infection and from then on only attached to the infection thread tip. This observation deviates from earlier observations and statements that the nucleus is first completely uncoupled from the root hair tip and subsequently recoupled to the point of invasion (Ridge 1992). No observations were made which show that rhizobia are able to induce the depolymerization of microtubules but apparently they can influence their structure and possibly also the way that they function.

The plant microtubular cytoskeleton clearly assists in the growth of the infection thread. Inside the thread there are no components available that might sustain polarized growth and without a plant structure which determines the site and the width of the thread only an isodiametrically growing structure can develop. The microtubular cytoskeleton can successfully be used to prevent invasion of non-pathogenic fungi (Kobayashi et al. 1997). Here the cytoskeleton plays the role of a counter force to the entry attempts of the fungus by forming a rigid structure, which is located just beneath the plasma membrane at the site of appressorium formation. Possibly, infection thread growth results from a basically similar mechanism. However, now the force generated by the cytoskeleton is not directed against but in favour of penetration. Thus the cytoskeleton functions as a mould for the infection thread form whose integrity is maintained by the synthesis of a very rigid and solid infection thread wall. Migration of the nucleus by means of the microfilament system (Lloyd et al. 1987) and maintenance of a fixed distance between nucleus and infection thread tip provoke progression of the infection thread.

Although the description of the changes in the microtubular cytoskeleton brings us a little bit closer to understanding the mechanism of infection thread growth in legume root hairs, many questions remain or arise:

1. How do root hairs curl? Microtubules change their organization during its formation but it remains unclear what role they play. Are they only involved in localized cell wall modifications or are they actively involved in root hair tip curling? If so, how is this force generated?
2. How are microtubules directed to the future infection point? Are they initiated and organized at microtubule organizing centres (MTOCs) only at the root hair nucleus or also at the future infection point? If so, do rhizobia have a role in this? Does a localized presence of Nod factors induce microtubule rearrangements?
3. Does the change in microtubular cytoskeleton structure involve the action of specific microtubule associated proteins or tubulin isotypes?
4. How is infection thread growth regulated to ensure the necessary co-ordinated secretion of matrix material, plant plasma membrane and cell wall components in line with bacterial proliferation?
5. How is infection thread growth directed? Apparently the nucleus plays an important role because even in the unusual growth of infection threads from the epidermal cell body toward the root hair tip the nucleus precedes it (Nutman 1959).
6. What is the role of the microfilament system and how is it connected to the microtubular system?

To find answers to these questions a more dynamic approach has to be developed, as is briefly described in the last part of this chapter.

5 Technical Advances Provide New Tools for Studying the Cytoskeleton During Infection with Rhizobia

Studying the microtubular cytoskeleton in root hairs during infection has been done only by conventional techniques based on chemical fixation and subsequent visualization either by electron microscopy or fluorescence microscopy. However, although very informative, this approach is limited if one wants to follow the dynamics of the changes in the organization of the cytoskeleton during all stages of the infection process. Fortunately, new techniques are available now based on labelling the cytoskeleton with fluorescent markers and thereby allowing real time in vivo observations by fluorescence microscopy.

5.1 Visualization After Fixation

In general the microtubular cytoskeleton is fixed using 4% formaldehyde in a microtubule stabilising buffer consisting of 50 mM PIPES at pH 6.9, 50 mM $MgSO_4$, 50 mM EGTA and 10% DMSO (Baluska et al. 1992). The fixed plant material is subsequently embedded and cut or treated with cell wall digesting enzymes to allow access of anti-tubulin antibodies. However, we found that this widely applicable method does not give satisfactory visualization of the microtubular cytoskeleton in alfalfa root hairs. Due to its high osmolarity the use of a PIPES-containing buffer leads to pronounced plasmolysis of alfalfa root hairs before they are fixed. By removing calcium from the root hair cell wall EGTA diminishes the rigidity of the root hair which in turn decreases the level of structural preservation after embedding and sectioning. Alfalfa root hair cell walls appear to be extremely difficult to digest. In most cases only at the tip of the root hair the wall is removed sufficiently to allow antibodies to penetrate the root hair. Long enzyme treatments, necessary to permit a total access of the antibodies, does not promote good structural preservation. Therefore, we tested several fixation protocols to find an optimal combination of structural preservation and microtubular cytoskeleton visualization. Very satisfactory results were obtained with 2% glutaraldehyde in 50 mM phosphate buffer at pH 7.0 with 5% DMSO and 0.1% triton X-100 followed by embedding in low melting point wax (Vitha et al. 1997). In combination with a spot inoculation technique changes in the organization of the microtubular cytoskeleton were studied during the whole sequence of nodule development from the first contact with the rhizobia till nodule meristem formation (Timmers et al. 1999).

5.2 In Vivo Real Time Observation of Microtubules

Recently in vivo methods for visualising cytoskeletal structures by fluorescence microscopy became available. These techniques are very promising for the study of microtubular cytoskeleton dynamics in root hairs during infection with rhizobia. Microtubules can be made fluorescent by fusing a green fluorescent protein (GFP) gene either directly to a tubulin gene or indirectly to a gene coding for a microtubule binding protein or microtubule binding domain (Ludin and Matus 1998). GFP has many characteristics which makes it ideal as an in vivo reporter protein in many applications and its primary sequence has been changed to improve its applicability in plants (Leffel et al. 1997; Tsien 1998). Observation by confocal laser microscopy is essential for a detailed fine structural analysis (Hepler and Gunning 1998). Probably due to the relatively large size of the protein (238 amino acids), the fusion of GFP directly to tubulin has not resulted in plants that stably express the fusion protein, although tubulin-GFP fusions work fine in animal cells (Ludin and Matus 1998). Microtubule binding proteins or their microtubule binding domains are being used more successfully for fusions to GFP to visualize microtubules. Recently, cortical microtubules were convincingly labelled in living epidermal cells of *Vicia faba* with a *GFP-MAP4* reporter gene (Marc et al. 1998). We obtained recently a similar result in onion epidermal cells with a tau-GFP fusion (A.C.J. Timmers, A. Niebel, C. Balagué, G. Truchet, 1999, unpublished results). These results were achieved in transient expression systems but it remains to be seen whether this approach will be useful for studying stably transformed plants. Possibly, the addition of such a large label has a severe effect on normal plant cell function, thereby preventing plant regeneration after transformation. To circumvent this problem the transgenes can be put under control of an inducible or tissue specific promotor (Aoyama and Chua 1997). Two alternatives have been developed recently to diminish the influence of an added fluorescent label to a protein. In the first one recombinant proteins with just a small receptor domain of as few as six amino acids can be made fluorescent by administration of a small membrane-permeant ligand (Griffin et al. 1998). In the second, a novel targeting approach based on noncovalent heterodimerization of GFP and cytoplasmic structural proteins is thought to prevent structural distortions of those protein components being studied (Katz et al. 1998). Both methods are also very promising for the study of cytoskeleton dynamics in living plant cells.

ACKNOWLEDGMENTS

I thank G. Truchet for his constructive comments regarding the manuscript and N. Grimsley for English editing. The author is supported by a grant from the European Community TMR Programme (FMRX-CT96-0039).

References

Aist JR (1976) Papillae and related wound plugs of plant cells. Annu Rev Phytopathol 14:145-163

Aoyama T, Chua N-H (1997) A glucocorticoid-mediated transcriptional induction system in transgenic plants. Plant J 11:605-612

Ardourel M, Demont N, Debellé F, Maillet F, De Billy F, Promé J-C, Dénarié J, Truchet G (1994) *Rhizobium meliloti* lipooligosaccharide nodulation factors: different structural requirements for bacterial entry into target root hair cells and induction of plant symbiotic developmental responses. Plant Cell 6:1357-1374

Bakhuizen R (1988) The plant cytoskeleton in the *Rhizobium*-legume symbiosis. PhD thesis. Leiden University, The Netherlands

Baluska F, Parker JS, Barlow PW (1992) Specific patterns of cortical and endoplasmic microtubules associated with cell growth and tissue differentiation in roots of maize (*Zea mays* L.). J Cell Sci 103:191-200

Belford DS, Preston RD (1961) The structure and growth of root hairs. J Exp Bot 12:157-168

Bénaben V, Duc G, Lefebvre V, Huguet T (1995) TE7, an inefficient symbiotic mutant of *Medicago truncatula* Gaertln. Cv *Jemalong*. Plant Physiol 107:53-62

Bhuvaneswari TV, Bhagwat AA, Bauer WD (1981) Transient susceptibility of root cells in four common legumes to nodulation by rhizobia. Plant Physiol 68:1144-1149
Bhuvaneswari TV, Turgeon G, Bauer W (1980) Early events in the infection of soybean (*Glycine max* L. Merr) by *Rhizobium japonicum*. Plant Physiol 66:1027-1031
Bibikova T, Zhigilei A, Gilroy S (1997) Root hair growth in *Arabidopsis thaliana* is directed by calcium and an endogenous polarity. Planta 203:495-505
Callaham DA, Torrey JG (1981) The structural basis for infection of root hairs of *Trifolium repens* by *Rhizobium*. Can J Bot 59:1647-1664
Calvert HE, Pence MK, Pierce M, Malik NSA, Bauer WD (1984) Anatomical analysis of the development and distribution of *Rhizobium* infections in soybean roots. Can J Bot 30:2375-2384
Cocking E (1985) Protoplasts from root hairs of crop plants. Bio/Technol 3:1104-1106
Cormack RGH (1962) Development of root hairs in angiosperms. II. Bot Rev 28:446-464
Dart PJ (1974) The infection process. In: Quispel A (ed) The biology of nitrogen fixation. North-Holland, Amsterdam, pp 381-429
Dazzo FB, Orgamide GG, Philip-Hollingsworth S, Hollingsworth RI, Ninke KO, Salzwedel JL (1996) Modulation of development, growth dynamics, wall crystallinity, and infection sites in white clover root hairs by membrane chitolipooligosaccharides from *Rhizobium leguminosarum* biovar *trifolii*. J Bacteriol 178:3621-3627
Dénarié J, Debellé F, Promé J-C (1996) *Rhizobium* lipochitooligosaccharide nodulation factors: Signaling molecules mediating recognition and morphogenesis. Annu Rev Biochem 65:503-535
Diaz CD, van Spronsen PC, Bakhuizen R, Logman GJJ, Lugtenberg EJJ, Kijne JW (1986) Correlation between infection by *Rhizobium leguminosarum* and lectin on the surface of *Pisum sativum* L. roots. Planta 168:350-359.
Djordjevic MA, Gabriel DW, Rolfe BG (1987) *Rhizobium*-The refined parasite of legumes. Annu Rev Phytopathol 25:145-168
Dolan L, Duckett C, Grierson C, Linstead P, Schneider K, Lawson E, Dean C, Poethig S, Roberts K (1994) Clonal relationship and cell patterning in the root epidermis of *Arabidopsis*. Development 120:2465-2474
Emons AMC, Derksen J, Sassen MMA (1992) Do microtubules control plant cell wall microfibrils? Physiol Plant 84:486-493
Felle HH, Hepler PK (1997) The cytosolic Ca^{2+} concentration gradient of *Sinapis alba* root hairs as revealed by Ca^{2+}-selective microelectrode tests and Fura-dextran ratio imaging. Plant Physiol 114:39-45
Gage DJ, Bobo T, Long SR (1996) Use of green fluorescent protein to visualize the early events of symbiosis between *Rhizobium meliloti* and Alfalfa (*Medicago sativa*). J Bacteriol 178:7159-7166
Goodchild DJ, Bergersen FJ (1966) Electron microscopy of the infection and subsequent development of soybean nodule cells. J Bacteriol 92:204-213
Griffin BA, Adams SR, Tsien RY (1998) Specific covalent labeling of recombinant protein molecules inside live cells. Science 281:269-272
Halverson LJ, Stacey G (1986) Signal exchange in plant-microbe interactions. Microbiol Rev 50:193-225
Hasezawa S, Sano T, Nagata T (1998) The role of microfilaments in the organization and orientation of microtubules during the cell cycle transition from M Phase to G1 phase in tobacco BY-2 cells. Protoplasma 202:105-114
Heidstra R, Geurts R, Franssen H, Spaink HP, van Kammen A, Bisseling T (1994) Root hair deformation activity of Nodulation factors and their fate on *Vicia sativa*. Plant Physiol 105:787-797

Hepler PK, Gunning BES (1998) Confocal fluorescence microscopy of plant cells. Protoplasma 201:121-157

Higashi S, Abe M (1980) Scanning electron microscopy of *Rhizobium trifolii* infection sites on root hairs of white clover. Appl Environ Microbiol 40:1094-1099

Higashi S, Kushiyama K, Abe M. (1987) Electron microscopic observations of infection threads in driselase treated nodules of *Astragalus sinicus*. Can. J. Microbiol. 32:947-952.

Hirsch AM, LaRue TA (1997) Is the legume nodule a modified root or stem or an organ *sui generis*? Crit Rev Plant Sci 16:361-392

Hubbel DH (1981) Legume infection by *Rhizobium*: A conceptual approach. BioScience 31:832-837

Katz B-Z, Krylov D, Aota S-i, Olive M, Vinson C, Yamada KM (1998) Green fluorescent protein labeling of cytoskeletal structures-Novel targeting approach based on leucine zippers. BioTechniques 25:298-304

Kijne JW (1992) The *Rhizobium* infection process. In: Stacey G, Burris RH, Evans HJ (Eds) Biological nitrogen fixation. Chapman and Hall, New York, pp 349-398

Kobayashi Y, Kobayashi I, Funaki Y, Fujimoto S, Takemoto T, Kunoh H (1997) Dynamic reorganization of microfilaments and microtubules is necessary for the expression of non-host resistance in barley coleoptile cells. Plant J 11:525-537

Leffel SM, Mabon SA, Stewart CN (1997) Applications of green fluorescent protein in plants. BioTechniques 23:912-918

Libbenga KR, Harkes PAA (1973) Initial proliferation of cortical cells in the formation of root nodules in *Pisum sativum* L. Planta 114:17-28

Libbenga KR, Bogers RJ (1974) Root-nodule morphogenesis. In: Quispel A (Ed) The biology of nitrogen fixation. North-Holland, Amsterdam, pp 430-472

Lillich TT, Elkan GH (1968) Evidence countering the role of polygalacturonase in invasion of root hairs of leguminous plants by *Rhizobium* spp. Can J Microbiol 14:617-25

Ljunggren H, Fahraeus G (1961) The role of polygalacturonase in root hair invasion by nodule bacteria. J Gen Microbiol 26:521-528

Lloyd CW, Pearce KJ, Rawlins DJ, Ridge RW, Shaw P (1987) Endoplasmic microtubules connect the advancing nucleus to the tip of legume root hairs, but F-actin is involved in basipetal migration. Cell Motil Cytoskeleton 8:27-36

Long SR (1996) *Rhizobium* symbiosis: Nod factors in perspective. Plant Cell 8:1885-1898

Ludin B, Matus A (1998) GFP illuminates the cytoskeleton. Trends Cell Biol 8:72-77

Marc J, Granger CL, Brincat J, Fisher DD, Kao T-h, McCubbin AG, Cyr RJ (1998) A *GFP-MAP4* reporter gene for visualizing cortical microtubule rearrangements in living epidermal cells. Plant Cell 10:1927-1939

Miller DD, de Ruijter NCA, Emons AMC (1997) From signal to form: aspects of the cytoskeleton-plasma membrane-cell wall continuum in root hair tips. J Exp Bot 48:1881-1896

Mort AJ, Grover PB (1988) Characterization of root hair cell walls as potential barriers to the infection of plants by rhizobia. Plant Physiol 82:787-794

Mylona P, Pawlowski K, Bisseling T (1995) Symbiotic nitrogen fixation. Plant Cell 7:869-885

Napoli CA, Hubbell DH (1975) Ultrastructure of *Rhizobium*-induced infection threads in clover root hairs. Appl Microbiol 30:1003-1009

Newcomb W (1981) Nodule morphogenesis and differentiation. In: Giles KL, Atherly AG (Eds) Biology of the Rhizobiaceae. Int Rev Cytol suppl 13. Academic Press, New York, pp 247-298

Newcomb W, Sippell D, Peterson RL (1979) The early morphogenesis of *Glycine max* and *Pisum sativum* root nodules. Can J Bot 57:2603-2616

Nutman PS (1959) Some observations on root hair infection by nodule rhizobia. J Exp Bot 10:250-263

Rae AL, Bonfante-Fasolo P, Brewin NJ (1992) Structure and growth of infection threads in the legume symbiosis with *Rhizobium leguminosarum*. Plant J 2:385-395

Ridge RW (1990) Cytochalasin-D causes abnormal wall-ingrowths and organelle-crowding in legume root hairs. Bot Magazine Tokyo 103:87-96

Ridge RW (1992) A model of legume root hair growth and *Rhizobium* infection. Symbiosis 14:359-373

Ridge RW (1995) Recent developments in the cell and molecular biology of root hairs. J Plant Res 108:399-405

Ridge RW, Rolfe BG (1985) *Rhizobium* sp. Degradation of legume root hair cell wall at the site of infection thread origin. Appl Environ Microbiol 50:717-720

Robertson JG, Lyttleton P (1982) Coated and smooth vesicles in the biogenesis of cell walls, plasma membranes, infection threads and peribacteroid membranes in root hairs and nodules of white clover. J Cell Sci 58:63-78

Row HC, Reeder JR (1957) Root hair development as evidence of relationship among genera of Gramineae. Am J Bot 44:596-601

Sagan M, Huguet T, Duc G (1994) Phenotypic characterization and classification of nodulation mutants of pea (*Pisum sativum* L.). Plant Sci 100:59-70

Sahlman K, Fahraeus G (1963) An electron microscope study of root-hair infection by *Rhizobium*. J Gen Microbiol 33:425-427

Schnall JA, Quatrano RS (1992) Abscisic acid elicits the water stress response in root hairs of *Arabidopsis thaliana*. Plant Physiol 100:216-218

Seagull RW (1983) The role of the cytoskeleton during oriented microfibril deposition. I Elucidation of the possible interaction between microtubules and cellulose synthetic complexes. J Ultrastruct Res 83:168-175

Sprent JI (1989) Which steps are essential for the formation of functional legume nodules? New Phytol 111:129-153

Sprent JI, Faria SM (1988) Mechanisms of infection of plants by nitrogen fixing organisms. Plant Soil 110:157-165

Timmers ACJ, Auriac M-C, de Billy F, Truchet G (1998) Nod factor internalization and microtubular cytoskeleton changes occur concomitantly during nodule differentiation in alfalfa. Development 125:339-349

Timmers ACJ, Auriac M-C, Truchet G (1999) Refined analysis of early symbiotic steps of the *Rhizobium-Medicago* interaction in relationship with microtubule cytsokeletal rearrangements. Development 126:3617-3628

Tominaga M, Morita K, Sonobe S, Yokota E, Shimmen T. (1997) Microtubules regulate the organization of actin filaments at the cortical region in root hair cells of *Hydrocharis*. Protoplasma 199:83-92

Traas JA, Braat P, Emons AMC, Meekes H, Derksen J (1985) Microtubules in root hairs. J Cell Sci 76:303-320

Tsien RY (1998) The green fluorescent protein. Annu Rev Biochem 67:509-544

Turgeon BG, Bauer WD (1985) Ultrastructure of infection-thread development during the infection of soybean by *Rhizobium japonicum*. Planta 163:328-349

Utrup LJ, Cary AJ, Norris JH (1993) Five nodulation mutants of white sweetclover (*Melilotus alba* Desr.) exhibit distinct phenotypes blocked at root hair curling, infection thread development, and nodule organogenesis. Plant Physiol 103:925-932

van Batenburg FHD, Jonker R, Kijne JW (1986) *Rhizobium* induces marked root hair curling by redirection of tip growth: a computer simulation. Physiol Plant 66:476-480 van Spronsen PC, Bakhuizen R, van Brussel AAN, Kijne JW (1994) Cell wall degradation during infection thread formation by the root nodule bacterium *Rhizobium leguminosarum* is a two-step process. Eur J Cell Biol 64:88-94

Vandenbosch KA, Bradley DJ, Knox JP, Perotto S, Butcher GW, Brewin NJ (1989) Common components of the infection thread matrix and the intercellular space identified by immunocytochemical analysis of pea nodules and uninfected roots. EMBO J 8:335-342

Vitha S, Baluska F, Mews M, Volkmann D (1997) Immunofluorescence detection of F-actin on low melting point wax sections from plant tissues. J Histochem Cytochem 45:89-95

Wood SM, Newcomb W (1989) Nodule morphogenesis: The early infection of alfalfa (*Medicago sativa*) root hairs by *Rhizobium meliloti*. Can J Bot 67:3108-3122

Wymer C, Lloyd C (1996) Dynamic microtubules: implications for cell wall patterns. Trends Plant Sci 1:222-228

Chapter 15
Nod-Factors in Symbiotic Development of Root Hairs

Helen R. Irving[1], Nawal M. Boukli[2], Marilyn N. Kelly[1], and William J. Broughton[2]

[1]Department of Pharmaceutical Biology & Pharmacology, Monash University, 381 Royal Parade, Parkville, VIC 3052, Australia
[2]L.B.M.P.S., Université de Genève, 1, ch. De l'Impératrice, 1292 Chambésy, Genève, Switzerland

1 Introduction

Generally, root hairs emerge at the apical end of some epidermal cells, and elongate by polar growth of the tip (Cormack 1962; see Chapters 1 to 3 this volume). Young, elongating root hairs are extensively colonized by soil-borne micro-organisms. Intimate associations between dissimilar organisms are common, especially in nutrient poor environments. Prevalent amongst these associations are nitrogen-fixing symbioses in which photosynthetic organisms provide carbohydrates in exchange for organic nitrogen exported by diazotrophs. Soil bacteria of the genera *Azorhizobium*, *Bradyrhizobium*, *Mesorhizobium*, *Rhizobium* and *Sinorhizobium* (collectively called rhizobia) enter the roots (and occasionally adventitious-roots on the stems) of legumes (family Leguminosae), and induce the formation of highly specialised organs, called nodules. Rhizobia present in the root nodules convert to an endo-symbiotic form, the bacteroids, in which dinitrogen is reduced to ammonia. Bacteroids within nodules contribute a disproportionately large portion of fixed nitrogen to the global pool.

Mutual attraction between the symbionts begins when legume roots secrete a variety of substances, including flavonoids that co-ordinate expression of rhizobial nodulation (*nod*) genes. In turn, the Nod-proteins direct the synthesis of lipochito-oligosaccharides (LCO's) known as Nod-factors. Although they are not known to be produced by plants, Nod-factors are in fact plant hormone-like substances (Relić et al. 1993) that provoke deformation of root hairs and induce cell division in inner or outer cortical cells of the root (Truchet et al. 1991). Exactly which cells are stimulated to divide depends on the type of nodule which in turn is dependent on the host-plant. Together, these changes result in nodule initiation and allow

Key words. actin, auxin, cytokinin, cytoskeleton, flavonoids, G-proteins, intra-cellular free calcium, phospholipase C, protein phosphorylation, *Rhizobium*, root-hair deformation, signal transduction rhizobia to enter the plant (Relić et al. 1994). Root hairs are the first "port-of-call" for rhizobia, and thus play an important role in symbiotic recognition.

2 Signal Exchange

Symbiotic interactions between rhizobia and legumes involve a complex interplay of signals culminating in co-ordinated gene expression. The temporal co-ordination of different signals controls specificity in many signal transduction pathways. Two signals operate early in the symbiosis – plant flavonoids and bacterial Nod-factors (see Hanin et al. 1999).

2.1 Flavonoids

Plants jettison surprisingly large amounts of organic matter into the soil, most of which support the growth of micro-organisms in the rhizosphere. Helal and Sauerbeck (1989) used $^{14}CO_2$, to label whole plants and showed that 19% of the photosynthate was released as organic material into the soil surrounding the roots. Amongst these compounds, flavonoids (2-phenyl-1,4-benzopyrone derivatives), are found throughout the plant kingdom (Kellis and Vickery 1984), but in the legume rhizosphere, these and related phenolic substances play an important symbiotic role. Flavonoids are synthesised via the phenylpropanoid biosynthetic pathway, and form the strongest inducers of bacterial nodulation (*nod*)-gene expression. Significant differences in the flavonoid compositions of seeds, roots and root exudates of *Glycine max* (Kosslak et al. 1987; Sutherland et al. 1990; Kape et al. 1992), *Phaseolus vulgaris* (Hungria et al. 1991a, b; Hungria et al. 1992), *Medicago sativa* (Maxwell and Philips 1990; Hartwig and Philips 1991), *Trifolium repens* (Redmond et al. 1986) and *Vicia sativa* (Zaat et al. 1989; Recourt et al. 1992) occur. Many flavonoids are stored and released as glycosides or related conjugates (Redmond et al. 1986). These conjugates are more soluble in water than aglycones but are diffusible and can be hydrolysed to aglycones which are more active at inducing *nod*-genes (Hartwig and Philips 1991). In at least one case however, isoflavone glucosides were shown to act as *nod*-gene inducers (Smit et al. 1992).

The initial steps of the symbiosis involve communication between both parties. Inoculation of *Vicia sativa* with *R. leguminosarum* bv. *viciae* results in an increase in *nod*-gene inducing activity of root exudates (Recourt et al. 1992; Table 1). Possibly synthesis is involved, as flavonoids have been observed in soya-bean root exudates following treatment with *B. japonicum* and NGR234 Nod-factors (Schmidt et al. 1994). Also, rhizobia are able to degrade a wide range of flavonoids (Rao and Cooper 1994, 1995). Degradation leads to the appearance first of chalcones, which can be more efficient inducers of *nod*-genes than other flavonoids (Hartwig et al. 1990).

Interestingly some, though not all, *nod*-gene inducing flavonoids also serve as chemo-attractants for rhizobia (Aguilar et al. 1988). At least in *B. japonicum* – *G. max* associations, they are not the principal attractants of root-exudates however. Aspartate, dicarboxylic acids, and glutamate play this role in the soya-bean

Table 1. Known biological and molecular responses of legume roots and root hairs to Nod-factors

Responses	Nod-factors[a]	Concentration	Plants	References
Accumulation of flavonoids	Nod*Rlv*(Ac)	10^{-9} M	*Vicia*	Recourt et al. 1991
	NodNGR(OH or Ac) NodBj		*Glycine*	Schmidt et al. 1994
Increase of $[Ca^{2+}]_i$	NodNGR(Ac,S), Nod*Rm*IV(S)	10^{-9} M	*Vigna*	Gehring et al. 1997
Depolarisation of root hairs	Nod*Rm*(S)	10^{-9} M	*Medicago*	Ehrhardt et al. 1992 Felle et al. 1995 Kurkdjian 1995
Calcium spiking in root hairs	Nod*Rm*IV(Ac, S)		*Medicago*	Ehrhardt et al. 1996
Root-hair deformation	Nod*Rm*IV(S)	10^{-9} - 10^{-12} M	Medicago *Melilotus*	Lerouge et al. 1990
	Nod*Rlv*, Nod*Rm*	10^{-9} - 10^{-12} M	*Vicia*	Spaink et al. 1991
	NodNGR(S)	10^{-7} - 10^{-13} M	*Vigna*	Price et al. 1992
	NodNGR(Ac)	10^{-8} - 10^{-11} M	*Macroptilium*	Schultze et al. 1992
	NodNGR(S)	10^{-7} - 10^{-10} M	*Medicago*	
Root-hair curling	*NodNGR*		*Macroptilium* *Vigna*	Relić et al. 1993 Relić et al. 1994
Penetration of bacteria	*NodNGR*(S)	10^{-6} - 10^{-7} M	*Calopogonium* *Glycine, Vigna*	Relić et al. 1994
Induction of pre-infection threads	*NodRlv*(Ac)		*Vicia*	van Brussel et al. 1992
Induction of nodule meristems	*NodRlv*(Ac) Nod*Rm*(S) Nod*Be*		*Vicia* *Medicago* *Glycine*	Spaink et al. 1991 Truchet et al. 1991 Stokkermans and Peters 1994
Induction of Nodulin genes	*NodRlv* (Ac)	10^{-12} - 10^{-13}M	*Pisum* *Medicago*	see Table 2[b]
Induction of other genes	*NodNGR*(Ac,S)	10^{-6} - 10^{-7} M	*Vigna*	see Table 2[b]

[a]Abbreviations used for Nod-factors (Nod*Be*, *Bradyrhizobium elkani*; Nod*Bj*, *B. japonicum;* NodNGR, *Rhizobium* sp. NGR234; Nod*Rlv*, *R. leguminosarum* bv. *vicieae*; Nod*Rm*, *R. meliloti*). IV after the abbreviation shows that *N*-acylated tetramers were used; Ac that they were acetylated; OH that they were hydroxylated; and S that they were sulphated. [b]Further details of these genes (including references) are given in Table 2

rhizosphere (Barbour et al. 1991), and in any event, the effect is only marginal. Moreover, chemotaxis and *nod* gene induction do not necessarily correlate. Amongst the flavonoid inducers of *R. leguminosarum* bv *phaseoli nod*-genes, apigenin and luteolin evoke similar chemotactic responses as the inhibitors of *nod*-gene induction, umbelliferone and acetosyringone. On the other hand, the inducer naringenin has almost no effect on chemotaxis (Aguilar et al. 1988). Nevertheless, chemotaxis, enhanced expression of *nod*-genes and symbiotic infection all appear to be focused at the same part of the root (Dharmatilake and Bauer 1992). This is the region containing the "elongating root hairs" (see Heidstra et al. 1994; Gehring et al. 1997; de Ruijter et al. 1998).

In other words, flavonoids are the first symbiotic signals encountered by soil-borne organisms such as rhizobia. The constitutively expressed NodD proteins of homologous rhizobia (i.e., rhizobia which are able to nodulate the particular host) sense these signals and activate Nod-factor production in the bacteria.

2.2 Nod-Factors

Binding of flavonoids to the NodD-proteins activates transcription of *nod*-genes. By attaching to conserved sequences in the upstream promoter-regions of many (though not all) *nod*-genes, the flavonoid-NodD protein complexes regulate transcription of genes involved in the synthesis of a class of lipo-oligosaccharides called Nod-factors (Hanin et al. 1999). Nod-factors are *N*-acylated oligomers of β-1,4-linked *N*-acetyl-D-glucosamine with various substitutions on the sugar residues (see Dénarié et al. 1996; Hanin et al. 1999). An acyl chain of various lengths (commonly 16-20 C) and degrees of saturation is attached to the non-reducing terminus of the chitin chain. Sulphur or acetate substitutions are common at the reducing end of the chain (Dénarié et al. 1996). These and other substitutions on the Nod-factors are important in refining the specificity of the interaction with root hairs and the physical responses induced (see Dénarié et al. 1996; Long 1996).

Nod-factors are remarkably important compounds in the symbiosis. Purified Nod-factors elicit many plant responses that are characteristic of the bacteria. Root hairs are the first "port of call" and respond specifically to Nod-factors but several responses are also detected in cortical cells. Nevertheless, the essential characteristic of Nod-factors is their ability to deform root hairs.

3 Root Hairs

Actively growing root hairs are most susceptible to rhizobial infection or Nod-factors (Heidstra et al. 1994; de Ruijter et al. 1998), and have high exocytotic and cell wall assembly activities (Ridge 1988). Growing root hairs have a vesicle-rich region at the tip (Miller et al. 1997) containing Golgi bodies (Bonnett and Newcomb 1966; Emons 1987; Ridge 1988; Sherrier and VandenBosch 1994; Galway et al. 1997) that is devoid of visible bundles of actin filaments (Miller et al. 1999). The vacuole in growing root hairs remains in the basal part of the cell, while the nucleus lies alongside the apical part of the vacuole (Galway et al. 1997). A characteristic pattern of cytoplasmic streaming directed by the actin cytoskeleton is present in elongating root hairs. The cytoplasm flows towards the tip in strands located near the plasma membrane and returns before reaching the tip, often in strands near the centre. Both the organelle distribution and the pattern of cytoplasmic streaming of elongating root hairs is distinct from that of mature root hairs (Miller et al. 1997). The microtubule cytoskeleton appears to determine the direction of growth in a cell (Baskin et al. 1994; Wymer et al. 1996) whereas the actin cytoskeleton seems to implement growth. However, both parts of the cytoskeleton are intimately associated and it appears that microtubules organise actin filaments in root hairs (Tominaga et al. 1997). The pattern of cytoplasmic streaming during root-hair tip growth suggests that actin filaments may actually move Golgi vesicles from their site of production (the Golgi bodies) to the base of the vesicle-rich region where they are released and inserted into the plasma membrane (Miller et al. 1999). The polar tip growth of root hairs is also characterized by the presence of a spectrin-like antigen which is absent from mature root hairs (de Ruijter et al. 1998). Spectrin is a large multifunctional protein that has a prominent role in the cytoskeleton with actin, calmodulin, and phosphatidylinositol bisphosphate binding sites (Viel and Branton 1996).

3.1 Root-Hair Deformation and Curling

Rhizobia (and their Nod-factors) stimulate and redirect growth of the root-hair cell wall (Dart 1977; Kijne 1992) resulting in deformed and curled root hairs. Although the mechanism of root-hair deformation is poorly understood, there are similarities to root-hair development (Heidstra et al. 1994; Miller et al. 1999). The tips of Nod-factor treated *Vicia sativa* root hairs swell (Heidstra et al. 1994) and this is a growth phenomenon (Vijn et al. 1995) independent of actin polymerisation (Miller et al. 1999). Induction of tip growth is preceded by local hydrolysis of the cell-wall (Heidstra et al. 1994) and the root-hair tips regain the characteristics of polar tip growth (de Ruijter et al. 1998). The actin cytoskeleton also resumes the features of polar tip growth (Miller et al. 1999). Schiefelbein and Somerville (1990) showed that the phenotype of certain root-hair specific mutants of *Arabidopsis thaliana* resemble root hair-deformations caused by rhizobia. Thus initiation and deformation of root hairs appear to be ontogenetically related.

Some but not all legume root hairs curl into a "shepherds crook" shape in response to rhizobia and also Nod-factors (Relić et al. 1993, 1994; D'Haeze et al. 1998). Although initially curling was thought to represent specificity in the symbiosis, it is now apparent that certain Nod-factor-legume combinations provoke deformation (or curling) irrespective of the nodulation ability of the rhizobia from which they were derived (Relić et al. 1993; Gehring et al. 1997). It is likely that curling of this type entraps rhizobia within the inner folds of the curl, and permits the extremely specific entry of rhizobia into the infection thread (Relić et al. 1994). The "shepherd's crook" type curling thus provides the opportunity for the homologous rhizobia to closely proximate themselves to the plant membranes and is probably necessary for rhizobial penetration into the host plant.

3.2 Rhizobial Entry

Within these curled root hairs, rhizobia promote the formation of infection threads (van Brussel et al. 1992; Table 1) and it is through these tubes that the rhizobia enter the plant. Nod-factors are also necessary signals for entry of rhizobia into legume roots. Co-inoculation of NodNGR factors with a *nodABC*$^-$ mutant of NGR234 or *B. japonicum* USDA110 permits these non-nodulating strains to nodulate and fix nitrogen on *Vig. unguiculata* and *G. max* respectively (Relić et al. 1994). NodNGR factors also allow entry of *R. fredii* USDA257 into the roots of *Calopogonium caeruleum*, a non-host (Relić et al. 1994), and the *nodABC*$^-$ mutant of NGR234 into *Macroptilium atropurpureum* (Relić et al. 1993). Simultaneous application of purified *Azorhizobium caulinodans* Nod-factors and a *nodA*$^-$ mutant of the same strain restored the formation of outer cortical infection pockets in *Sesbania rostrata* (D'Haeze et al. 1998).

4 Hormonal Changes

All plant hormones influence root-hair growth and appearance. Although not produced by plants, Nod-factors are in fact "hormone-like" substances. Allen et al. (1953) were the first to show that substituted benzoic acids induce nodule-like structures on the roots of various legumes. Direct application of cytokinins (e.g., BAP, 2iP, or kinetin), or inhibitors of auxin transport (NPA, TIBA, etc) produce the same effect on *Mac. atropurpureum* (Relić et al. 1993) and *M. sativa* roots (Hirsch et al. 1989). Nod-factors, at 10 to 100 times lower concentrations, elicit pseudo-nodules on *M. sativa* (Truchet et al. 1991; Schultze et al. 1992), *Sesbania caulidulans* (Mergaert et al. 1993) and *Mac. atropurpureum* (Relić et al. 1993). In other words, Nod-factors and increases in the cytokinin/auxin balance have the same effect on the formation of "pseudo"-nodules. As MN1008, a Nod$^-$ mutant of *M. sativa* is insensitive to Nod-factor application, but responds to physiological concentrations of cytokinin (as measured by accumulation of transcripts of *MsEnod2* and *MsEnod40*), the changes in hormone balance are probably downstream of Nod-factor perception (Hirsch et al. 1997).

Although the evidence is incomplete, it is possible that rhizobia provoke this apparent shift in hormone balance by stimulating the production (Mathesius et al. 1998a) and secretion of additional flavonoid compounds in legume roots (van Brussel et al. 1990; Recourt et al. 1991; Schmidt et al. 1994). Since certain flavonoids (Jacobs and Rubery 1988), LCO's and derivatives of chitin oligosaccharides can act as natural inhibitors of auxin transport, they may trigger a transient down-regulation of auxin synthesis at the nodule initiation site (Mathesius et al. 1998b). It is not known if this effect is direct or indirect, but in the case of chitin-based molecules, it is probably the later. Assuming that cytokinin levels remain constant in the same area of the root, this could explain why a relative increase in cytokinin levels stimulates nodule formation (Allen et al, 1953; Hirsch et al. 1989; Hirsch 1992; Hirsch and Fang 1994).

Other hormonal changes may also be involved. Abscisic acid and gibberellins have been implicated in nodule initiation (Williams and de Mallorca 1982; Dobert et al. 1992). An exopolysaccharide-deficient mutant of *R. leguminosarum* bv. *viciae* failed to nodulate its host unless ethylene levels were suppressed by growing the plant-roots in the dark or by endogenous application of aminoethoxyvinylglycine (AVG) (van Workum et al. 1995). In an ethylene insensitive (sickle) mutant of *Medicago truncatulata*, a much larger percentage of infection threads proceeded through to the cortex and initiated a higher number of nodule primordia in the inner cortex than occurred in wild-type plants (Penmetsa and Cook 1997). Accumulation of transcripts encoding 1-aminocyclopropane (ACC) oxidase (and therefore a possible site of ethylene production) was shown in cells opposite the phloem poles (Heidstra et al. 1997). Since nodule primordia predominate opposite the proto-xylem poles, it is possible that cell division leading to nodule primordia is negatively regulated by ethylene gradients. Consistent with this hypothesis is the observation that increased numbers of nodule primordia are found opposite the phloem poles following external application of compounds that inhibit the production and perception of ethylene, namely AVG or Ag^+. In contrast, ethylene is required for the formation of root hairs as well as their subsequent growth in *V. sativa* (Heidstra et al. 1997). Ethylene thus plays different, contradictory roles. On the one hand it promotes the formation of root hairs while on the other it inhibits nodulation. At the present moment no satisfactory explanation for this conundrum exists.

5 Specific Effects of Nod-Factors on Root Hairs

The ability of Nod-factors to induce pseudo nodules on many but not all legumes (*M. truncatulata* is an exception – AM Hirsch, personal communication, 1999) suggests that they may be regarded as "hormone-like" molecules (Relić et al. 1993). These "hormone-like" properties of Nod-factors are further exemplified by their ability to induce root-hair deformation (Had). When Nod-factors are applied to roots growing between glass slides, a certain developmental stage of root hairs responds to the Nod-factor (Heidstra et al. 1994). These are the hairs that are terminating growth (de Ruijter et al. 1998) and the response is a reinitiation of tip growth (de Ruijter et al. 1998, Miller et al. 1999) which involves re-direction of tip-growth. The nodulation competent zone of the root is the region of "elongating root hairs" and it is this area that contains root hairs responsive to Nod-factors (Heidstra et al. 1994, Gehring et al. 1997, de Ruijter et al. 1998; Miller et al. 1999).

Purified Nod-factors provoke root-hair deformation at extremely low concentrations (e.g., from 10^{-13} M) in nodulation-competent roots (Lerouge et al. 1990; Spaink et al. 1991; Price et al. 1992; Sanjuan et al. 1992; Schultze et al. 1992; Mergaert et al. 1993; Heidstra et al. 1994; Table 1). Within one to two hours, bulbous swellings at the tip of the root-hair are clearly visible (Heidstra et al. 1994; de Ruijter et al. 1998). New tip growth results from this swelling, and in some instances branch points form in the root hair (e.g., Lerouge et al. 1990; Relić et al. 1993; Gehring et al. 1997), as polar tip growth is reinitiated (de Ruijter et al. 1998; Miller et al. 1999). Brief exposures to Nod-factors are difficult to interpret since little is known about their accumulation and attachment to root-surfaces. Nevertheless, very brief treatment (e.g., one min) with NodNGR factors stimulates root-hair deformation over the following 24 h (Gehring et al. 1997). Longer exposures (> 4 min) provoke the more classic deformation type responses

(Gehring et al. 1997). These results indicate that perception of Nod-factors and activation of the signal transduction cascade occurs within one minute of application of Nod-factors.

6 Nod-Factor Signalling

The molecular mechanisms underlying the physiological changes in root hairs such as deformation are only now being explored. Since Nod-factors act specifically at very low concentrations, it is probable that Nod-factor signal transduction processes are mediated by specific receptors present in root hairs (see Cohn et al. 1998).

6.1 Nod-Factor Receptors

A Nod-factor signal receptor should be able to respond to Nod-factors at sub-nano-molar concentrations with high specificity, and possess the means of tranducing this binding signal to other cellular components. Nod-factor binding sites have been characterized from microsomal membranes of *Medicago* spp. (Bono et al. 1995, Niebel et al. 1997). The first site identified in *M. truncului* root extracts and *M. varia* cell suspensions (NFBS1) has a low affinity for Nod-factors (dissociation constant of 70 to 90 nM). A second site with higher affinity (dissociation constant of 1.9 nM)(NFBS2) has also been found in the microsomal fraction of *M. varia* cell suspension cultures (Niebel et al. 1997). This site appears to be specific for Nod-factors as the chitin backbone, *N-N'-N"-N"'*-tetracetylchitotetraose (TACT), was a poor competitor for the site (Niebel et al. 1997). A homologue of NFBS1 was also found in tomato roots however, suggesting that Nod-factor binding proteins of this type play only general roles in the symbiosis (Bono et al. 1995). In contrast, NFBS2, which was isolated from a plasma-membrane enriched fraction taken from *M. varia* cell suspension cultures, requires specific Nod*Rm* factor substitutions (e.g., the O-acetate group, the acyl chain, etc) of the non-reducing sugar for binding activity. Yet, the sulphate group that is necessary for all biological activities on *Medicago*, is not discriminated by NFBS2 (Gressent et al. 1999). It thus seems that a specific Nod*Rm* factor receptor has yet to be isolated.

Other experiments suggest that independent reception systems may be necessary to trigger symbiotic development. Felle et al. (1996) analysed the ability of Nod-factors to induce either cytosolic alkalisation or membrane depolarisation in root hairs (see below). Nod*Rm*IV(S) factors which are morphogenic on *M. sativa*, are able to induce membrane depolarization and rapid intracellular alkalization of root hairs. In contrast, non-sulphated Nod-factors (Nod*Rm*IV) elicit intracellular pH changes in the absence of membrane depolarization. Together, these data point to the existence of two Nod-factor recognition events. Whether this recognition is perceived by two separate receptors or by another mechanism, remains to be determined but the data accords well with the suggestion made by Ardourel et al. (1994). Their model postulated the existence of a low-stringency signal receptor that would mediate such events as root-hair deformation, and a second, high-affinity receptor necessary for triggering rhizobial invasion.

Other evidence supporting the existence of more than one functional receptor for Nod-factors has been obtained from studies with symbiotically induced genes. The early nodulin, *Enod40* is usually induced within 24 h of inoculation. The mRNA of *Enod40* in *G. max* is transiently induced by a simple, non-acylated chitin pentamer (equivalent to the sugar backbone of Nod-factors) whereas sustained expression required the application of soya-bean specific Nod-factors (Minami et al. 1996b). In a related study, Minami et al. (1996a) examined the ability of specific chitin and Nod-factor molecules to induce expression of another early nodulin *Enod2* in *G. max*. No single chitin-containing molecule was able to induce *Enod2* mRNA expression. Rather, a mixture of at least two chitin containing molecules was necessary, only one of which had to be a soya-bean specific Nod-factor.

At the present moment, it is difficult to summarise Nod-factor perception. In part this is because various combinations of legumes and Nod-factors have been used making commonalities hard to find. Also, in the search for simpler and more reproducible conditions for studying Nod-factor perception, workers have resorted to cell-suspension cultures and chitin fragments. Although it is possible that these systems accurately mirror Nod-factor reception *in planta*, it should not be forgotten that non-symbiotic tomato cell-suspension cultures have well developed mechanisms for perceiving both Nod-factors and chitin-fragments (Staehelin et al. 1994). As an added complication, Philip-Hollingsworth et al. (1997) suggest that Nod-factors may be internalized within root hairs, an idea which is consistent with the presence of a Nod-factor inducible lipid-transfer protein in root hairs (Krause et al. 1994).

6.2 Sequential Temporal Changes

The sequential changes observed in root hairs in response to Nod-factors are presented in Fig. 1. Felle et al. (1998) used ion selective microelectrodes in the Nod-factor responsive zone of *M. sativa* roots to measure extra-cellular ions. They detected an influx of Ca^{2+} evoked by Nod-factors that was inhibited by the Ca^{2+} channel antagonist nifedipine indicating that Ca^{2+} entry was via a Ca^{2+} channel. It was followed by a Cl^- efflux from the roots (Felle et al. 1998) and accompanied by root-hair membrane depolarisation (Ehrhardt et al. 1992; Felle et al. 1995; Kurkdjian 1995; Felle et al. 1998). It is thought that the Cl^- ion efflux triggers the membrane depolarisation (Felle et al. 1998). Concomitantly with the membrane depolarisation, Nod-factors induce an intracellular alkalization in root hairs of about 0.2 pH units (over a period of five min) that only returns to basal levels upon removal of Nod-factors (Felle et al. 1996).

Several changes in intracellular free calcium ($[Ca^{2+}]_i$) following the initial Ca^{2+} influx (Felle et al. 1998) occur. Nod-factors evoke a rapid (within one min) plateau type increase in $[Ca^{2+}]_i$ at the tip of *Vig. unguiculata* root hairs that lasts for minutes before declining to basal levels (Gehring et al. 1997). The $[Ca^{2+}]_i$ increase appears to be a prerequisite for root-hair deformation as it is induced by

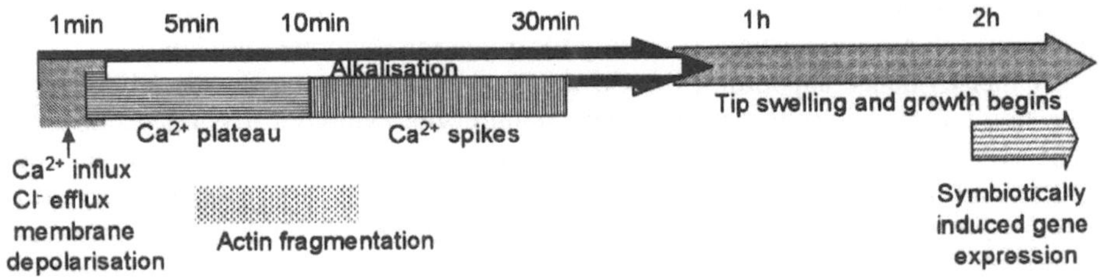

Fig. 1. A schematic time line of the early sequential responses of root hairs to Nod-factors. The details are discussed in the text

Nod-factors (homologous or heterologous) that provoke deformation but not by the chitin backbone, TACT, of Nod-factors. Whether such an increase is due to a series of very rapid spikes or a sustained elevation of Ca^{2+} has not been ascertained. Transient Ca^{2+} influxes have been reported in *M. sativa* (Allen et al. 1994) which could contribute to this Ca^{2+} increase. The plateau type Ca^{2+} response returns to basal levels by 10 min (Gehring et al. 1997), but further changes in Ca^{2+} occur. Periodic spiking of $[Ca^{2+}]_i$ in the region proximal to the nucleus of *M. sativa* root hairs occurs and is often delayed by up to 10 min following Nod-factor application (Ehrhardt et al. 1996). The regular spiking occurs for up to an hour and is partially dependent upon external Ca^{2+} (Ehrhardt et al. 1996) presumably to replenish internal Ca^{2+} stores.

During this period, dramatic changes also occur in the cytoskeleton in preparation for renewed root-hair growth. The actin cytoskeleton of root hairs is organised in long bundles of filamentous actin parallel to the axis of the hair (Miller et al. 1997, 1999). Five to 10 min after Nod-factor application to *P. vulgaris* root hairs, these bundles become fragmented (Cárdenas et al. 1998). Within an hour of these changes, the root-hair tip swells (Heidstra et al. 1994; Gehring et al. 1997; de Ruijter et al. 1998; Miller et al. 1999) and the actin cytoskeleton partially recovers (Cárdenas et al. 1998). Root-hair tip growth is stimulated as the root hairs take on the characteristic cytoplasmic streaming of immature growing hairs with a Ca^{2+} gradient at the tip and accumulation of spectrin-like compounds at the tip (de Ruijter et al. 1998). These growing root hairs proceed to form the deformed and curled structures characteristically induced by Nod-factors or rhizobia (Lerouge et al. 1990; Relić et al. 1993; Heidstra et al. 1994; Relić et al. 1994; Gehring et al. 1997; de Ruijter et al. 1998; Table 1). Such curled root hairs contain fragmented actin bundles (Ridge 1992; Allen et al. 1994). Expression of symbiotically induced genes usually begins two to three hours after exposure to Nod-factors (see below).

6.3 Signalling Mechanisms

Changes in $[Ca^{2+}]_i$ appear critical to the process and the question of how these changes are regulated arises. Increases in $[Ca^{2+}]_i$ can be due to Ca^{2+} entry from the exterior as initially occurs (Felle et al. 1998) and also from Ca^{2+} released from internal stores. A variety of evidence suggests that both sources of Ca^{2+} are important in generating Nod-factor induced responses (e.g., Ehrhardt et al. 1996; Gehring et al. 1997; Pingret et al. 1998). If Ca^{2+} is released from internal stores it may be triggered by production of inositol trisphosphate (IP_3). All the components of the IP_3 signal cascade exist in plants (Coté and Crain 1993) including intracellular IP_3 receptors (e.g., Biswas et al. 1995) although IP_3 has yet to be shown to be operative in Nod-factor signalling. Phospholipase C hydrolyses phosphatidylinositol (4,5)-bisphosphate to produce IP_3 and diacylglycerol (Singer et al. 1997).

An *in vitro* phospholipase C assay using [^{3}H]phosphatidylinositol-(4,5)-bisphosphate was developed for *Vig. unguiculata* root plasma membranes isolated from the Nod-factor responsive zone (see methods). Phospholipase C activity is specifically induced by Nod-factors but not by the Nod-factor chitin backbone TACT (Fig. 2). As the phospholipase C inhibitor, neomycin, reduces Nod-factor induced phospholipase C activity to basal levels, it is probable that phospholipase C activity is indeed being measured (Fig. 2). Interestingly, preliminary experiments indicate that Nod-factor induction of phospholipase C is restricted to the Nod-factor responsive zone of the root as plasma membranes isolated from other zones are not stimulated (MN Kelly and HR Irving unpublished results, 1998). These data demonstrate that Nod-factors can utilise phospholipase C in

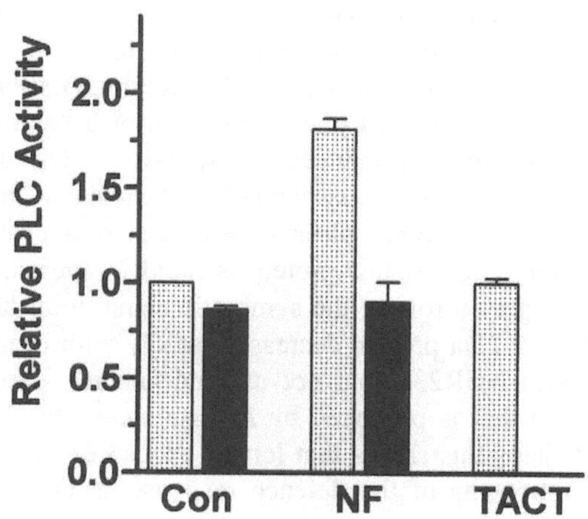

Fig. 2. Effect of *Rhizobium* sp NGR234 Nod-factors on induction of phospholipase C (*PLC*) activity in plasma-membranes isolated from *Vigna unguiculata* zone II roots. PLC activity of all treatments is expressed relative to the untreated control sample. Plasma-membranes were incubated with NodNGR[Ac] (*NF*) or TACT at 10^{-7} M in the absence (*dotted bars*) or presence (*black bars*) of 15 μM neomycin. Values are presented as mean ± sem of at least three experiments

their signalling process as recently proposed by Pingret et al. (1998) who demonstrated that expression of a Nod-factor induced gene, *pMtENod*12-GUS, was stimulated by mastoparan (a G-protein activator) in the absence of Nod-factor and that neomycin inhibits gene expression in the presence of Nod-factor in transgenic *M. sativa* plants. Moreover, Pingret et al. (1998) found that pertussis toxin inhibited expression of the Nod-factor regulated gene, *pMtENod*12-GUS. This suggests that Nod-factor signalling involves a heterotrimeric G-protein (GTP-binding protein) that activates phospholipase C.

G-proteins are well known in vertebrates where they couple the signal from ligand activated receptors to downstream effectors. Evidence is steadily accumulating that G-proteins are also present in plants (see Ma 1994) and heterotrimeric G-proteins are implicated in plant defence responses (see Blumwald et al. 1998). In preliminary experiments using plasma-membranes isolated from the Nod-factor responsive zone of *Vig. unguiculata* roots, GTPase activity increased in response to both Nod-factors and TACT (HR Irving unpublished results 1998) which is indicative of G-protein activation. As it is believed that the initial stages of the symbiosis resemble plant defence responses (Djordjevic et al. 1987), it is possible that chitin receptors recognise TACT and also use G-protein mediated signalling processes. However, the phospholipase C appears to be specific to Nod-factors as TACT has no effect on phospholipase C activity (Fig. 2).

The prolonged changes in $[Ca^{2+}]_i$ (and also the alkalization) will have pronounced effects on calcium dependent proteins such as calmodulin, protein kinases and phosphatases. Thus we expect that particular kinases will be activated to generate a phosphorylation signal cascade leading to further cellular activation. To test whether changes in the patterns of phosphorylation follow the induction of nodulation, plasma-membrane proteins were isolated from roots of *Vig, unguiculata.* Treatment of *Vig. unguiculata* roots for 2 h with NGR234, NodNGR-factors, pentameric chitin, *Ralstonia* (*Pseudomonas*) *syringae* and water (data not shown) resulted in enhanced phosphorylation of a band of 33 kDa that could be a stress kinase (Fig. 3) induced by spraying. On the other hand, treatment with NGR234 as well as *Ra. syringae* but particularly NodNGR-factors, enhanced phosphorylation of another plasma-membrane root protein of 65 kDa (Fig. 3). Since phosphorylation of this protein is rapidly (approx. 2 h) enhanced by Nod-factors, it could play a role in the symbiotic signal transduction pathway. Phosphorylation of this 65 kDa protein decreases rapidly with time. Twenty-four hours after spraying with NGR234 (but not its Nod-factors), phosphorylation of the same 62 kDa protein that is provoked by *Ra. syringae* after 2 h of treatment becomes apparent. Perhaps this shows that longer exposures to NGR234 induces some of the same components of the defence response as those provoked by a pathogen.

A schematic model incorporating some of the signalling events likely to occur in root hairs during the initiation of symbiosis is presented in Fig. 4. This model is based on vertebrate paradigms and the evidence reviewed above. Nod-factors are perceived by receptors (R) found in cytoskeleton or plasma membrane (e.g.,

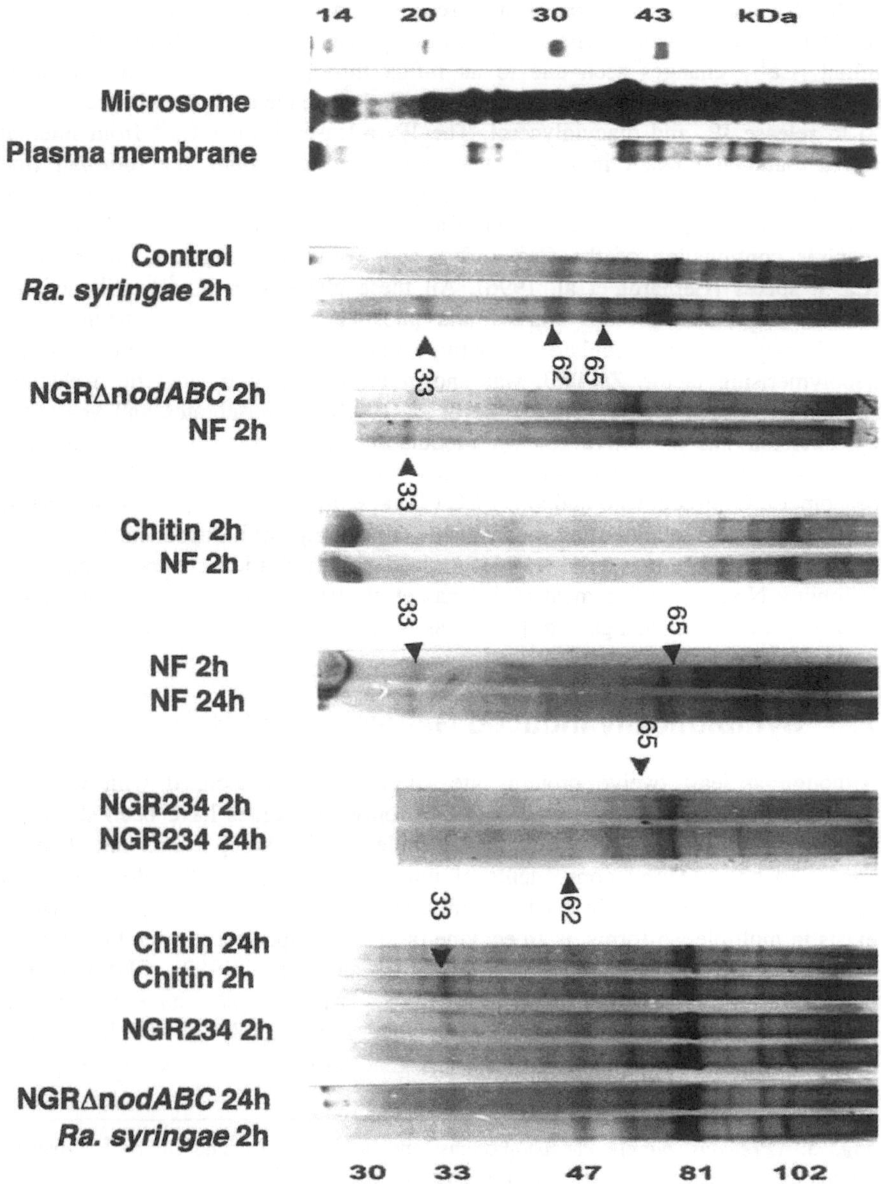
14
20
30
43
kDa
Microsome
Plasma membrane
Control
Ra. syringae 2h
33
62
65
NGRΔn*odABC* 2h
NF 2h
33
Chitin 2h
NF 2h
33
65
NF 2h
NF 24h
65
NGR234 2h
NGR234 24h
33
62
Chitin 24h
Chitin 2h
NGR234 2h
NGRΔn*odABC* 24h
Ra. syringae 2h
30
33
47
81
102

Niebel et al. 1997). Activation of these receptors involves a G-protein that directly modulates a Ca^{2+} channel and later activates phospholipase C. This results in the initial $[Ca^{2+}]_i$ changes seen due to an influx through membrane Ca^{2+} channels (Felle et al. 1998) which in turn stimulates Ca^{2+}-dependent phospholipase C (Fig. 2) to release IP_3 and diacylglycerol. The IP_3 releases further Ca^{2+} from internal stores that could result in a Ca^{2+} wave propagating through the root-hair (the plateau type response observed by Gehring et al. (1997) could represent a wave that was not resolved). The internal Ca^{2+} stores need to be replenished which requires continuation of the Ca^{2+} influx and eventually results in the periodic $[Ca^{2+}]_i$ spikes (Ehrhardt et al. 1996). All these changes in $[Ca^{2+}]_i$ will activate calcium dependent protein kinases and phosphatases and these enzymes and $[Ca^{2+}]_i$ may in turn modulate the actin cytoskeleton. Recently, a maize actin-depolymerising factor, *ZmAdf3*, was shown to be phosphorylated by a calcium stimulated protein kinase (Smertenko et al. 1998). In vertebrates and yeast, the cytoskeleton has an active role in modulating signal transduction events. For example, actin depolymerisation occurs in response to activation of the monomeric G-protein Rho and also PLC (Takai et al. 1995). A Rho GTPase, Rop1 may be involved in signalling mechanisms controlling actin-dependent tip-growth of pea pollen tubes (Lin et al. 1996). Actin depolymerisation occurs in root hairs following Nod-factor treatment (Cárdenas et al. 1998) and this depolymerisation may depend upon changes in $[Ca^{2+}]_i$ as well as activation of monomeric G-proteins and phospholipase C.

7 Symbiotically Induced Genes in Root Hairs

Although at least twelve proteins are known to be associated with root-hair deformation (Krause and Broughton 1992), only four genes have been shown to be specifically expressed in root hairs (Table 2). Of these, *Enod5* and *Enod12* appear to be structural components of infection threads and therefore probably play only passive roles in bacterial invasion. Chalcone synthase (CHS) which exists in multiple iso-forms, is an enzyme of central importance in the biosynthetic pathway of all classes of flavonoids (Halbrock and Scheel 1989). CHS may also play a role in nodulation. Extensive studies with *G. max* have shown that inoculation with compatible *Bradyrhizobium* spp. rapidly (≤ 24 h) induces expression of specific members of the *chs* gene family (Estabrook and Sengupta-

Fig. 3. (*Previous page*). Phosphoproteins and *in vitro* phosphorylation of plasma membranes of variously treated *Vigna unguiculata* roots. Typical results from a number of different experiments are shown with phosphorylated proteins (*arrows*). Treatments included 10^{-7} M NodNGR[Ac] factors, 10^{-7} M chitin pentamers, 10^7 cfu of the wild-type *Rhizobium* sp. NGR234, 10^7 cfu of the NodABC$^-$ mutant NGR*ΔnodABC*, and 10^7 cfu of the pathogen *Ralstonia* (*Pseudomonas*) *syringae*. In each case, the treatments were sprayed onto seedlings held in large glass Petri dishes as described by Krause et al. (1994). Roots were harvested by plunging into liquid nitrogen either 2 h or 24 h after treatment. The molecular weights indicated at the bottom correspond to the last six lanes. *A* silver-stained microsomal protein fraction. *B* Silver-stained plasma membrane protein fraction

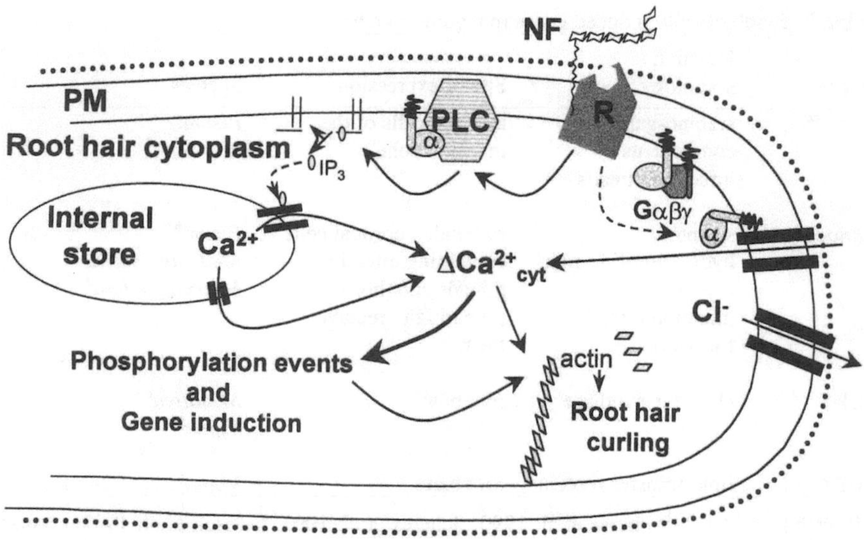

Fig. 4. Schematic representation of some of the signalling pathways activated by Nod-factors in root hairs during the early stages of symbiosis. This model is based upon analogous signal transduction pathways present in vertebrates and results reviewed in this article. Nod-factors (*NF*) bind to the plasma membrane (*PM*) located receptor (*R*) which activates a heterotrimeric G-protein (*Gαβγ*) cascade that stimulates inward calcium channels and the enzyme phospholipase C (*PLC*) to release inositol trisphosphate (IP_3). It should be mentioned however that the γ-sub-unit has not so far been found in plants

Gopalan 1991). *R. leguminosarum* bv. *trifolii* induces CHS expression in the roots of *Trifolium subterraneum* before bacterial penetration begins (Lawson et al. 1994). *Rhizobium* sp. NGR234 up-regulates expression of the *chs* gene family in the roots and root hairs of *Vig. unguiculata* (Krause et al. 1997). Transcripts of one member of the family, *VuChs*1 are induced within 24 h of inoculation in root hairs but not roots, although in *G. max* (Estabrook and Sengupta-Gopalan, 1991) and *M. sativa* (Junghans et al. 1993; McKhann and Hirsch 1994; McKhann et al. 1998) induction is less marked. Nevertheless, the timing of CHS expression, as well as its regulation by Nod-factors suggests that modulation of expression of the *chs*-gene family is associated with root-hair deformation.

Expression of a lipid-transfer protein (LTP) mimics that of a gene expected to play an important role early in the symbiosis. Although also present in aerial parts of *Vig. unguiculata* plants, the number of LTP-like transcripts increases dramatically in root hairs but not in nodules following treatment with NodNGR factors (Krause et al. 1994). As this rapid rise in LTP-like transcripts corresponds

Table 2. Symbiotically induced genes in legume root hairs

Gene	Function / homology	Site of expression	Species
Enod5	arabino-galactan components of infection threads	infected cells of the invasion zone	*Pisum*[a]
Enod12	secretory hydroxyproline-rich glycoprotein, infection thread formation	root hairs, cortical cells containing infection threads and those preparing to receive them	*Pisum*[b] *Medicago sativa*[c] *M. truncatulata*[d]
CHS	chalcone synthase	root hairs	*M. sativa*[e] *Vigna*[f]
LTP	lipid-transfer protein	root hairs	*Vigna*[g]

[a]Horvath et al. 1993, [b]Scheres et al. 1990; Scheres et al. 1992, [c]Allison et al. 1993; Csanádi et al. 1994, [d]Pichon et al. 1992, [e]McKhann et al. 1997, [f]Krause et al. 1997, [g]Krause et al. 1994

to deformation of *Vig. unguiculata* root hairs (Krause and Broughton 1992), it appears that the LTP-like gene is transiently expressed during nodule development. Whether it has a role in transporting Nod-factors remains speculative.

Extensins are a class of cell-wall proteins. Levels of transcripts corresponding to extensins decrease in *Vig. unguiculata* root hairs within one day of inoculation. These decreases contrast with the observation that extensins are synthesized at higher rates during stress responses (see Arsenijević -Maksimović et al. 1997). Perhaps extensins allow the creation of a more impenetrable barrier or they could serve to immobilize potential pathogens in cell walls by their microbial agglutination activity (Zhang et al. 1993). Reduced amounts of transcripts following inoculation with rhizobia may mean that fewer proteins are synthesized and that root-hair cell walls change composition. If one assumes that extensins give rigidity to plant cell walls, such a change in composition might loosen the cell wall structure and facilitate the penetration of symbiotic bacteria into the host.

8 Conclusions

Small, tantalizing pieces of the symbiotic signal-transduction puzzle are now being found. Cellular transduction of hormonal signals typically involves chains of consecutive events including rapid perturbation of cytoplasmic calcium levels and phosphorylation cascades (Fluhr 1998). Direct biochemical evidence that both events play a role in symbiotic recognition within root hairs is only just beginning to accumulate. Changes in cytosolic calcium levels are indeed rapid and precede morphological responses (Table 1; Fig. 1). It is thus possible that perception of Nod-factors in legume roots resembles that which occurs in other plant hormone signal-transduction pathways. Yet it should not be forgotten that entry of bacteria into the root-hair does not automatically lead to nodules. Abortion of infection threads is common and varies with both the host and the bacterium. In some associations, perhaps only 1.5% of infection threads lead to nodules, while in other cases it can be much higher (see Broughton 1978). Thus, other signals (and presumably signal transduction pathways) are probably involved.

Perhaps the best analogy for the regulation of infection-thread initiation and development is in comparison with pollen-tube growth in self-incompatible plants. After initiation of pollen-tube formation, continuation of growth of the pollen tube is regulated by a self-incompatibility system (see Hiscock et al. 1996). Legume-*Rhizobium* symbioses also invite comparison with the associations between vesicular-arbuscular mycorrhizae and higher plants. It is possible that nitrogen-fixing nodules have evolved from them (Hirsch and Kapulnik, 1998). Most probably the incompatibility mechanism(s) controlling infection thread growth in legumes involve recognition of various extra-cellular components of rhizobia (Albrecht et al. 1999). Since transcriptional activation of *Rhizobium* spp. NGR234 *nod*-genes occurs within the first hour of contact with flavonoids and declines rapidly thereafter (Perret et al. 1999), these extra-cellular components are unlikely to be Nod-factors. Rather, other rhizobial exports including extra-cellular polysaccharides (see Perret et al. 2000) and/or proteins excreted by the type III secretion pathway (Viprey et al. 1998, 2000) probably play this role.

9 Methods

To test whether Nod-factors induce phospholipase C activity in the root hair plasma membrane, microsomes from *Vig. unguiculata* were prepared. Plasma membranes were purified using an aqueous two-phase system consisting of 5.7% Dextran T500 and 5.7% PEG 3350 as described (Briskin et al. 1987; Widdell and Larsson 1987). The *in vitro* phospholipase C assay was based on procedures described by Drøbak et al. (1994). In brief, 100mM phosphoinositides dissolved in chloroform-methanol [2:1 (v/v)], 50mM Tris-maleate buffer (pH 6.0) with 10μM $CaCl_2$, and [H^3]-phosphatidylinsitol-4,5-bisphosphate were sonicated for 3 min to form a micellar suspension. Plasma membranes (4–6μg protein) were added to micro-centrifuge tubes together with tritiated phosphoinositide mixture, and Tris-maleate buffer to make a total reaction volume of 50μL. After incubation at 25°C for 10min, the assay was stopped by the addition of ice cold chloroform-methanol (2:1 [v/v]) and 0.6M HCl. The mixture was vortexed and centrifuged at 12 000g, after which the radioactivity present in duplicate aliquots (100μL) of the top phase was determined using scintillation spectroscopy.

To test whether changes in the patterns of phosphorylation follow the induction of nodulation, plasma membrane proteins were isolated from roots of *Vig. unguiculata* by zonal centrifugation. Subsequently, plasma membranes were purified from the microsomal fraction (refer to Fig. 3a) using a dextran-polyethylene glycol two-phase partition method (Yoshida et al. 1983) with both polymers at a concentration of 5.9% (see Fig. 3b). Thiophosphorylation of the plasma membranes was performed *in vitro* in screw cap micro-centrifuge tubes as described by Farmer et al. (1991). To do so, 2μg of the plasma-membrane protein was added to 8.7 μl phosphorylation buffer (20mM MES pH 7.6 containing 5mM 2-mercaptoethanol, 10% (vol/vol) glycerol, 5mM $MgCl_2$, 1mM $MnSO_4$, 0.1mM $CaCl_2$ and 0.4% (wt/vol) octaethylene glycododecylether) containing 12.5 μCi adenosine γ-35S-(thio)triphosphate. The reaction was initiated by the addition of [γ-^{35}S]ATP (1000Ci/mmol) to a final concentration of 55 nM, followed by vortex mixing for 5 s. After incubation at 30°C for 10 min, the proteins were separated on 10% polyacrylamide-SDS gels which were then dried and autoradiographed.

ACKNOWLEDGMENTS

We wish to thank A.A.N. van Brussel, J.E. Cooper, M.A. Djordjevic, A.M. Hirsch, X. Perret, R.W. Ridge, and G. Stacey for their helpful comments on this manuscript. D. Gerber and S. Relić are gratefully thanked for their generous support. Financial assistance was provided by the Fonds National Suisse de la Recherche Scientifique (Grant No. 31-45921.95), the Université de Genève and Monash University.

References

Aguilar J, Ashby A, Richards J, Loake G, Watson M, Shaw C (1988) Chemotaxis of *Rhizobium leguminosarum* biovar *phaseoli* towards flavonoids inducers of the symbiotic nodulation genes. J Gen Microbiol 134:2741-2746

Albrecht C, Guerts R, Bisseling T (1999) Legume nodulation and mycorrhizae formation: two extremes in host specificity meet. EMBO J 18:281-288

Allen E, Allen O, Newman A (1953) Pseudonodulation of leguminous plants induced by 2-bromo-3,5-dichlorobenzoic acid. Am J Bot 40:429-435

Allen NS, Bennett MS, Cox D, N, Shipley A, Ehrhardt DW, Long SR (1994) Effects of Nod-factors on alfalfa root hair Ca^{++} and H^{+} currents and on cytoskeletal behavior. In: Daniels M (Ed) Advances in molecular genetics of plant-microbe interactions. Kluwer Academic Publishers, Dordecht, pp 107-113

Allison L, Kiss G, Bauer P, Poiret M, Pierre M, Savouré A, Kondorosi E, Kondorosi A (1993) Identification of two alfalfa early nodulin genes with homology to members of the pea Enod12 gene family. Plant Mol Biol 21:375-380

Ardourel M, Demont N, Debellé F, Maillet F, de Billy F, Promé J-C, Dénarié J, Truchet G (1994) *Rhizobium meliloti* lipooligosaccharide nodulation factors: different structural requirements for bacterial entry into target root hair cells and induction of plant symbiotic developmental responses. Plant Cell 6:1357-1374

Arsenijević -Maksimović I, Broughton W, Krause A (1997) Rhizobia modulate root-hair specific expression of extensin genes. Mol Plant-Micr Inter 10:95-101

Barbour WM, Hattermann DR, Stacey G (1991) Chemotaxis of *Bradyrhizobium* to soybean exudates. Appl Environ Microbiol 57:2635-2639.

Baskin T, Wilson J, Cork A, Williamson R (1994) Morphology and microtubule organization in *Arabidopsis* roots exposed to oryzalin or taxol. Plant Cell Physiol 35:935-942

Biswas S, Dalal B, Sen M, Biswas B (1995) Receptor for *myo*-inositol trisphosphate from the microsomal fraction of *Vigna radiata*. Biochem J 306:631-636

Blumwald E, Aharon GS, Lam BC-H (1998) Early signal transduction pathways in plant-pathogen interactions. Trends Plant Sci 3:342-346

Bonnett H, Newcomb E (1966) Coated vesicles and other cytoplasmic components of root hairs of radish. Protoplasma 62:59-75

Bono J-J, Riond J, Nicolaou K, Bockovich NJ, Estevez VA, Cullimore JV, Ranjeva R (1995) Characterization of a binding site for chemically synthesized lipo-oligosaccharidic NodRm factors in particulate fractions prepared from roots. Plant J 7:253-260

Briskin DP, Leonard RT, Hodges TK (1987) Isolation of the plasma membrane: Membrane markers and general principles. Meth Enz 148:542-575

Broughton W (1978) Control of specificity in legume-*Rhizobium* associations. J Appl Bact 45:165-194

Cárdenas L, Vidali L, Dominguez J, Peréz H, Sánchez F, Hepler P, Quinto C (1998) Rearrangement of actin microfilaments in plant root hairs responding to *Rhizobium etli* nodulation signals. Plant Physiol 116:871-877

Cohn J, Day BR, Stacey G (1998) Legume nodule organogenesis. Trends Plant Sci 3:105-110

Cormack R (1962) Development of root hairs in angiosperms. II. Bot Rev 28:446-464

Coté GG, Crain RC (1993) Biochemistry of phosphoinositides. Annu Rev Plant Physiol Plant Mol Biol 44:333-356

Csanádi G, Szécsi J, Kaló P, Kiss P, Endre G, Kondorosi A, Kondorosi E, Kiss G (1994) *ENOD*12, an early nodulin gene, is not required for nodule formation and efficient nitrogen fixation in alfalfa. Plant Cell 6: 201-213

Dart P (1977) Infection and development of leguminous nodules. In Hardy R, Silver W (Eds) A treatise on dinitrogen fixation, Sect. III. Cambridge University Press, London, pp 367-472

de Ruijter N, Rook M, Bisseling T, Emons A (1998) Lipochito-oligosaccharides re-initiate root hair tip growth in *Vicia sativa* with high calcium and spectrin-like antigen at the tip. Plant J 13:341-350

Dénarié J, Debellé F, Promé J-C (1996) *Rhizobium* lipo-chitooligosaccharide nodulation factors: signaling molecules mediating recognition and morphogenesis. Annu Rev Biochem 65:503-535

D'Haeze W, Gao M-S, de Rycke R, van Monatagu M, Engler G, Holsters M (1998) Roles for azorhizobial Nod-factors and surface polysaccharides in intercellular invasion and nodule penetration, respectively. Mol Plant-Micr Inter 11:999-1008

Dharmatilake A, Bauer W (1992) Chemotaxis of *Rhizobium meliloti* towards nodulation gene-inducing compounds from alfalfa roots. Appl Environ Microbiol 58:1153-1158

Djordjevic MA, Gabriel DW, Rolfe BG (1987) *Rhizobium* - the refined parasite of legumes. Annu Rev Phytopath 25:145-168

Dobert R, Rood S, Blevins D (1992) Gibberellins and the legume-*Rhizobium* symbiosis. Plant Physiol 98:221-224

Drøbak BK, Watkins PAC, Valenta R, Dove SK, Lloyd CW, Staiger CJ (1994) Inhibition of plant plasma membrane phosphoinositide phospholipase C by the actin-binding protein, profilin. Plant J 6:389-400

Ehrhardt DW, Atkinson EM, Long SR (1992) Depolarization of alfalfa root hair membrane potential by *Rhizobium meliloti* Nod-factors. Science 256:998-1000

Ehrhardt DW, Wais R, Long SR (1996) Calcium spiking in plant root hairs responding to *Rhizobium* nodulation signals. Cell 85:673-681

Emons A (1987) The cytoskeleton and secretory vesicles in root hairs of *Equisetum* and *Linnobium* and cytoplasmic streaming in root hairs of *Equisetum*. Ann Bot 60:625-632

Estabrook EM, Sengupta-Gopalan C (1991) Differential expression of phenylalanine ammonia-lyase and chalcone synthase during soybean nodule development. Plant Cell 3: 299-308

Farmer EE, Moloshok TD, Saxton MJ, Ryan CA (1991) Oligosaccharide signalling in Plants. Specificity of oligouronide-enhanced plasma membrane phosphorylation. J Biol Chem 266:3140-3145

Felle HH, Kondorosi E, Kondorosi A, Schultze M (1995) Nod signal-induced plasma membrane potential changes in alfalfa root hairs are differentially sensitive to structural modifications of the lipochitooligosaccharide. Plant J 7:939-947

Felle HH, Kondorosi E, Kondorosi A, Schultze M (1996) Rapid alkalinization in alfalfa root hairs in response to rhizobial lipochitooligosaccharide signals. Plant J 10:295-301

Felle HH, Kondorosi E, Kondorosi A, Schultze M (1998) The role of ion fluxes in Nod-factor signalling in *Medicago sativa*. Plant J 13:455-463

Fluhr R (1998) Ethylene perception: from two-component signal tranducers to gene induction. Trends Plant Sci 3:141-146

Galway M, Heckman J, Schiefelbein J (1997) Growth and ultrastructure of *Arabidopsis* root hairs: the rhd3 mutation alters vacuole enlargement and tip growth. Planta 201:209-218

Gehring CA, Irving HR, Kabbara AA, Parish RW, Boukli NM, Broughton WJ (1997) Rapid, plateau-like increases in intracellular free calcium are associated with Nod-factor-induced root hair deformation. Mol Plant-Micr Inter 10:791-802

Gressent F, Drouillard GF, Mantegazza S, Samain E, Geremia RA, Niebel A, Driguez H, Ranjeva R, Cullimore J, Bono JJ (1999) Ligand specificity of a high-affinity binding site for lipo-chitooligosaccharide Nod-factors in *Medicago* suspension cultures Proc Natl Acad Sci 96:4704-4709.

Halbrock K, Scheel D (1989) Physiology and molecular biology of phenylpropanoid metabolism. Annu Rev Plant Physiol Plant Mol Biol 40:347-369

Hanin M, Jabbouri S, Broughton W, Fellay R, Quesada-Vincens D (1999) Molecular aspects of host-specific nodulation. In: Stacey G, Keen N (Eds) Plant-microbe interactions. American Phytopathological Society, St Paul MN, pp 1-37

Hartwig U, Maxwell C, Joseph C, Philips D (1990) Effects of alfalfa *nod* gene-inducing flavonoids on *nodABC* transcription in *Rhizobium meliloti* strains containing different *nodD* genes. J Bacteriol 172:2769-2773

Hartwig U, Philips D (1991) Release and modification of *nod*-gene-inducing flavonoids from alfalfa seeds. Plant Physiol 95:804-807

Heidstra R, Guerts R, Franssen H, Spaink H, van Kammen A, Bisseling T (1994) Root hair deformation activity of nodulation factors and their fate on *Vicia sativa*. Plant Physiol 105:787-797

Heidstra R, Yang W, Yalcin Y, Peck S, Emons A, van Kammen A, Bisseling T (1997) Ethylene provides positional information on cortical cell division but is not involved in Nod-factor-induced root hair tip growth in *Rhizobium*-legume interaction. Development 124:1781-1787

Helal H, Sauerbeck D (1989) Carbon turnover in the rhizosphere. Z Pflanzenernahr Bodenk 152:211-216

Hirsch, AM (1992) Tansley Review No. 40. Developmental biology of legume nodulation. New Phytol 122:211-237

Hirsch AM, Bhuvaneswari TV, Torrey JG, Bisseling T (1989) Early nodulin genes are induced in alfalfa root outgrowths elicited by auxin transport inhibitors Proc Natl Acad Sci 86:1244-1248.

Hirsch A, Fang Y (1994) Plant hormones and nodulation: what's the connection? Plant Mol Biol:5-9

Hirsch AM, Fang Y, Asad S, Kapulink Y (1997) The role of phytohormones in plant-microbe symbioses. Plant Soil 194:171-184

Hirsch AM, Kapulnik Y (1998) Signal transduction pathways in mycorrhizal associations: Comparisons with the *Rhizobium*-legume symbiosis. Fungal Genet. Biol. 23:205-212

Hiscock S, Kues U, Dickinson H (1996) Molecular mechanisms of self-incompatibility in flowering plants and fungi - different means to the same end. Trends Cell Biol 6:421-428

Horvath B, Heidstra R, Lados M, Moerman M, Spaink H, Promé J-C, van Kammen A, Bisseling T (1993) Lipo-chitoligosaccharides of *Rhizobium* induce infection-related early nodulin gene expression in pea root hairs. Plant J 4:723-764

Hungria M, Johnston A, Philips D (1992) Effects of flavonoids released naturally from bean (*Phaseolus vulgaris*) on *nodD*-regulated gene transcription in *Rhizobium leguminosarum* bv. *viciae*. Mol Plant-Micr Inter 5:199-203

Hungria M, Joseph C, Philips D (1991a) Anthocyanidins and flavonols: major *nod* gene inducers from seeds of a black-seeded common bean (*Phaseolus vulagaris L*). Plant Physiol 97:751-758

Hungria M, Joseph C, Philips D (1991b) *Rhizobium nod* gene inducers exuded naturally from roots of common bean (*Phaseolus vulgaris* L). Plant Physiol 97:759-764

Jacobs M, Rubery PH (1988) Naturally occurring auxin transport regulators. Science 241:346-349

Junghans H, Dalkin K, Dixon RA (1993) Stress responses in alfalfa (*Medicago sativa* L.). 15. Characterizations and expression patterns of members of a subset of the chalcone synthase multigene family. Plant Mol Biol 22: 239-253

Kape R, Parniske M, Brandt S, Werner D (1992) Isoliquiritigenin, a strong *nod* gene- and glyceollin resistance-inducing flavonoid from soybean root exudate. Appl Environ Microbiol 58:1705-1710

Kellis J, Vickery L (1984) Inhibition of human estrogen synthetase (aromatase) by flavones. Science 225:1032-1034

Kijne J (1992) The *Rhizobium* infection process. In Stacey G, Burris R Evans H (Eds) Biological nitrogen fixation. Chapman and Hall, New York, pp 349-398

Kosslak R, Bookland R, Barkei J, Paaren H, Appelbaum E (1987) Induction of *Bradyrhizobium japonicum* common *nod* genes by isoflavones isolated from *Glycine max*. Proc Natl Acad Sci USA 84:7428-7432

Krause A, Broughton WJ (1992) Proteins associated with root-hair deformation and nodule initiation in *Vigna unguiculata*. Mol Plant-Micr Inter 5:96-103

Krause A, Lan VT, Broughton WJ (1997) Induction of chalcone synthase expression by rhizobia and Nod-factors in root hairs and roots. Mol Plant-Micr Inter 10:388-393

Krause A, Sigrist CJ, Dehning I, Sommer H, Broughton WJ (1994) Accumulation of transcripts encoding a lipid transfer-like protein during deformation of nodulation-competent *Vigna unguiculata* root hairs. Mol Plant-Micr Inter 7:411-418

Kurkdjian AC (1995) Role of the differentiation of root epidermal cells in Nod-factor (from *Rhizobium meliloti*)-induced root-hair depolarisation of *Medicago sativa*. Plant Physiol 107:783-790

Lawson CGR, Djordjevic MA, Weinman JJ, Rolfe BG (1994) *Rhizobium* inoculation and physical wounding result in the rapid induction of the same chalcone synthase copy in *Trifolium subterraneum*. Mol Plant-Microbe Interact 7: 498-507

Lerouge P, Roche P, Faucher C, Maillet F, Truchet G, Promé JC, Dénarié J (1990) Symbiotic host-specificity of *Rhizobium meliloti* is determined by a sulphated and acylated glucosamine oligosaccharide signal. Nature 344:781-784

Lin Y, Wang Y, Zhu J-k, Yang Z (1996) Localization of a Rho GTPase implies a role in tip growth and movement of the generative cell in pollen tubes. Plant Cell 8:293-303

Long SR (1996) *Rhizobium* symbiosis: Nod-factors in perspective. Plant Cell 8:1885-1898

Ma H (1994) GTP-binding proteins in plants: new members of an old family. Plant Mol Biol 26:1611-1636

Mathesius U, Bayliss C, Weinman JJ, Schlaman HRM, Spaink HP, Rolfe BG, McCully ME, Djordjevic MA (1998a) Flavonoids synthesized in cortical cells during nodule initiation are early developmental markers in white clover Mol Plant-Microbe Interact 11:1223-1232

Mathesius U, Schlaman H, Spaink H, Sautter C, Rolfe B, Djordjevic M (1998b) Auxin transport inhibition precedes root nodule formation in white clover roots and is regulated by flavonoids and derivatives of chitin oligosaccharides. Plant J 14:23-34

Maxwell C, Philips D (1990) Concurrent synthesis and release of *nod*-gene-inducing flavonoids from alfalfa roots. Plant Physiol 93:1552-1558

McKhann HI, Hirsch AM (1994) Isolation of chalcone synthase and chalcone isomerase cDNAs from alfalfa (*Medicago sativa* L.): highest transcript levels occur in young roots and root tips. Plant Mol Biol 24: 767-777

McKhann H, Paiva N, Dixon R, Hirsch A (1997) Chalcone synthase transcripts are detected in alfalfa root hairs following inoculation with wild-type *Rhizobium meliloti*. Mol Plant-Micr Inter 10:50-58

McKhann HI, Paiva NL, Dixon RA, Hirsch AM (1998) Expression of genes for enzymes of the flavonoid biosynthetic pathway in the early stages of the *Rhizobium*-legume symbiosis. In Flavonoids in the living system. J. Manthey and B. Buslig (Eds), Plenum Press, New York. pp. 45-54.

Mergaert P, van Montagu M, Promé J-C, Holsters M (1993) Three unusual modifications, a D-arabinosyl, an N-methyl, and a carbamoyl group, are present on the Nod-factors of *Azorhizobium caulinodans* strain ORS571. Proc Natl Acad Sci USA 90:1551-1555

Miller DD, de Ruijter NC, Bisseling T, Emons AMC (1999) The role of actin in root hair morphogenesis: studies with lipochito-oligosaccharide as a growth stimulator and cytochalasin as an actin perturbing drug. Plant J 17:141-154

Miller DD, de Ruijter NC, Emons AMC (1997) From signal to form: aspects of the cytoskeleton-plasma membrane-cell wall continuum in root hair tips. J Exp Bot 48:1881-1896

Minami E, Kouchi H, Carlson R, Cohn J, Kolli V, Day R, Ogawa T, Stacey G (1996a) Cooperative action of lipo-chitin nodulation signals on the induction of the early nodulin, ENOD2, in soybean roots. Mol Plant-Micr Inter 9:574-583

Minami E, Kouchi H, Cohn J, Ogawa T, Stacey G (1996b) Expression of the early nodulin, ENOD40, in soybean roots in response to various lipo-chitin signal molecules. Plant J 10:23-32

Niebel A, Bono J, Ranjeva R, Cullimore J (1997) Identification of a high affinity binding site for lipo-oligosaccharidic NodRm factors in the microsomal fraction of *Medicago* cell suspension cultures. Mol Plant-Micr Inter 10:132-134

Penmetsa R, Cook D (1997) A legume ethylene-insensitive mutant hyperinfected by its rhizobial symbiont. Science 275:527-530

Perret X, Freiberg C, Rosenthal A, Broughton W, Fellay R (1999) High-resolution transcriptional analysis of the symbiotic replicon of *Rhizobium* sp. NGR234. Mol Microbiol 32:415-425.

Perret X, Staehelin C, Broughton W (2000) Molecular basis of symbiotic promiscuity. Microbiol Mol Biol Rev 64:180-201

Pichon M, Journet E-P, Dedieu A, de Billy F, Truchet G, Barker D (1992) *Rhizobium meliloti* elicits transient expression of the early nodulin gene *ENOD*12 in the differentiating root epidermis of transgenic alfalfa. Plant Cell 4:1199-1211

Philip-Hollingsworth S, Dazzo FB, Hollingsworth RI (1997) Structural requirements of *Rhizobium* chitooligosaccharides for uptake and bioactivity in legume roots as revealed by synthetic analogs and fluorescent probes. J Lipid Res 38:1229-1241

Pingret J-L, Journet E-P, Barker DG (1998) *Rhizobium* Nod-factor signaling: Evidence for a G protein-mediated transduction mechanism. Plant Cell 10:659-671

Price N, Relić B, Talmont F, Lewin A, Promé D, Pueppke S, Maillet F, Dénarié J, Promé J-C, Broughton W (1992) Broad-host-range *Rhizobium* species strain NGR234 secretes a family of carbamoylated, and fucosylated nodulation signals that are *O*-acetylated or sulphated. Mol Microbiol 6:3575-3584

Rao J, Cooper J (1994) Rhizobia catabolize *nod* gene inducing flavonoids via C-ring fission mechanisms. J Bacteriol 176:5409-5413

Rao J, Cooper J (1995) Soybean nodulating rhizobia modify *nod* gene inducers daidzein and genistein to yield aromatic products that can influence gene inducing activity. Mol Plant-Micr Inter 8:855-862

Recourt K, Schripsema J, Kijne J, van Brussel A, Lugtenberg B (1991) Inoculation of *Viciae sativa* subsp. *nigra* roots with *Rhizobium leguminosarum* bv. *viciae* results in release of nod gene activating flavanones and chalcones. Plant Mol Biol 16:841-852

Recourt K, Verkerke M, Schripsema J, van Brussel A, Lugtenberg B, Kijne J (1992) Major flavonoids in uninoculated and inoculated roots of *Viciae sativa* subsp. *nigra* are four conjugates of the nodulation gene inhibitor kaempferol. Plant Mol Biol 18:505-513

Redmond J, Batley M, Djordjevic M, Innes R, Kuempel P, Rolfe B (1986) Flavones induce expression of nodulation genes in *Rhizobium*. Nature 323:632-635

Relić B, Perret X, Estrada-Garcia MT, Kopcinska J, Golinowski W, Krishnan HB, Pueppke SG, Broughton WJ (1994) Nod-factors of *Rhizobium* are a key to the legume door. Mol Microbiol 13:171-178

Relić B, Talmont F, Kopcinska J, Golinowski W, Promé J-C, Broughton WJ (1993) Biological activity of *Rhizobium* sp. NGR234 Nod-factors on *Macroptilium atropurpureum*. Mol Plant-Micr Inter 6:764-774

Ridge R (1988) Freeze substitution improves the ultrastructural preservation of legume root hairs. Bot Mag Tokyo 101:427-441

Ridge RW (1992) A model of legume root hair growth and *Rhizobium* infection. Symbiosis 14:359-373

Sanjuan J, Carlson R, Spaink H, Bhat U, Barbour W, Glushka J, Stacey G (1992) A 2-*O*-methylfucose moiety is present in the lipo-oligosaccharide nodulation signal of *Bradyrhizobium japonicum*. Proc Natl Acad Sci USA 89:8789-8793

Scheres B, McKhan H, Zalensky A, Löbler M, Bisseling T, Hirsch A (1992) The *PsENOD12* gene is expressed at two different sites in Afghanistan pea pseudonodules induced by auxin transport inhibitors. Plant Physiol 100:1649-1655

Scheres B, van de Wiel C, Zalenski A, Horvath B, Spaink H, van Eck H, Zwartkruis F, Wolters A-M, Gloudemans T, van Kammen A, Bisseling T (1990) The *Enod12* gene product is involved in the infection process during the pea-*Rhizobium* interaction. Cell 60:281-294

Schiefelbein J, Somerville C (1990) Genetic control of root hair development in *Arabidopsis thaliana*. Plant Cell 2:235-243

Schmidt P, Broughton W, Werner D (1994) Nod-factors of *Bradyrhizobium japonicum* and *Rhizobia* sp. NGR234 induce flavonoid accumulation in soybean exudate. Mol Plant-Micr Inter 7:384-390

Schultze M, Quiclet-Sire B, Kondorosi E, Virelizier H, Glushka N, Endre G, Géro D, Kondorosi A (1992) *Rhizobium meliloti* produces a family of sulfated lipo-oligosaccharides exhibiting different degrees of plant host specificity. Proc Natl Acad Sci USA 89:92-196

Sherrier D, VandenBosch K (1994) Secretion of cell wall polysaccharides in *Vicia* root hairs. Plant J 5:185-195

Singer WD, Brown HA, Sternweis PC (1997) Regulation of eukaryotic phosphatidylinositol-specific phospholipase C and phospholipase D. Annu Rev Biochem 66:475-509

Smertenko A, Jiang C-J, Simmons N, Weeds A, Davies D, Hussey P (1998) Ser6 in the maize actin-depolymerizing factor, ZmADF3, is phosphorylated by a calcium-stimulated protein kinase and is essential for the control of functional activity. Plant J 14:187-193

Smit G, Puvanesarajah V, Carlson RW, Barbour WM, Stacey G (1992) *Bradyrhizobium japonicum nodD*1 can be specifically induced by soybean flavonoids that do not induce the *nodYABCSUIJ* operon. J biol Chem 267:310.

Spaink H, Sheeley D, van Brussel A, Glushka J, York W, Tak T, Geiger O, Kennedy E, Reinhold V, Lugtenberg B (1991) A novel highly unsaturated fatty acid moiety of lipo-oligosaccharide signals determines host specificity of *Rhizobium.* Nature 354:125-130

Staehelin C, Granado J, Müller J, Wiemken A, Mellor B, Felix G, Regenass M, Broughton WJ, and Boller (1994) Perception of *Rhizobium* nodulation factors by tomato cells and inactivation by root chitinases Proc Natl Acad Sci 91: 2196-2200

Stokkermans J, Peters N (1994) *Bradyrhizobium elkanii* lipo-oligosaccharide signals induce complete nodule structures on *Glycine soja* Siebold & Zucc. Planta 193:413-420

Sutherland T, Bassam B, Schuller L, Gresshoff P (1990) Early nodulation signals of the wild type and symbiotic mutants of soybean (*Glycine max*). Mol Plant-Micr Inter 3:122-128

Takai T, Sasaki T, Tanaka K, Nakanishi H (1995) Rho as a regulator of the cytoskeleton. Trends Biochem Sci 20:227-231

Tominaga M, Morita K, Sonobe S, Yokota E, Shimmen T (1997) Microtubules regulate the organization of actin filaments at the cortical region in root hair cells of *Hydrocharis.* Protoplasma 199:83-92

Truchet G, Roche P, Lerouge P, Vasse J, Camut S, de Billy F, Promé J-C, Dénarié J (1991) Sulphated lipo-oligosaccharide signals of *Rhizobium meliloti* elicit root nodule organogenesis in alfalfa. Nature 351:670-673

van Brussel A, Bakhuizen R, van Spronsen P, Spaink H, Tak T, Lugtenberg B, Kijne J (1992) Induction of pre-infection thread structures in the leguminous host plant by mitogenic lipo-oligosaccharides of *Rhizobium.* Science 257:70-72

van Brussel A, Recourt K, Pees E, Spaink H, Tak T, Wijffelman C, Kijne J, Lugtenberg B (1990) A biovar-specific signal of *Rhizobium leguminosarum* bv. *viciae* induces increased nodulation gene-inducing activity in root exudate of *Vicia sativa* subsp. *nigra.* J Bacteriol 172:5394-5401

van Workum W, van Brussel A, Tak T, Wijffelman C, Kijne J (1995) Ethylene prevents nodulation of *Vicia sativa* ssp. *nigra* by exopolysaccharide-deficient mutants of *Rhizobium leguminosarum* bv. *viciae.* Mol Plant-Micr Inter 8:111-119

Viel A, Branton D (1996) Spectrin: on the path from structure to function. Curr Opin Cell Biol 8:49-55

Vijn I, Martinez-Abarca F, Yang W, Das Neves L, van Brussel A, van Kammen A, Bisseling T (1995) Early nodulin gene expression during Nod-factor-induced processes in *Vicia sativa.* Plant J 8:111-119

Viprey V, Del Greco A, Golinowski W, Broughton WJ, Perret X (1998) Symbiotic implications of type III protein secretion machinery in *Rhizobium.* Mol Microbiol 28:1381-1389

Viprey V, Perret X, Broughton W (2000) Host-plant invasion by Rhizobia. In Hacker J, Olschlager T (Eds) Subcellular biochemistry. Bacterial invasion into eukaryotic cells. Plenum Publishing Corp, London, (in press)

Widdell S, Larsson C (1987) Plasma membrane purification. In Senger H (Ed) Blue light responses: Phenomena and occurrence in plants and microorganisms. CRC Press, Boca Raton, Florida, pp 99-107

Williams P, de Mallorca S (1982) Abscisic acid and gibberellin-like substances in root and root nodules of *Glycine max*. Plant Soil 65:19-26

Wymer C, Wymer S, Cosgrove D, Cyr R (1996) Plant cell growth responds to external forces and the response requires intact microtubules. Plant Physiol 110:425-430

Yoshida S, Uemura M, Niki T, Sakai A, Gusta L (1983) Partition of membrane particles in aqueous two-polymer phase system and its practical use for purification of plasma membranes from plants. Plant Physiol 72:105-114

Zaat S, Schripsema J, Wijffelman C, van Brussel A, Lugtenberg B (1989) Analysis of the major inducers of the *Rhizobium nodA* promoter from *Viciae sativa* root exudate and their activity with different *nodD* genes. Plant Mol Biol 13:175-188

Zhang S, Sheng J, Liu Y, Mehdy M (1993) Fungal elicitor-induced bean proline-rich protein mRNA down-regulation is due to destabilization that is transcription and translation dependent. Plant Cell 5:1089-1099

Chapter 16

Rhizobium-Induced Plant Gene Expression in Root Hairs

Andrea Jahraus and Ton Bisseling

Laboratory of Molecular Biology, Department of Plant Sciences, Wageningen University, Dreijenlaan 3, 6703 HA Wageningen, The Netherlands

1 Introduction

Root hairs, seen as an extension and surface enlargement of root epidermal cells, so-called trichoblasts, serve for the uptake of nutrients and provide anchorage into the soil. In addition, the young emerging root hairs are the first cells of legumes with which symbiotic soil bacteria of several genera, collectively called rhizobia, interact (Bhuvaneswari et al. 1980; Broughton and Perret 1999; Turgeon and Bauer 1985). This interaction leads to root nodules, new organs that are formed within the cortex of the root. Here, the bacteria are hosted in order to fix atmospheric nitrogen into ammonia for the benefit of the plant. In return, the host plant pays back by supplying the bacteria with organic nutrients. In several areas soil nitrogen is a limiting factor, and there symbiosis makes leguminous species significantly more competitive than other species.

During the establishment of the symbiosis, the interaction of rhizobia with epidermal cells induces deformation of root hairs (Heidstra and Bisseling 1996; Kannenberg and Brewin 1994; see also this volume for further details). Some hairs form a "shepherd's crook"-like curl in which the rhizobia are captured (Brewin 1991). Within the pocket of the curl, the bacteria modify and partly dissolve the plant cell wall (Ridge and Rolfe 1985) and enter actively the hair by invagination of the plant plasma membrane. The host deposits new cell wall material at the site of infection, forming the so-called infection thread. Most strikingly, this infection thread enclosing the bacteria is able to "grow" directionally towards the base of the trichoblast and subsequently into the root interior towards the nodule primordia in the inner cortex (e.g., Bauer 1981).

Root hair deformation is initiated upon an exchange of signals between bacteria and the legume host. The communication involves the secretion of flavonoids and other molecules (e.g., sugars or amino acids) by the plant into the

Key words. early nodulins, *ENOD12*, *ENOD5*, extensins, gene expression, *Rhizobium* symbiosis, *rip1*, root epidermis, root hairs soil which chemotactically attract the rhizobia to the root system (Phillips and Kapulnik 1995; Snehal et al. 1999). Flavonoids were shown to induce the transcription of bacterial nodulation genes (nod genes: Caetano-Anolles et al. 1988; Long and Staskawicz 1993; Peters et al. 1986). These genes encode proteins that are involved in the production and export of the so-called Nod-factors, signal molecules consisting of a chito-oligosaccharide backbone with a fatty acid moiety at the non-reducing terminal sugar residue. Dependent on the bacterial species, they further can possess different modifications, such as a sulfate group, which restricts the symbiotic interaction to certain host plant species (for more details on Nod-factor structure and host specificity see Fisher and Long 1992; Roche et al. 1992; Schultze and Kondorosi 1995). The bacterial Nod-factors are secreted or at least transferred to the outer lipid layer of the gram-negative bacteria and thus exposed to the root epidermis. They are essential and also sufficient on their own to induce depolarization of root hair membranes (Ehrhardt et al. 1992), root hair deformation (Heidstra et al. 1994) and root cortical cell division (Lerouge et al. 1990; Truchet et al. 1991). In addition to Nod-factors, exopolysaccharides and other unknown molecules play a role in certain steps of nodulation, especially in infection processes such as infection thread formation and growth (Bladergroen and Spaink 1998; Downie and Johnston 1986; Finan et al. 1985; Kannenberg and Brewin 1994; Leigh and Walker 1994). As it will be shown below, the use of rhizobial mutants affected either in their outer polysaccharide-constitution (exopoly-saccharide- or lipopolysaccharide-mutants) or in their Nod-factor production (mutants producing no or modified Nod-factors) helped to distinguish if e.g., an altered host gene expression is specifically caused by Nod-factors or if other bacterial factors are involved. Furthermore, the use of EPS- or LPS-mutants that can block infection in specific stages, can contribute to the classification of certain gene patterns into the hierarchy of events during infection of susceptible root hairs (Cheng and Walker 1998; Kannenberg and Brewin 1994; Leigh and Walker 1994).

A general consensus starts to arise that root hair deformation occurs within the first 3 h after Nod-factor application (Heidstra et al. 1994; Sieberer and Emons, 2000; R. Mirabella, pers. communication). In principle, bacteria can also induce root hair deformation in such a short time period but production of sufficient Nod-factors depends on growth parameters of the bacterial culture (Bhuvaneswari et al. 1983). Infection thread formation seems to occur after one to two days (Cook et al. 1995; Kijne 1991; Timmers et al. 1999). The morphological changes in epidermal cells are preceded by several physiological changes (for reviews see this book; Heidstra and Bisseling 1996; Schultze and Kondorosi 1998) and are accompanied by the expression of "symbiosis-specific" plant genes. The so-called early nodulin genes (*ENOD*s) are by definition plant genes that are specifically expressed during the interaction with rhizobia and activated during early stages of the symbiosis prior to the onset of nitrogen fixation (Nap and Bisseling 1990). In this chapter, we will give an overview of presently known early nodulin genes and we will especially focus on those expressed in the root epidermis.

2 Approaches to Identify Early Nodulin Genes

Before the early nodulins are described, we would like to give a short survey of the different approaches used to search for these genes. The principle of most of the studies is to compare the protein or mRNA composition of *Rhizobium*-inoculated or Nod-factor treated roots/root hairs with the same plant source from untreated plants or plants inoculated with a mutant bacterial strain or non-host/modified Nod-factors. Earlier approaches were based on differential screening of cDNA libraries, often in combination with subtractive hybridization in order to enrich for differentially expressed genes. Subtractive hybridization has been used either to make subtractive libraries enriched for the clones of interest or to produce more specific probes to screen libraries. Most of the early nodulin genes were isolated from nodule libraries and some of them were shown to be also activated in the root epidermis by bacteria. In addition to the screening of cDNA libraries, early nodulins were identified by two-dimensional protein gel (2D-gel) electrophoresis (Galat et al. 1996; Jungblut et al. 1994; Wilkins et al. 1996) and mRNA display methods such as differential display (Kozian et al. 1999; Liang et al. 1998; Martin et al. 1999). The latter method can cope with rather small amounts of material and is therefore ideal to study gene expression in root hairs. Whereas 2D-gel electrophoresis only provided information about the 1000 most abundant proteins in root hairs (of a total of about 20,000), the RNA display methods proved to be useful to study mRNAs with a very low copy number per cell.

3 Early Nodulins Identified by 2D-Gel Electrophoresis

Earlier attempts using 2-D protein gel analysis in order to identify proteins that specifically accumulate in root hairs upon inoculation with rhizobia had only a very limited outcome (Gloudemans et al. 1989; Krause and Broughton 1992). Since it was not possible to separate the microbes from root hairs after inoculation and due to the lack of purified Nod-factors at the beginning of the 1990's, 2-D gel analysis did not allow the discriminatation between proteins from the bacteria and proteins from the host plant. Thus, it was not possible to determine whether the observed changes in protein pattern after bacterial inoculation were related to plant proteins. The effect of *Rhizobium* infection on gene expression in root hairs could, however, be determined by in vitro translation of root hair mRNAs since bacterial RNA is not translated in such a eukaryotic cell-free translation system (Govers et al. 1985; Horst et al. 1980). The translation products are then separated by 2-D-gel electrophoresis.

Gloudemans et al. (1989) searched for early nodulin mRNAs by comparing the mRNA population of root hairs of inoculated and non-inoculated pea plants. They observed several differences in the in vitro translated protein populations but showed that most of the corresponding pea genes appeared to be also expressed at a high level in an early stage of root hair development. Some root hair specific mRNAs decreased during root hair development, while others remained constant (Gloudemans et al. 1989). Thus, they concluded that *Rhizobium* could influence the developmental stage of root hairs by either stimulating formation of new root hairs or transferring existing root hairs to a younger stage of development. The latter is actually in agreement with today's view of the re-initiation of tip growth during root hair deformation (Heidstra et al. 1994; De Ruijter et al. 1998). In conclusion, only two proteins occurred at markedly elevated levels after inoculation with *R. leguminosarum* bv *viciae*, that were not detected during early stages of root hair development. According to their molecular weight they were named RH-42, which was newly formed as judged by 2D-gel electrophoresis, and RH-44 which showed a considerably increased expression level. Neither RH-42 nor RH-44 are activated by an exopolysaccharide-mutant (EPS-mutant) of *Rhizobium*, which has a strikingly reduced ability to infect pea roots, or by mutants not able to produce Nod-factors (Nod$^-$ mutant) and thus not able to trigger root hair deformation any longer. Hence, it is unclear which bacterial signal molecules are involved in the activation of these host genes. However, it is probable that RH-42 and RH-44 play a role in infection since these genes are first up-regulated after about one day post inoculation, and bacteria unable to infect do not up-regulate their expression.

A by-product of these root hair studies was the identification of a group of small proteins of around 14 kDa which occurred at elevated levels in root hair fractions. One of them, termed RH2, was sequenced and subsequently cloned (Mylona et al. 1994). In situ hybridization demonstrated that RH2 is specifically expressed in the epidermal root cell layer. Although RH2 does not play a specific role in symbiosis, this gene can be used as an important epidermal marker during root hair isolation procedures. In addition, its promoter showed to be very useful in directing epidermal specific expression in transformed plants (J. Ramons and R. Mirabella, pers. communication).

Another attempt to characterize symbiosis-specific changes in the protein composition of root hairs with the help of 2D-gel analysis came from the laboratory of Broughton and colleagues. These authors used isolated root hairs of *Vigna unguiculata* which had been inoculated with either a wild type host strain or with a Nod$^-$ mutant (Krause and Broughton 1992). In principle, this allows differentiation between bacterial and plant proteins although, for example, possible discrepancies in wild type and mutant bacterial gene expression could lead to misinterpretation. Furthermore, various bacterial mutants can alter plant gene expression specifically (Arsenijevic Maksimovic et al. 1997). Nonetheless, protein extracts of root hair fractions can so be analyzed directly, circumventing the in vitro translation. This approach led to the discovery of changes in protein quantity post inoculation. Three of the proteins were present at reduced levels after four days of inoculation with the wild type strain which was not seen with the mutant strain. Thus, rhizobia seem to be able to repress genes specifically but only at a time point beyond root hair infection. On the other hand, nine proteins were present at elevated levels that could vary in a time-dependent manner during infection. The earliest changes in protein pattern were first detectable after one day of bacterial inoculation as observed also in pea (Gloudemans et al. 1989). In conclusion, studies on the protein level are not very suitable to follow rapid changes in expression during symbiotic interactions.

Unfortunately, for none of these "symbiosis-specific" polypeptides, detected by 2D-gel analysis, corresponding genes were cloned nor have the proteins been analyzed. Therefore, the knowledge of the function of these host proteins in the *Rhizobium*-plant interaction is still rather limited. Recent technical advances in the 2D-gel electrophoresis resolution, the sequencing of model legumes (*Medicago truncatula/Lotus japonicum*) genomes in the near future and the possibility to sequence subpicomolar amounts of proteins with the mass spectrometry technology make a reconsideration of this approach worthwhile. In addition, proteome-analysis has the advantage of being closer to the biological consequences of altered gene expression. Current investigations actually utilize this technique to compare root protein extracts before and after inoculation with rhizobia/purified Nod-factors at successive stages of root nodule development (Dumas et al. 1999, *Medicago truncatula* workshop, Amsterdam, The Netherlands).

4 Cloned Hyroxyproline-Rich Early Nodulins Involved in Symbiosis

Most of the *ENOD* genes cloned up to now were identified by differential screening of nodule or root hair cDNA libraries as well as homology-based cloning (Allison et al. 1993; Govers et al. 1991; Greene et al. 1998; Krause et al. 1994, 1997; Pichon et al. 1992; Scheres et al. 1990b; Scheres et al. 1990a; Vijn et al. 1995c). We first want to focus on *ENOD*s belonging to the group of hydroxyproline-rich glycoproteins (HPRG-proteins) that comprises some nodulins but also many proteins found in non-symbiotic tissue such as extensins, arabinogalactan proteins and lectins (Cassab 1998; Showalter 1993). Their sequences all contain several characteristic (hydroxy)proline-rich domains and they are thought to be cell wall proteins which are extensively post-translationally modified. Within the nodulins displaying this characteristic structure, *ENOD5*, *ENOD11* and *ENOD12* are the only members which have been demonstrated to be also expressed in the root epidermis.

4.1 ENOD5

ENOD5 was originally cloned from pea (*Pisum sativum*; Scheres et al. 1990a). Later, an *ENOD5* homolog was isolated from *Vicia sativa* showing 76% identity at the amino acid level to Ps*ENOD5* (Vijn et al. 1995c). The predicted molecular weight of the polypeptide is 14 kDa which is probably encoded by a gene with a single or low copy number. *ENOD5* contains a putative amino-terminal plasma membrane translocation signal peptide. The amino acid composition of *ENOD5* is reminiscent of that of arabinogalactan proteins with a high percentage of proline, serine, glycine and alanine and a short sequence of five amino acids consisting of alternating prolines and alanines (Scheres et al. 1990a). Greene et al. (1998) used Ps*ENOD5* sequences to search for cross-hybridizing genes in *Medicago truncatula*. The two new *ENOD* clones (Mt*ENOD16* and Mt*ENOD20*) are homologous to *ENOD5* but have an additional region that is not present in

ENOD5. This additional region encodes a copper-binding domain and reveals that Mt*ENOD16* and Mt*ENOD20* are phytocyanin-like proteins. Whether *ENOD5* and *ENOD16*/*ENOD20* have a similar function is unknown.

In situ hybridization studies on infected pea or vetch roots showed that the *ENOD5* gene is only expressed in root cortical cells containing a growing infection thread and in infected cells of nodules (Scheres et al. 1990a; Vijn et al. 1995b). However, by RT-PCR (reverse transcriptase PCR) it was shown that Nod-factors induce the expression of this gene also in pea root hairs (Horvath et al. 1993). Hence, the presence of infection threads or bacteria is not required to trigger *ENOD5* expression but might enhance it since only very sensitive methods such as RT-PCR could detect it in the absence of bacteria. RT-PCR experiments also revealed that Vs*ENOD5* is already induced within 3 h after Nod-factor application to *Vicia sativa* plants (Heidstra et al. 1997a). In addition, these authors demonstrated that there is a low level of *ENOD5* expression in untreated roots using whole root segments as starting material. More details about *ENOD5* expression in root hairs will be described in combination with *ENOD12* expression (see below).

4.2 ENOD12

ENOD12 is the most detailed studied early nodulin gene. It had been first isolated by differential screening of a pea cDNA nodule library with a subtractive cDNA probes from inoculated and non-inoculated plants (Scheres et al. 1990b). *PsENOD12* contains a typical hydrophobic N-terminus of 24 amino acids which probably functions as a plasma membrane translocation signal. *ENOD12* is mainly composed of repeating pentapeptide units of the PPXXX-type, a structural feature that has also been found in a family of hydroxyproline-rich cell wall proteins of soybean known as SbPRPs (Hong et al. 1990). The copy number of the *ENOD12* genes varies between legume species. In *Pisum sativum* (Govers et al. 1991; Scheres et al. 1990b) and *Medicago sativa* (Allison et al. 1993; Bauer et al. 1994) two genes exist, while in *Medicago truncatula* and *Vicia sativa* only a single copy of *ENOD12* is present (Pichon et al. 1992; Vijn et al. 1995c). Especially the signal peptide sequence and the proline-rich regions are highly conserved in the different *ENOD12* polypeptides that vary in length between 80 and 140 amino acids. Csanádi et al. (1994) summarized the evolutionary relations between the different *ENOD12* genes. Interestingly, *Medicago* species have been found which carry a null allele in heterologous configuration. These plants were used to generate *Medicago* plants lacking the *ENOD12* gene (see below; Csanádi et al. 1994).

A comparison of the expression patterns of the different *ENOD12* genes revealed some differences not only between the two different isoforms but also between plant species. The two pea isoforms *ENOD12*A and B, the Mt*ENOD12* gene and one of the alfalfa *ENOD12* genes, Ms*ENOD12*B, are expressed in a similar way (Scheres et al. 1990b; Govers et al. 1991; Pichon et al. 1992; Allison et al. 1993; Bauer et al. 1994). Within a few hours after inoculation, *ENOD12* genes are transcribed in root hairs (see below). During the infection process they are active in clusters of cortical cells where the infection thread is present.

Contrary to Ps*ENOD5*, *ENOD12* is expressed several cell layers in front of the infection thread tip (Scheres et al. 1990a,b). In nodules, they are detected in a zone where bacteria are released into the plant cytoplasm. Again, in contrast to Ps*ENOD5*, the infected as well as the non-infected cells of the central tissue of this zone express *ENOD12*. However, the situation is different in *Vicia sativa* and with *ENOD12*A in alfalfa. In *Vicia*, Vs*ENOD12* is also induced in a more mature nodule part where the rhizobia enclosed in the peribacteroid membrane proliferate (Vijn et al. 1995c). In *Medicago sativa*, the two *ENOD12* genes are differentially expressed in contrast to the two isoforms in pea (Allison et al. 1993; Bauer et al. 1994). Here, expression of MsENDO12A was first detected in young nodules (Bauer et al. 1994). Moreover, this gene was also expressed in nodules formed under certain conditions in absence of bacteria, lateral root primordia and in root tips. Thus, its expression does not require the presence of Nod-factors or rhizobia and Bauer et al. (1996, 1997) proposed a correlation between Ms*ENOD12*A expression and meristematic activity in alfalfa and *Lotus japonicum*.

A good impression of ENOD12 gene activation in the root epidermis was achieved using transgenic *Medicago* plants carrying a Mt*ENOD12*-promoter-ß-glucuronidase (GUS) reporter gene (Pichon et al. 1992; Journet et al. 1994). This transgene has a similar expression as the endogenous *ENOD12* gene and hence can be used to study Nod-factor signaling in the epidermis. *ENOD12*-GUS is activated in the epidermis within 2-3 hours and its expression is induced in a region that begins in the root elongation zone just above the root tip, continues throughout the zone of root hair emergence and development, and terminates in the region where root hairs have matured. Rhizobia are generally initiating nodulation in a part of the root that lies between the elongating root tip and the zone of the root hair emergence (Bhuvaneswari et al. 1981; Caetano-Anolles and Gresshoff 1991) but the studies with the *ENOD12* transgene reveals that the zone of epidermal cells able to respond to Nod-factors is much broader. However, within the very early induction period, GUS-staining is restricted to epidermal cells prior to and during the earliest stages of root hair emergence, obviously the most responsive zone in the root. After 48-72 h of inoculation a number of discrete, dark blue-colored loci within the mature root hair region were observed. At this time, staining can be detected in the interior region of roots, while with the exception of these intensely staining loci, all the surrounding epidermal cells (including root hairs) were without detectable GUS activity. The persistent expression in epidermal foci could reflect the sites of bacterial entry but more work needs to be done here. In conclusion, it seems that the early transcription of the ENOD12 gene is transient for the majority of differentiating root epidermal cells, a phenomenon also found in pea root hairs and alfalfa roots (Horvath et al. 1993; Bauer et al. 1994).

4.3 ENOD11

Using *ENOD12* as a probe to screen a *Medicago truncatula* library, Barker et al. (1999) recently isolated Mt*ENOD11*. It is a single copy gene closely related to *ENOD12* but lacking the tyrosine residues in the proline-rich pentapeptide motif.

By using the Mt*ENOD11*-promoter-GUS construct it was shown that the *ENOD11* promoter is activated by Nod-factors in the root epidermis similar to *ENOD12* (Barker et al. 1999, *Medicago truncatula* workshop, Amsterdam, The Netherlands; Gamas et al. 1996; Cook et al. 1995). The advantages of *ENOD11* over *ENOD12* as reporter gene will be mentioned below.

4.4 Extensins

Recent data from several laboratories demonstrated that extensins play a role in symbiosis (Brewin et al. 1999, MPMI-meeting, Amsterdam, The Netherlands; Perlick and Puhler 1993; Arsenijevic-Maksimovic et al. 1997). These cell wall proteins are also rich in hydroxyproline displaying the typical repetitive motif of Ser-(Hyp4) and are supposed to function in strengthening the cell wall (Cassab 1998; Cosgrove 1997). The Brewin group identified with a specific antibody a major protein component of the infection thread matrix that additionally was secreted at elevated levels from epidermal cells as an early response to *Rhizobium* infection. Immunopurification of the antigen and peptide sequencing provided enough information to clone the corresponding gene in pea, which turned out to be an extensin like glycoprotein (MPG). It shows homology to extensin genes of *Vicia faba* and *Medicago truncatula* that were shown to be expressed at enhanced levels during symbiotic events (Perlick and Puhler 1993; Gamas et al. 1996). Brewin's data demonstrated for the first time the cellular localization of an early nodulin protein. Similar to other HPRGs it is also detected in other tissues, for example root tips are able to secrete MGP. It seems probable that also the other HPRG-like nodulins are part of the cell wall matrix and could be part of the infection thread matrix. In contrast to the general view that extensin like genes are up-regulated during symbiosis, Arsenijevic-Maksimovic et al. (1997) studied three extensins from a root hair library (Krause et al. 1994) which surprisingly showed a reduced expression one day after inoculation with rhizobia and which were enriched in root hairs (see also Chapter 15 this volume). The role of these extensin isoforms in symbiosis needs further investigation.

5 Other Cloned Early Nodulin Genes Involved in Symbiosis

5.1 Rip1

Mt*rip1* (*Rhizobium*-induced peroxidase) has been isolated from a nodule cDNA library from *Medicago truncatula* using subtractive hybridization and PCR to enrich the probe for symbiotic-related cDNAs (Cook et al. 1995). *Rip1* transcripts were not only found in nodules but were also present in high amounts in the root epidermis shortly after inoculation with rhizobia. Prior to infection, low levels of *rip1* transcripts could be detected in roots. *Rip1* showed high homology (max. 59% on the amino acid level) to several plant peroxidases with the conserved distal and proximal heme group ligands. The protein is probably located extracellularly since it contains a putative membrane translocation signal sequence. Similar to *ENOD12*, the copy number of this gene varies between different ecotypes: *Medicago truncatula* A-68 probably contains one or two copies of *rip1* while ecotype A-17 has multiple loci.

Upon inoculation, but prior to infection, the expression of *rip1* was strongly enhanced in the differentiating root epidermis adjacent to the root tip, reaching its maximal level 1 d post inoculation and then it declined (Cook et al. 1995; Peng et al. 1996). Further, Nod-factors on their own were sufficient and necessary for the full induction of *rip1* since Nod$^-$ mutants did not trigger any response. Nod-factors were able to enhance the expression in the epidermis within 3 h and the expression level is markedly higher than that of *ENOD12*. During the period of nodule morphogenesis, *rip1* declined in the epidermis to levels below those observed in non-inoculated roots (Peng et al. 1996). This suppression occurs in the zone of the epidermis just above the growing root tip where nodule formation is suppressed in the cortex due to the presence of nodules in the older parts of the roots (Bhuvaneswari et al. 1981; Caetano-Anollés and Gresshoff 1991). When this zone regains its susceptibility after a couple of days in order to form more nodules on the host plant, *rip1* transcripts are again detected at levels comparable to those associated with non-inoculated roots. Thus, the absence of *rip1* transcripts in the differentiating epidermis subsequent to primary nodulation events is temporally correlated with a feedback inhibition of nodulation and hence, the expression in this zone could be essential for the competence of this zone for infection.

5.2 Chalcone Synthase

The two following genes are also discussed in Chapter 15 this volume, therefore these genes will not be described in detail here.

Chalcone synthase (CHS) is involved in the biosynthetic pathway of flavonoids and its expression is increased by several environmental stimuli but also during different stages of symbiotic interaction (Djordjevic et al. 1997; Estabrook and Sengupta Gopalan 1991; Grosskopf et al. 1993; McKhann et al. 1997; Wingender et al. 1989). Krause et al. (1997) studied CHS expression in root hairs of *Vigna unguivulata* and isolated one CHS gene from a root hair library (Krause et al. 1994) with high homology to other chalcone synthases. The kinetics of Vu*CHS1* transcripts, tested with a specific 5' end probe from this clone, varied within tissues. In root hairs, transcript levels were induced within one day post inoculation with rhizobia and subsequently declined. Nod-factors could induce Vu*CHS1* expression in root hairs in the same way, but surprisingly Nod$^-$ mutants also induced Vu*CHS1* in root hairs. Thus, it seems likely that this chalcone synthase gene is regulated by several rhizobial signals.

5.3 Lipid Transfer-Like Protein

Krause et al. (1994) screened a root hair library of *Vigna unguivulata* using a heterologous lipid transfer-like protein as probe hypothesizing that these proteins might play a role in Nod-factor perception. They isolated a cDNA with homology to nonspecific lipid transfer proteins (LTP) from Barley and *Arabidopsis* (41%

and 48% identity on the amino acid level). Eight cysteine residues, characteristic for LTPs, are highly conserved. Only one copy of this gene is present in the *Vigna unguivulata* genome, which was expressed in almost all tissues, being most abundant in young leaves. Inoculation with *Rhizobium* or Nod-factors led to an increase of LTP-transcript levels in root hairs after one day, while a Nod$^-$ mutant did not have any effect.

6 Early Nodulin Genes as Marker Genes to Study the Mode of Action of Nod-factors

6.1 Suitability of ENOD12 as a Molecular Marker

As described in previous pages, our knowledge about the function of early nodulins is rather limited. However, early nodulin genes have been very successfully used as marker genes to study early steps of the interaction. In principle, all *ENOD*s with a suitable expression pattern can serve as marker genes to study Nod-factor signaling but *ENOD12* has been most frequently used to study symbiosis specific responses in the epidermis. However, like other early nodulin genes *ENOD12* is first activated after 2-3 h, which suggests that the activation of this gene requires de novo protein synthesis, which is in fact confirmed by studies with cycloheximide (Heidstra et al. 1997b; Vijn et al. 1995b). Since other protein(s) have to be made before *ENOD12* is activated, it is probable that the signal transduction cascade leading to its activation is rather complex. The recently cloned *MtENOD11* promoter will be slightly better than the *ENOD12* promoter since its gene is expressed at higher levels. Furthermore, *ENOD11* is constitutively expressed in the root cap and this can serve as an internal control to check, for example, the toxicity of applied pharmaceutical drugs. However, since *ENOD11* is also first activated after ca. 2 h of Nod-factor application, it seems probable that the activation of this gene also involves de novo protein synthesis. Therefore it will be useful to obtain additional marker genes that are induced preferably within minutes after Nod-factor addition.

A transcription factor, ENBP1, has been identified in *Vicia sativa* (Christiansen et al. 1996) that binds to the 200 base pairs upstream of the start codon of the pea *ENOD12* promoters which are sufficient for Nod-factor induced expression (Vijn et al. 1995a). However, it is unknown whether ENBP is de novo synthesized before the *ENOD12* promoter is activated in the epidermis.

Moreover, the fact that *ENOD*12 induction is independent of root hair deformation, as was shown by using the exocytosis blocker Brefeldin A (Heidstra et al. 1997b), gives the opportunity to dissect the diversity and hierarchy of signal transduction pathways leading to gene expression and/or root hair deformation. In the following paragraph several examples will be given that illustrate the use of early nodulin genes as markers to study the mode of action of Nod-factors.

6.2 Induction of ENOD12 in the Epidermis Shows that Both Trichoblasts and Atrichoblasts Respond to Nod-factors

Transgenic *Medicago varia* A2 plants carrying an *ENOD12-GUS* fusion were used to study which epidermal cells respond to Nod-factors of *S. meliloti*. Within the epidermal zone that responds to Nod-factors, *ENOD12-GUS* is activated in atrichoblasts as well as trichoblasts (Pichon et al. 1992; Journet et al. 1994). This result is consistent with the depolarization response that can also be triggered by Nod-factors in both cell types (see Chapter 10 this volume; Ehrhardt et al. 1992; Felle et al. 1995, 1996). Symplastic communication between root epidermal cells in *Arabidopsis* becomes restricted as they differentiate into trichoblasts and atrichoblasts (Duckett et al. 1994). If this is also the case in legumes then it would imply that trichoblasts and atrichoblasts both recognize Nod-factors. Whether this cell to cell communication does not occur between the trichoblasts and atrichoblasts remains to be studied in legumes but Journet et al. (1994) showed that Nod-factor induced *ENOD12* expression in the epidermis is at least not a systemic reaction. Application of an agarose-block containing Nod-factors caused only *ENOD12-GUS* activity in the region of the epidermis that had been in direct contact with the agarose block. The high affinity of Nod-factors for membranes as well as their immobility in epidermal cell walls support the idea of a locally restricted response of epidermal cells to this bacterial signal which could explain how the plant host "senses" the presence of bacteria (see Chapter 5 this volume).

A couple of hours after the application of Nod-factors *ENOD12-GUS* expression is restricted to the epidermis and it is not active in the hypodermal or cortical cell layers (Journet et al. 1994). In *Egeria* and *Arabidopsis* the epidermis was shown to be symplastically isolated from the hypodermal layer (Duckett et al. 1994; Erwee and Goodwin 1985). If this is also the case in legumes this could explain why the response is restricted to the epidermis.

6.3 Early Nodulin Gene Induction to Study Nod-factor Structure-Function Relationship

In pea and *Medicago* the relation between Nod-factor structure and its ability to activate *ENOD12* has been studied (Horvath et al. 1993; Journet et al. 1994). It was shown that saturation of the fatty acid chain or the presence of the O-acetyl group at the non-reducing sugar had only a slight effect on the *ENOD12*-inducing activity of Nod-factors. However, in the case of *Medicago* the non-sulphated NodRm factors were almost 1000-fold less active in eliciting *ENOD12* expression as compared with the sulphated NodRm factor. These results demonstrate that the induction of the *ENOD12* promoter is a quantitative assay. This is in contrast to an assay that is based on a growth response such as the root hair deformation assay.

6.4 Signal Transduction

The power of early nodulin based reporter systems to study Nod-factor signaling was demonstrated by Pingret et al. (1998). By a pharmacological approach they obtained strong indications that heterotrimeric G-proteins play a role in Nod-factor signal transduction. Their work further suggested that phospholipase C activation as well as Ca^{2+} influx/release are involved in Nod-factor induced signaling leading to *ENOD12* expression (for further details see Chapter 15 this volume).

7 Concluding Remarks

In this chapter we described several genes whose expression in root hairs is enhanced or decreased by rhizobia. In most cases, these genes have been cloned and their expression has only been studied at the RNA level. Such studies provided the first indications about the role of these genes in the early steps of symbiosis but by using more sensitive detection systems it has now been shown that "early nodulin" genes are also transcribed in non-symbiotic parts of the plants (Bauer et al. 1997; Heidstra et al. 1997b; Scheres et al. 1992). This implies that *ENOD*s could have multiple functions in different processes or one general function used in different tissues. Only biochemical and cell biological studies as well as reverse genetics can supply a full understanding of their function. Although almost no such studies have been performed it is striking that several of the up-regulated genes are HPRG-like proteins that are most likely cell-wall matrix proteins. It seems probable that their function is at least in part similar. This assumption is consistent with the observation that *Medicago* cultivars that lack *ENOD12* are not affected in their ability to interaction with rhizobia (Csanádi et al. 1994). Brewin and colleagues showed that at least one of these HRPG-like proteins (an extensin like protein) is part of the infection thread matrix (Brewin et al. 1999; MPMI meeting, Amsterdam, The Netherlands). Cross-linking of cell wall extensins has been widely reported as a rapid response to plant pathogens both in vivo (Otte and Barz 1996) and in vitro (Wojtaszek et al. 1997). Extracellular located peroxidases are known to be responsible for this H_2O_2-dependent insolubilization of extensins and thus contribute to the stiffening of the extracellular matrix. Intriguingly, the *rip1* peroxidase was identified as *ENOD* and found to be strongly expressed in the root epidermis. Brewin et al. proposed that HRPG in the infection thread matrix could be differentially cross-linked. For example a tight plug of cell wall matrix could be formed at the site where the infection thread enters the root hairs.

In addition to the root hair genes whose expression is changed during the interaction with rhizobia, several other genes will play a pivotal role in the early steps of symbiosis. Among these genes will be receptors for the rhizobial signals as well as proteins that function as molecular switches. It is very probable that these genes are constitutively expressed in root hairs. This assumption is supported by the notion that several cell physiological changes as well as an alteration of the actin cytoskeleton configuration occur within minutes after Nod-factor addition (see Chapter 3 this volume). Hence, to obtain a full understanding of the role of plant genes in early steps of symbiosis it is important to study the proteins encoded by different classes of plant genes at a biochemical, cell biological and genetic level.

References

Allison LA, Kiss GB, Bauer P, Poiret M, Pierre M, Savoure A, Kondorosi E, Kondorosi A (1993) Identification of two alfalfa early nodulin genes with homology to members of the pea Enod12 gene family. Plant Mol Biol 21:375-80

Arsenijevic Maksimovic I, Broughton WJ, Krause A (1997) Rhizobia modulate root-hair-specific expression of extensin genes. Mol Plant Microbe Interact 10:95-101

Bauer P, Crespi MD, Szecsi J, Allison LA, Schultze M, Ratet P, Kondorosi E, Kondorosi A (1994) Alfalfa Enod12 genes are differentially regulated during nodule development by Nod-factors and *Rhizobium* invasion. Plant Physiol 105:585-92

Bauer P, Poirier S, Ratet P, Kondorosi A (1997) MsEnod12A expression is linked to meristematic activity during development of indeterminate and determinate nodules and roots. Mol Plant Microbe Interact 10:39-49

Bauer P, Ratet P, Crespi MD, Schultze M, Kondorosi A (1996) Nod-factors and cytokinins induce similar cortical cell division, amyloplast deposition and MsEnod12A expression patterns in alfalfa roots. Plant J 10:91-105

Bauer WD (1981) Infection of legumes by rhizobia. Annu Rev Plant Physiol 32:407-449

Bhuvaneswari TV, Bhagwat AA, Bauer WD (1981) Transient susceptibility of root cells in four common legumes to nodulation by rhizobia. Plant Physiol 68:1144-1149

Bhuvaneswari TV, Mills KK, Crist DK, Evans WR, Bauer WD (1983) Effects of culture age on symbiotic infectivity of *Rhizobium japonicum*. J Bacteriol 153:443-51

Bhuvaneswari TV, Turgeon BG, Bauer WD (1980) Early events in the infection of soybean *Glycine max* (L.) Merr. by *Rhizobium japonicum*. I. Localization of infectible root cells. Plant Physiol 66:1027-1031

Bladergroen MR, Spaink HP (1998) Genes and signal molecules involved in the rhizobia-Leguminoseae symbiosis. Curr Opin Plant Biol 1:353-9

Brewin NJ (1991) Development of the legume root nodule. Annu Rev Cell Biol 7:191-226

Broughton WJ, Perret X (1999) Genealogy of legume-*Rhizobium* symbioses. Curr Opin Plant Biol 2:305-11

Caetano-Anollés G, Crist-Estes DK, Bauer WD (1988) Chemotaxis of *Rhizobium meliloti* to the plant flavone luteolin requires functional nodulation genes. J Bacteriol 170:3164-9

Caetano-Anollés G, Gresshoff PM (1991) Plant genetic control of nodulation. Annu Rev Microbiol 45:345-82

Caetano-Anollés G, Gresshoff PM (1991) Alfalfa controls nodulation during the onset of *Rhizobium*-induced cortical cell division. Plant Physiol 95:366-373

Cassab GI (1998) Plant cell wall proteins. Annu Rev Plant Physiol Plant Mol Biol 49:281-309

Cheng HP, Walker GC (1998) Succinoglycan is required for initiation and elongation of infection threads during nodulation of alfalfa by *Rhizobium meliloti*. J Bacteriol 180:5183-91

Christiansen H, Hansen AC, Vijn I, Pallisgaard N, Larsen K, Yang WC, Bisseling T, Marcker KA, Jensen EO (1996) A novel type of DNA-binding protein interacts with a conserved sequence in an early nodulin ENOD12 promoter. Plant Mol Biol 32:809-21

Cook D, Dreyer D, Bonnet D, Howell M, Nony E, VandenBosch K (1995) Transient induction of a peroxidase gene in Medicago truncatula precedes infection by *Rhizobium meliloti*. Plant Cell 7:43-55

Cosgrove DJ (1997) Assembly and enlargement of the primary cell wall in plants. Annu Rev Cell Dev Biol 13:171-201

Csanádi G, Szecsi J, Kalo P, Kiss P, Endre G, Kondorosi A, Kondorosi E, Kiss GB (1994) ENOD12, an early nodulin gene, is not required for nodule formation and efficient nitrogen fixation in alfalfa. Plant Cell 6:201-13

De Ruijter NCA, Rook M, Bisseling T, Emons AMC (1998) Lipochito-oligosaccharides re-initiate root hair tip growth in *Vicia sativa* with high calcium and spectrin-like antigen at the tip. Plant J 13:341-350

Djordjevic MA, Mathesius U, Arioli T, Weinman JJ, Gaertner E (1997) Chalcone synthase gene expression in transgenic subterranean clover correlates with localised accumulation of flavonoids. Australian Journal of Plant Physiology 24:119-132

Downie JA, Johnston AW (1986) Nodulation of legumes by *Rhizobium*: the recognized root? Cell 47:153-4

Duckett CM, Oparka KJ, Prior DAM, Dolan L, Roberts K (1994) Dye-coupling in the root epidermis of *Arabidopsis* is progressively reduced during development. Development 120:3247-3255

Ehrhardt DW, Atkinson EM, Long SR (1992) Depolarization of alfalfa root hair membrane potential by *Rhizobium meliloti* Nod-factors. Science 256:998-1000

Erwee MG, Goodwin PB (1985) Symplastic domains in extrastellar tissues of *Egeria densa Planch.* Planta 163:9-19

Estabrook EM, Sengupta Gopalan C (1991) Differential expression of phenylalanine ammonia-lyase and chalcone synthase during soybean nodule development. Plant Cell 3:299-308

Felle HH, Kondorosi E, Kondorosi A and Schultze M (1995) Nod signal-induced plasma membrane potential changes in alfalfa root hairs are differentially sensitive to structural modifications of the lipochitooligosaccharide. Plant J 7:939-947

Felle HH, Kondorosi E, Kondorosi A, Schultze M (1996) Rapid alkalinization in alfalfa root hairs in response to rhizobial lipochitooligosaccharide signals. Plant J 10:295-301

Finan TM, Hirsch AM, Leigh JA, Johansen E, Kuldau GA, Deegan S, Walker GC, Signer ER (1985) Symbiotic mutants of *Rhizobium meliloti* that uncouple plant from bacterial differentiation. Cell 40:869-77

Fisher RF, Long SR (1992) *Rhizobium*-plant signal exchange. Nature 357:655-60

Galat A, Gerbod M, Bouet F, Riviere S (1996) Proteins and their amino acid compositions: uniqueness, variability, and applications. Arch Biochem Biophys 330:229-37

Gamas P, Niebel FdC, Lescure N, Cullimore J (1996) Use of a subtractive hybridization approach to identify new *Medicago truncatula* genes induced during root nodule development. Mol Plant Microbe Interact 9:233-42

Gloudemans T, Bhuvaneswari TV, Moerman M, Van Brussel T, Van Kammen A, Bisseling T (1989) Involvement of *Rhizobium leguminosarum* nodulation genes in gene expression in pea root hairs. Plant Mol Biol 12:157-168

Govers F, Gloudemans T, Moerman M, Van Kammen A, Bisseling T (1985) Expression of plant genes during the development of pea root nodules. EMBO J 4:861-867

Govers F, Harmsen H, Heidstra R, Michielsen P, Prins M, van Kammen A, Bisseling T (1991) Characterization of the pea ENOD12B gene and expression analyses of the two ENOD12 genes in nodule, stem and flower tissue. Mol Gen Genet 228:160-6

Greene EA, Erard M, Dedieu A, Barker DG (1998) MtENOD16 and 20 are members of a family of phytocyanin-related early nodulins. Plant Mol Biol 36:775-83

Grosskopf E, Ha DTC, Wingender R, Roehrig H, Szecsi J, Kondorosi E, Schell J, Kondorosi A (1993) Enhanced levels of chalcone synthase in alfalfa nodules induced by a Fix negative mutant of *Rhizobium meliloti*. Mol Plant Microbe Interact 6:173-181

Heidstra R, Bisseling T (1996) Nod-factor-induced host responses and mechanisms of Nod-factor perception. New Phytologist 133:25-43

Heidstra R, Geurts R, Franssen H, Spaink HP, Van Kammen A, Bisseling T (1994) Root hair deformation activity of nodulation factors and their fate on *Vicia sativa*. Plant Physiol 105:787-797

Heidstra R, Nilsen G, Martinez-Abarca F, van Kammen A, Bisseling T (1997a) Nod-factor-induced expression of leghemoglobin to study the mechanism of NH4NO3 inhibition on root hair deformation. Mol Plant Microbe Interact 10:215-20

Heidstra R, Yang WC, Yalcin Y, Peck S, Emons AM, van Kammen A , Bisseling T (1997b) Ethylene provides positional information on cortical cell division but is not involved in Nod-factor-induced root hair tip growth in *Rhizobium*- legume interaction. Development 124:1781-7

Hong, JC, Nagao, RT, Key, JL (1990) - Characterization of a proline-rich cell wall protein gene family of soybean. A comparative analysis. J Biol Chem 265:2470-5

Horst J, Stanbro H, Merril CR (1980) On procaryotic gene expression in eucaryotic systems. Hum-Genet 54:289-302

Horvath B, Heidstra R, Lados M, Moerman M, Spaink HP, Prome JC, van Kammen A, Bisseling T (1993) Lipo-oligosaccharides of *Rhizobium* induce infection-related early nodulin gene expression in pea root hairs. Plant J 4:727-33

Journet EP, Pichon M, Dedieu A, de Billy F, Truchet G, Barker DG (1994) *Rhizobium meliloti* Nod-factors elicit cell-specific transcription of the ENOD12 gene in transgenic alfalfa. Plant J 6:241-9

Jungblut P, Otto A, Zeindl-Eberhart E, Plessner KP, Knecht M, Regitz-Zagrosek V, Fleck E, Wittmann-Liebold B (1994) Protein composition of the human heart: the construction of a myocardial two-dimensional electrophoresis database. Electrophoresis 15:685-707

Kannenberg EL, Brewin NJ (1994) Host-plant invasion by *Rhizobium*: the role of cell-surface components. Trends Microbiol 2:277-83

Kijne JW (1991) The rhizobium infection process. In: Stacey G, Burris RH and Evans HJ (Eds) Biological Nitrogen Fixation. Chapman and Hall, London, pp 348-397

Kozian, Dh, Kirschbaum, Bj (1999) Comparative gene-expression analysis. Trends Biotechnol 17:73-8

Krause A, Broughton WJ (1992) Proteins associated with root-hair deformation and nodule initiation in *Vigna unguiculata*. Mol Plant Microbe Interact 5:96-103

Krause A, Lan VTT , Broughton WJ (1997) Induction of chalcone synthase expression by rhizobia and Nod-factors in root hairs and roots. Mol Plant-Microbe Interact 10:388-393

Krause A, Sigrist CJA, Dehning I, Sommer H, Broughton WJ (1994) Accumulation of transcripts encoding a lipid transfer-like protein during deformation of nodulation-competent *Vigna unguiculata* root hairs. Mol Plant-Microbe Interact 7:411-418

Leigh JA, Walker GC (1994) Exopolysaccharides of *Rhizobium*: synthesis, regulation and symbiotic function. Trends Genet 10:63-7

Lerouge P, Roche P, Faucher C, Maillet F, Truchet G, Prome JC, Denarie J (1990) Symbiotic host-specificity of *Rhizobium meliloti* is determined by a sulphated and acylated glucosamine oligosaccharide signal. Nature 344:781-4

Liang P, Pardee A, (1998) Differential display. A general protocol. Mol Biotechnol 10:261-7

Long SR, Staskawicz BJ (1993) Prokaryotic plant parasites. Cell 73:921-35

Martin, Kj, Pardee, Ab (1999) - Principles of differential display. Methods Enzymol 303:234-58

McKhann HI, Pavia NL, Dixon RA, Hirsch AM (1997) Chalcone synthase transcripts are detected in Alfalfa root hairs following inoculation with wild-type *Rhizobium meliloti*. Mol Plant Microbe Interact 10:50-58

Mylona P, Moerman M, Yang WC, Gloudemans T, Van de Kerckhove J, van Kammen A, Bisseling T, Franssen HJ (1994) The root epidermis-specific pea gene RH2 is homologous to a pathogenesis-related gene. Plant Mol Biol 26:39-50

Nap JP, Bisseling T (1990) Nodulin function and nodulin gene regulation in root nodule development. In: Gresshoff PM (Ed). Molecular biology of symbiotic nitrogen fixation. CRC Press, Inc., Boca Raton, Florida, pp 181-229

Otte O, Barz W (1996) The elicitor-induced oxidative burst in cultured chickpea cells drives the rapid insolubilization of two cell wall structural proteins. Planta 200:238-246

Peng H-M, Dreyer DA, VandenBosch KA, Cook D (1996) Gene structure and differential regulation of the *Rhizobium*-induced peroxidase gene *rip1*. Plant Physiol 112:1437-1446

Perlick AM, Puhler A (1993) A survey of transcripts expressed specifically in root nodules of broadbean (*Vicia faba* L.). Plant Mol Biol 22:957-70

Peters NK, Frost JW, Long SR (1986) A plant flavone, luteolin, induces expression of *Rhizobium meliloti* nodulation genes. Science 233:977-80

Phillips DA, Kapulnik Y (1995) Plant isoflavonoids, pathogens and symbionts. Trends Microbiol 3:58-64

Pichon M, Journet EP, Dedieu A, de Billy F, Truchet G, Barker DG (1992) *Rhizobium meliloti* elicits transient expression of the early nodulin gene ENOD12 in the differentiating root epidermis of transgenic alfalfa. Plant Cell 4:1199-211

Pingret J-L, Journet E-P, Barker DG (1998) *Rhizobium* Nod-factor Signaling: Evidence for a G-Protein–Mediated Transduction Mechanism. Plant Cell 10:659-671

Ridge RW, Rolfe BG (1985) *Rhizobium* sp. degradation of legume root hair cell wall at the site of infection thread origin. Appl Envtl Micro 50:717-720

Roche P, Debelle F, Lerouge P, Vasse J, Truchet G, Prome JC, Denarie J (1992) The lipo-oligosaccharidic symbiotic signals of *Rhizobium meliloti*. Biochem Soc Trans 20:288-91

Scheres B, McKhann HI, Zalensky A, Loebler M, Bisseling T, Hirsch AM (1992) The PsENOD12 gene is expressed at two different sites in Afghanistan pea pseudonodules induced by auxin transport inhibitors. Plant physiol 100:1649-1655

Scheres B, Van De Wiel C, Zalensky A, Horvath B, Spaink H, Van Eck H, Zwartkruis F, Wolters AM, Gloudemans T et al. (1990a) The ENOD12 gene product is involved in the infection process during the pea *Rhizobium* interaction. Cell 60:281-294

Scheres B, Van Engelen F, Van Der Knaap E, Van De Wiel C, Van Kammen A, Bisseling T (1990b) Sequential induction of nodulin gene expression in the developing pea nodule. Plant Cell 2:687-700

Schultze M, Kondorosi A (1995) What makes nodulation signals host-plant specific? Trends Microbiol 3:370-2

Schultze M, Kondorosi A (1998) Regulation of symbiotic root nodule development. Annu Rev Genet 32:33-57

Showalter AM (1993) Structure and function of plant cell wall proteins. Plant Cell 5:9-23

Sieberer B and AMC Emons (2000) Cytoarchitecture and pattern of cytoplasmic streaming in developing root hairs of *Medicago truncatula* and during deformation by Nod-factors. Protoplasma, submitted

Snehal P, Poornima I, Vaishali G, Trilok P, Anjana D, Pandya S, Iyer P, Gaitonde V, Parekh T, Desai A (1999) Chemotaxis of *Rhizobium* SP.S2 towards *Cajanus cajan* root exudate and its major components. Current Microbiology 38:205-209

Timmers AC, Auriac MC, Truchet G (1999) Refined analysis of early symbiotic steps of the *Rhizobium-Medicago* interaction in relationship with microtubular cytoskeleton rearrangements. Development 126:3617-3628

Truchet G, Roche P, Lerouge P, Vasse J, Camut S, De Billy F, Promé J-C, Dénarié J (1991) Sulfated lipo-oligosaccharide signals of *Rhizobium meliloti* elicit root nodule organogenesis in alfalfa. Nature 351:670-673

Turgeon BG, Bauer WD (1985) Ultrastructure of infection-thread development during the infection of soybean by *Rhizobium japonicum*. Planta 163:328-349

Vijn I, Christiansen H, Lauridsen P, Kardailsky I, Quandt HJ, Broer I, Drenth J, Ostergaard Jensen E, van Kammen A, Bisseling T (1995a) A 200 bp region of the pea ENOD12 promoter is sufficient for nodule- specific and Nod-factor induced expression. Plant Mol Biol 28:1103-10

Vijn I, Martinez-Abarca F, Yang WC, das Neves L, van Brussel A, van Kammen A, Bisseling T (1995b) Early nodulin gene expression during Nod-factor-induced processes in *Vicia sativa*. Plant J 8:111-9

Vijn I, Yang WC, Pallisgard N, Ostergaard Jensen E, van Kammen A, Bisseling T (1995c) VsENOD5, VsENOD12 and VsENOD40 expression during *Rhizobium*-induced nodule formation on *Vicia sativa* roots. Plant Mol Biol 28:1111-9

Wilkins MR, Pasquali C, Appel RD, Ou K, Golaz O, Sanchez JC, Yan JX, Gooley AA, Hughes G, Humphery-Smith I, Williams KL, Hochstrasser DF (1996) From proteins to proteomes: large scale protein identification by two- dimensional electrophoresis and amino acid analysis. Biotechnology 14:61-5

Wingender R, Rohrig H, Horicke C, Wing D, Schell J (1989) Differential regulation of soybean chalcone synthase genes in plant defence, symbiosis and upon environmental stimuli. Mol Gen Genet 18:315-22

Wojtaszek P, Trethowan J, Bolwell GP (1997) Reconstitution in vitro of the components and conditions required for the oxidative cross-linking of extracellular proteins in French bean (*Phaseolus vulgaris* L.). FEBS Letters 405:95-98

Chapter 17
The Involvement of Root Hairs in Mycorrhizal Associations

Frédérique C. Guinel[1] and Ann M. Hirsch[2]

[1]Department of Biology, Wilfrid Laurier University, Waterloo, Ontario N2L 3C5, Canada
[2]Department of Molecular, Cell and Developmental Biology and Molecular Biology Institute, 405 Hilgard Avenue, University of California, Los Angeles, CA 90095-1606, USA

1 Introduction

The involvement of root hairs in the establishment of the *Rhizobium*-legume symbiosis is well known although the root hair is by no means the only point of entry of the bacteria into the plant root. A number of legumes are invaded by the so-called "crack entry" mode of penetration (Sprent 1989) whereby rhizobia enter the root either between epidermal cells or through breaks made through the root cortex by elongating lateral roots. In root hair entry, rhizobia dock on the root hairs, which deform in response to secreted Nod factors, and then the rhizobia enter the curled root hair by means of an infection thread (see Chapter 15 this volume).

Evidence, which indicates that the signal transduction pathways used by the *Rhizobium*-legume nitrogen fixation symbiosis and the fungal-plant mycorrhizal interaction may be conserved, has been accumulating for some time (Hirsch and Kapulnik 1998). Because of this potential conservation at the molecular level, we were interested in learning whether the modes of entry for mycorrhizal fungi were similar or distinct from those of rhizobia. Intuitively, it would seem that root hairs would be the first plant cells that the fungal hyphae encounter, and entry via root hairs would suggest conservation at the cellular level between the nitrogen-fixing *Rhizobium*-legume and phosphate-acquiring fungus-plant symbioses. Arbuscular-mycorrhizal (AM) fungi are known to enter the root both intracellularly and intercellularly, often after forming appressoria. Intercellular penetration by AM fungi is a mode of entry that superficially resembles crack entry. However, very little has been described about the involvement of root hairs in establishing mycorrhizal associations. Peterson and Farquhar (1996) dealt with some of these

---

Key words. ectomycorrhizae, endomycorrhizae, evolution, fungal entry, mycotrophy, root hairs issues in their review. Even in Smith and Read (1997), surprisingly little is said about mechanisms allowing the entry of the arbuscular mycorrhizal fungal hyphae into the plant root and no mention is made about root hair penetration. We thought it timely to reevaluate the involvement of root hairs in the establishment of either ectomycorrhizal or AM symbioses, especially because several text books (see Fig. 7.22 in Mauseth 1995 and Fig. 5.11 in Taiz and Zeiger 1998) imply that this is a common mode of entry for AM fungi.

In this review, we will concentrate mainly on ectomycorrhizae and arbuscular mycorrhizae, neglecting the details of the other types of mycorrhizal symbioses due to lack of space. There is some evidence for epidermal hair entry for orchid mycorrhizal fungi and the reader is directed to references in Smith and Read (1997) for more information.

2 Ectomycorrhizae

Ectomycorrhizae are fungal-plant interactions that occur mainly on woody shrubs and trees. The fungal partner is usually an ascomycete or a basidiomycete and the plant partner can be either a gymnosperm or an angiosperm (see Smith and Read 1997). The fungus forms a mantle or sheath around the root, and becomes intimately associated with root cells in what is known as the Hartig net, whereby the fungal hyphae branch and become interdigitated between epidermal (most angiosperms) or between epidermal and cortical cells (gymnosperms and a few angiosperms). The mycelium does not penetrate living cells of the plant root.

Ectomycorrhiza formation is very complex, and often difficult to study in nature. We focus on the process as studied in the laboratory, whereby axenic seedlings are colonized either in growth pouches or in Petri dishes using a known inoculum. This approach provides the opportunity to simplify this complex relationship and ask specific questions about chemical signalling in the establishment of the symbiosis.

2.1 Precolonization Signalling

Previous investigations have implicated a number of different compounds that trigger germination of spores of ectomycorrhizal fungi, many of which are of root origin (Horan and Chilvers 1990; Gogala 1991). Abietic acid, extracted from pine roots, stimulates spore germination at very low concentrations (10^{-7} M) (Fries et al. 1987). However, little is known about other germination factors and whether or not there is any specificity in any of these molecules.

Chemoattractants, which attract the fungus, are also secreted by the host plant, possibly by the plant root apex (Horan and Chilvers 1990). Analyses of the responses of eucalypt mycorrhizal fungi grown on membranes that separated them from their hosts indicated that hyphae penetrated the membranes to make contact with the host roots. Roots of non-host plants were not able to elicit this fungal response nor were non-eucalypt mycorrhizal fungi able to grow towards *Eucalyptus* roots. The chemical nature of the host signal molecules is not known, but several candidates, including plant hormones and volatiles, have been suggested (see Smith and Read 1997).

Under field conditions, signalling molecules may also come from sources other than the plant or the fungus. Garbaye and Bowen (1989) found that bacteria, the majority of them Gram-negative species, living in association with mycorrhizal fungi, stimulate mycorrhizal development as "helpers". Whether they do this by secreting molecules that improve the efficiency of mycorrhizal formation either as attractants or nutrients is not known.

Signal molecules have also been implicated in very early plant responses. Flavonoids are induced in roots that are colonized by ectomycorrhizal fungi (see references in Martin and Tagu 1995). However, an increase in phenolics is often indicative of an incompatible response (Malajczuk et al. 1982). Phenylpropanoids accumulate in a tissue-specific manner in conifer mycorrhizae where they may be involved in restricting hyphal growth (Weiss et al. 1999).

Elicitors composed of *N*-acetyl-glucosamine oligomers, which were isolated from culture medium of the fungus *Hebeloma crustuliniforme*, induce a number of plant responses, including extracellular alkalinization, efflux of Cl^- and K^+, influx of Ca^{2+}, phosphorylation of a 63 kDa protein and dephosphorylation of a 65-kDa protein, as well as synthesis of hydrogen peroxide, in cell suspension cultures of spruce (*Picea abies*) (Salzer et al. 1996). These responses are very similar to those elicited in legume root hairs to the application of rhizobial Nod factor (see Chapters 14-16 this volume). In addition, a G-protein activated signal transduction pathway is likely to be utilized in ectomycorrhiza formation based on studies using specific activators and inhibitors (Hebe et al. 1999). Mastoparan, an activator of trimeric G-proteins, induced ion release from cultured spruce cells and alkalinization of the medium in the absence of the elicitor. Phosphorylation of a 63-kDa protein also occurred upon addition of mastoparan as well as of cantharidin and calyculin A, both of which function as protein phosphatase inhibitors. Protein kinase inhibitors such as staurosporine prevented elicitor-induced effluxes of Cl^- and K^+ (Hebe et al. 1999). These results parallel those found for the *Rhizobium*-legume symbiosis whereby Pingret et al. (1998) observed that a Nod-factor induced gene, *MsENOD12*-GUS, was stimulated in transgenic alfalfa by mastoparan in the absence of Nod factor. However, in contrast to the Nod factor-responding cells, the identity of target genes or of the responding cells in planta to the *H. crustuliniforme* signal molecules is unknown. Do root hair or epidermal cells respond? Are the cellular responses induced by *N*-glucosamine oligomers part of the signal transduction cascade for the development of ectomycorrhizae?

Hyphae of several fungi also change color upon making contact with their host root apices, suggesting that some chemical within the root exudate triggers the change (Nylund and Unestam 1982; Massicotte et al. 1989, 1990). However, the identity of the inducer is unknown.

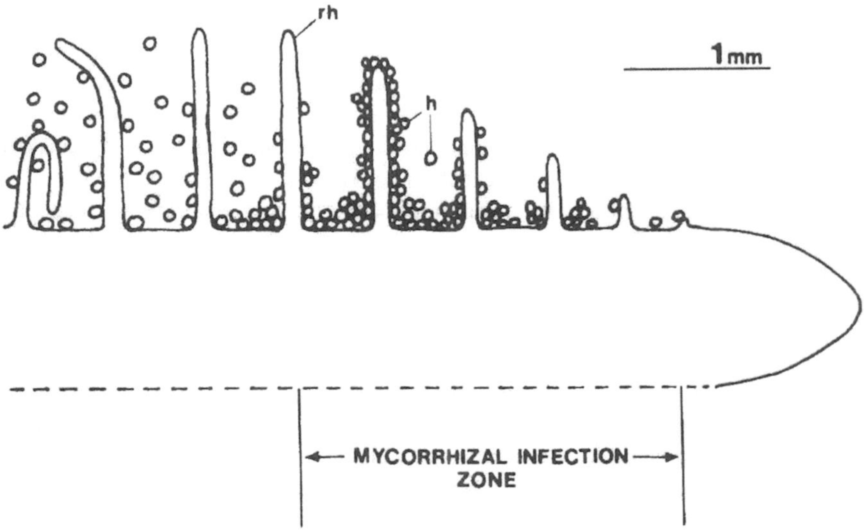

Fig. 1. Diagram of the colonization gradient relative to the position of root hairs in a second order lateral root. *h*, hyphae; *rh*, root hair. Reprinted with permission from Thomson et al. (1989)

2.2 Root Hair Involvement

A distinct zone of the lateral root, 0.5-4.0 mm from the root apex, is responsive to ectomycorrhizal fungi (Fig. 1). [However, see Horan and Chilvers (1990) for comments about the significance of root cap cells for colonization.] This zone overlaps in part with the susceptible zone for infection in the *Rhizobium*-legume symbiosis (see Chapter 15 this volume). Fungal hyphae first make contact with the root at a position immediately proximal to the root cap; in some roots, this region swells. The fungal hyphae proliferate forming a mantle or sheath consisting of multiple layers. Hyphae derived from the inner mantle are highly branched, and embedded in an extracellular matrix (see review by Tagu and Martin 1996). Eventually the hyphae penetrate between epidermal cells and form the Hartig net.

If root hairs are present, a different pattern of growth may occur. Upon making contact with root hairs, ectomycorrhizal fungal hyphae alter their growth and orientation to the root surface, becoming more intimately associated with the root surface (Jacobs et al. 1989). The hyphae partially envelop the root hairs and further root hair growth may be suppressed, or alternatively, the root hairs become incorporated into the mantle (Figs. 2-5) (Massicotte et al. 1987, 1990; Thomson et al. 1989). If root hair growth is suppressed, the hair becomes highly vacuolated, and the hyphae appear to aggregate at the base of the root hair (Thomson et al. 1989). There is no evidence for entry of the hyphae into root hair cells contrary to earlier reports (see references in Thomson et al. 1989). After extensive hyphal proliferation, the root hairs are no longer visible because a compact mantle that is comprised of fungal hyphae completely covers the root (Fig. 5).

Root hair inhibition or suppression may be a consequence of accumulation of the tryptophan betaine hypaphorine. This compound, found at high concentrations in hyphal mats of the fungus *Pisolithus tinctorius*, has been reported to have auxin-like activity. In studies on the *P. tinctorius-Eucalyptus* interaction, the hypaphorine concentration in the colonizing fungal hyphae was found to be 3 to 5 times higher than in adjacent parts of the fungus (Béguiristain and Lapeyrie 1997). However, hypaphorine does not always exactly mimic auxin responses on *Eucalyptus* root development in spite of the fact that it acts much like IAA or 2,4-D in inducing the expression of glutathione-*S*-transferase (Nehls et al. 1998). In contrast to exogenous IAA, which stimulated root hair formation (Horan 1991 cited in Béguiristain and Lapeyrie 1997), Béguiristain and Lapeyrie (1997) found that root hair length was significantly reduced following hypaphorine treatment. Hypaphorine accumulation appears to be controlled by molecules released from host roots (Béguiristain and Lapeyrie 1997).

2.3 Binding to the Root Surface

A number of investigators have noted a tight adherence of the fungal hyphae to the root cells (Bonfante-Fasolo 1988; Piché et al. 1988; Lapeyrie et al. 1989). A fibrillar substance between the root hair surface and the surrounding hyphae appears to be correlated with this tight adherence (Thomson et al. 1989; Lei et al. 1991). Root hairs as well as the cells at the root surface secrete a considerable amount of proteins and carbohydrates, which together form an extracellular matrix in which the fungal hyphae are embedded. However, binding of fungal hyphae occurs whether root hairs are present or not.

It is possible that the plant exudates function as a carbon source for the fungi, but it is also likely that these various polysaccharides and glycoproteins act as adhesins and facilitate the recognition of specific proteins on the symbiotic partner. The fungus too secretes polysaccharides and glycoproteins, and these may make up part of the "glue" that holds the two organisms together. The exact chemical nature of these adhesins is unknown, but Lei et al. (1991) observed that the lectin concanavalin A recognized glycoproteins on the root surface.

Lectins are carbohydrate-binding proteins that have been proposed to be important for host specificity in the *Rhizobium*-legume symbiosis. Numerous studies have confirmed their importance in host recognition in that symbiosis (see Hirsch 1999). Studies on lectin involvement in the mycorrhizal interaction are much more limited, but Giollant et al. (1993) characterized a lectin from *Lactarius deterrimus* and found that antibodies localized this lectin not only on cultured mycelium but also on young root hairs of spruce. In contrast, older root hairs lack the antigen. It would be of interest to determine whether lectins are universally associated with ectomycorrhiza formation, and also whether lectin localization correlates directly with the susceptible root zone pictured in Fig. 1.

Hydrophobins, small proteins produced by the fungus, have also been implicated in hyphal-root adhesion. These are cysteine-rich proteins that exist normally as soluble monomers, but spontaneously polymerize upon reaching a hydrophilic/hydrophobic interface. In *Schizophyllum commune*, hydrophobins govern fruiting body morphogenesis by bringing about hyphal aggregation (Wessels 1993). As *P. tinctorius* colonizes *Eucalyptus*, the relative levels of two different hydrophobin transcripts increased 5-fold (Tagu and Martin 1996). However, neither hydrophobin transcripts nor protein have been localized during the early steps of root colonization, so it is not clear which stage of ectomycorrhiza formation is correlated with the increase in hydrophobin expression.

In conclusion, it is unclear as to how important root hairs are for the establishment of ectomycorrhizae. A priori, it appears that they are not very significant because the entire root surface is competent for fungal colonization. In some ectomycorrhizal associations, however, there is a region of increased susceptibility to fungal colonization along the root where there are developing root hairs. The presence of root hairs may increase the root's surface area for recognition and/or binding of fungal hyphae. In the mycorrhizal symbiosis established between *Eucalyptus* and *Pisolithus*, root hair development is suppressed, and the fungal alkaloid hypaphorine appears to be the causative agent. It is of interest to see if similar molecules function in mycorrhizal formation in addition to the *Eucalyptus* and *Pisolithus* interaction. It would also be of interest to determine whether or not hypaphorine or lectin production is critical for ectomycorrhiza development. In addition, a tantalizing parallel with the *Rhizobium*-legume symbiosis is indicated by the discovery that fungal-produced *N*-glucosamine oligomers function as signal molecules and operate through a G-protein mediated signal transduction pathway. Again, it is unknown whether or not these responses are essential for the formation of ectomycorrhizae or are merely interesting coincidences.

3 Arbuscular Mycorrhizae

3.1 Recognition and Infection

The details of recognition and infection have been reviewed recently by Bonfante and Perotto (1995) and Smith and Read (1997). Only the highlights will be presented in this review.

Prepenetration stimuli exist before the contact between the fungus and the host plant occurs. In a manner similar to that of the *Rhizobium*-legume system, flavonoids exuded by the host plant have been suggested to stimulate the growth of the fungi. CO_2 also appears to be responsible for an increase in extramatrical hyphal length; such an increase would allow the fungus to reach the root surface (see references in Smith and Read 1997). Stimulation of hyphal growth is not the only influence that plant exudates could exhibit. Recently, water-soluble root exudates of bean were shown to act as fungal chemoattractants (Vierheilig et al. 1998).

Once the fungal mycelium touches the surface of the root, a specialized structure known as the appressorium differentiates on the epidermis. This structure

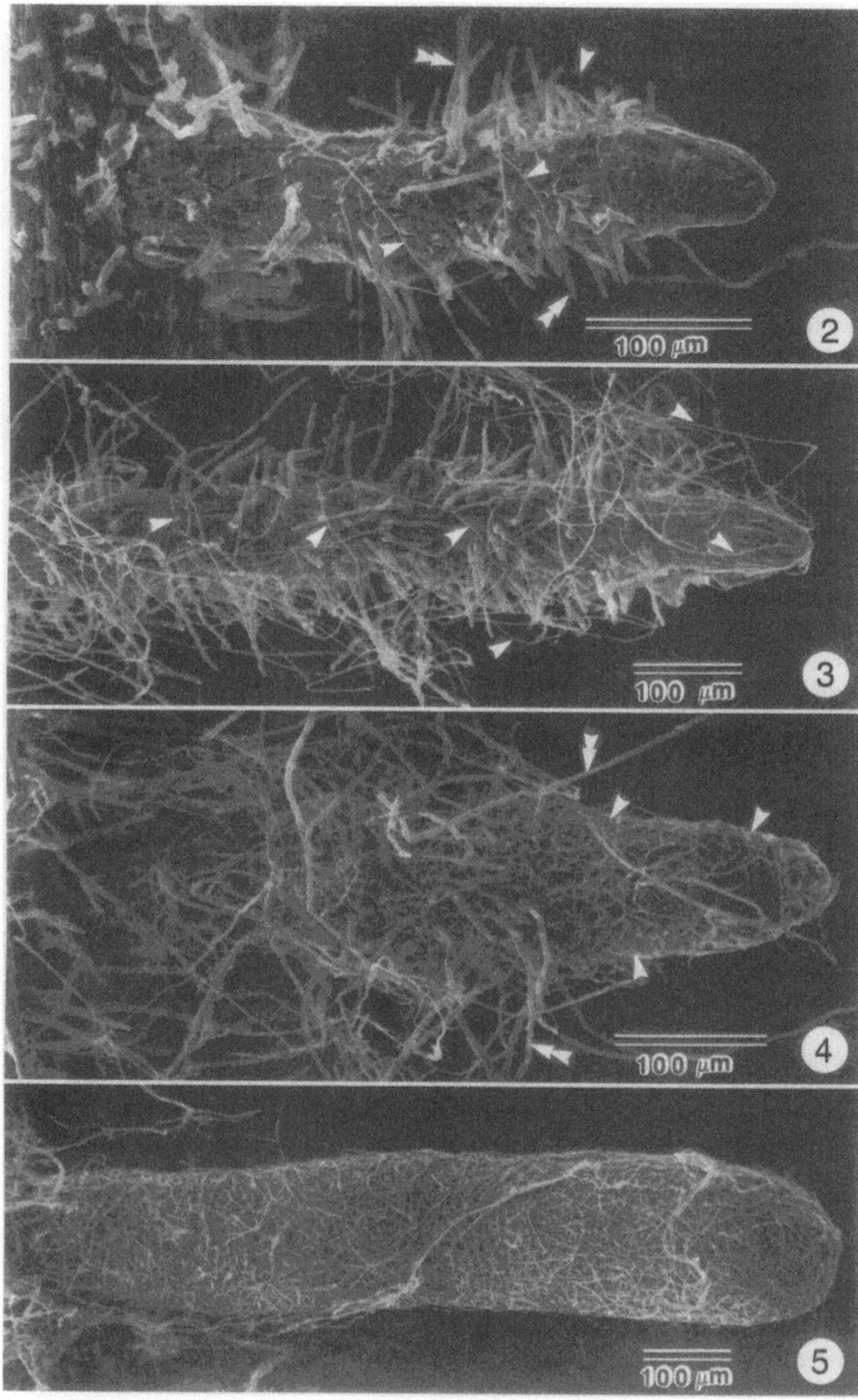
100 μm
2
100 μm
3
100 μm
4
100 μm
5

permits the entry of the fungal hypha within the epidermal cell (Garriock et al. 1989). In her 1984 review, Bonfante-Fasolo mentioned that there were three ways by which hyphae enter the host root. 1) The hyphae enter through the wall of a root hair or of an epidermal cell. 2) A fungal hypha, originating from an appressorium, could also find its way through empty spaces occurring between the outermost layers of older cells; it would then enter healthy cortical cells and form an intracellular coil. 3) A hypha could progress intercellularly from the entry point between two epidermal cells to the inner cortex where it would become intracellular. In all cases, the anatomy of the root and the chemistry of the outer (epidermal) cell wall will have an effect on the type of entry the fungus uses (Bonfante-Fasolo 1984). When penetrating the epidermal wall to enter within the root intracellularly, the fungal hypha reduces its diameter before enlarging it again when the process of entry is completed (Garriock et al. 1989; McArthur and Knowles 1992; Smith and Read 1997).

Once within the epidermal cells, the hyphae grow intraradically following either one of the two patterns (*Arum* and *Paris* types) described by Gallaud (1905), or in a form intermediate between the two. Gallaud (1905) did not write extensively on the entry per se of the fungus, although of interest is his observation that very few cells of the epidermis of either *Arum* or *Paris* differentiate into root hairs. Depending on the host species, the fungal hyphae once within the root either continue to grow intracellularly (in the *Paris* type) or change to intercellular growth and eventually intracellular proliferation (in the *Arum* type).

3.2 Extent of VA-Fungal Penetration Through Root Hairs

Although root hair entry is normally considered to be rare (see Harley 1969, Peterson and Farquhar 1996), a recent survey of roots from a wooded area in Ontario proved the contrary (F.C. Guinel, unpublished results). Randomly collected roots were cleared and then stained with chlorazol black E (Brundrett et al. 1984). What was surprising was the large number of fungal hyphae that had entered the roots through the hairs (Figs. 6, 7).

Root hairs of pea (*Pisum sativum* L.), inoculated with *Glomus aggregatum,* are also frequently penetrated by fungal hyphae (Figs. 8-11). Of the thirteen secondary roots of 18 day-old-plants that were analyzed, on the average, per root, there were 24 successful root penetrations and on average 6 of these were through

---

Figs. 2–5. (*Previous page*). Scanning electron micrographs showing the changes in the morphology of *Betula alleghaniensis* root and its root hairs upon inoculation with *Pisolithus tinctorius*. *Fig. 2*. Root hairs (*double arrowheads*) and a few fungal hyphae (*arrowheads*) are present. *Fig. 3*. The hyphae (*arrowheads*) proliferate and interact with the root hairs. *Fig. 4*. A thin mantle (*arrowheads*) has formed on the root surface and the root hairs have become covered with hyphae (*double arrowheads*). *Fig. 5*. The root is covered with a dense mantle and the root hairs are no longer evident. Reprinted with permission from Massicotte et al. (1990)

root hairs. Of these 6, an average of 1.2 had passed the epidermal layer and entered the cortex (Fig. 11).

Fungal penetration through the root hair has also been occasionally mentioned in the literature (e.g., in ginseng, Whitbread et al. 1995; in sweetclover, Hirsch and Kapulnik 1998), although most of the time, it is not reported. It is known that the plant may have a strong influence on mycorrhiza formation and on the extent of fungal root colonization. Smith and Smith, in their excellent 1997 review, gave strong arguments in favor of a correlation between anatomical features of a root and the type (Gallaud's *Arum*- or *Paris*-type) of mycorrhiza that is formed. Bonfante-Fasolo (1984) proposed that root anatomy and cell wall chemistry have an effect on fungal entry. Could root hair entry be another feature imposed by the plant? Also, the absence or presence of root hairs seems to be of importance for the degree of mycorrhizal dependency (Baylis 1975).

We have reviewed the literature on AM-mycorrhizae with the following questions in mind. Is AM fungal entry through root hairs a species-specific characteristic? Are root hairs and mycotrophy evolutionarily linked?

3.3 Species Where Root Hair Penetration by the Fungus Appears to be the Norm

Several reports in the literature point out that in certain species the hyphae enter mainly through root hairs. These observations support Bonfante-Fasolo's (1984) proposal that the plant could be responsible for the mechanism of fungal entry. The angiosperms, which exhibit root hair entry, belong to four taxonomical orders (Table 1). Two orders in the Dicotyledoneae (Fabales and Apiales) and three in the Monocotyledoneae (Dioscorales and Liliales in the superorder Liliiflorae, and Poales) frequently exhibit root hair entry. One family, that of the Taxaceae, in the gymnosperms also shows fungal penetration into root hairs. The frequency with which the fungal hyphae penetrate the hairs appears to be species-dependent. Thus, in *Taxus* (Prat 1926) and oat (Winter 1951), the event occurs most of the time whereas in wheat (Winter 1951) or sweetclover (Hirsch and Kapulnik 1998), it appears to be infrequent.

3.3.1 Gymnosperms

Prat (1926) studied in detail the infection of *Taxus baccata*, a member of the Taxaceae where two types of roots, the pioneer and the feeder roots, coexist. Whereas the former are fat, straight with practically no branches, and grow quite rapidly, the latter are very thin, sinuous, highly branched, and short-lived. These two types of roots also differ in their root hairs. The pioneer roots have few hairs that are generally grouped in certain areas. In contrast, the feeder roots have numerous hairs that are often deformed or incompletely developed. Prat (1926) reported that whereas the pioneer roots were rarely infected by fungal endophytes, the feeder roots were almost always infected. According to Prat (1926), this is possibly because the walls of the hairs on these roots are made exclusively of cellulose. Prat (1926) noted that the fungal hyphae infect either totally developed

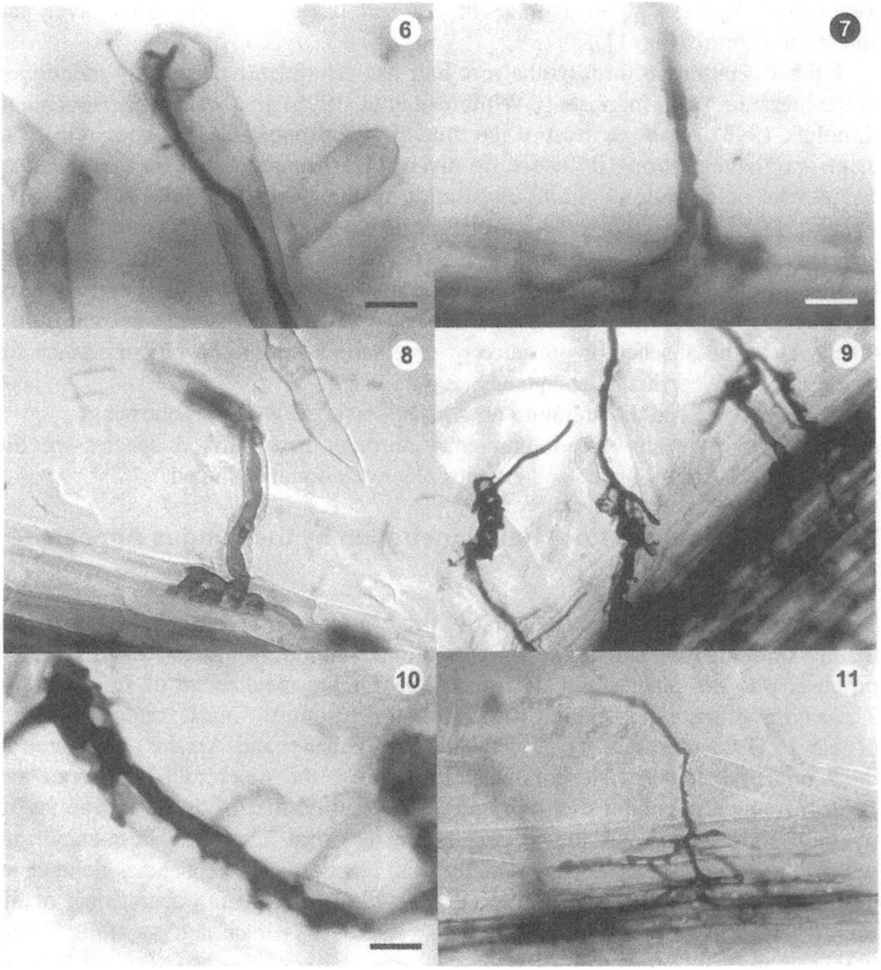

Figs. 6-7. Fungal entry through the root hair of an unknown species found in a wooded area of Opinicon, Canada. *Fig. 6.* The entry occurred at the tip of the hair without an appressorium. Bar 125 μm. *Fig. 7.* The fungal hypha grows straight towards the root epidermis, swelling and branching at the base of the hair. Bar 100 μm. **Figs. 8-11.** Pea root inoculated with *Glomus aggregatum*. *Fig. 8.* The hypha penetrates the root hair below the tip, without forming an appressorium. It enters the base of the root hair cell and then branches. x350. *Fig. 9.* Three root hairs contain fungal hyphae. On the left, the fungal hypha, after contacting the hair surface, grows by coiling around the hair, finally penetrating it in an area about 1/3 down its length. x216. *Fig. 10.* The hypha entered the root hair through an appressorium-like structure. Its topography inside the hair is rough. Bar 144 μm. *Fig. 11.* Having entered the root hair, the fungal hypha branched and infected cells of the inner root cortex. Intercellular hyphae are observed in the inner cortex. x150

hairs or hairs that have not yet finished developing. In the latter case, once infected, the root hair stops growing and appears atrophied. The infected hairs (about 10% of the total root hairs) are clustered in areas that are in contact with degrading wood. The fungal hypha penetrates at the tip or near the tip of the hair (Prat 1926, Fig. 12). Prat (1926) suggested that the host has an influence on the hypha because the fungus has a different topography when inside the root; it develops outgrowths that do not exist in the extramatrical phase. The internalized fungi grow in such a way that they are appressed against the interior wall of the root hair (Fig. 12).

The hypha occasionally branches before reaching the base of the hair and continuing to grow towards the center of the root. Once it reaches the inner periclinal wall of the basal epidermal cell, it swells and forms an appressorium-like structure that helps the fungal hypha cross this wall (Prat 1926). The fungus eventually forms a *Paris*-type association with large coils in the outer cortical cells.

To test whether the same observations could be made on a different member of the family, one of us (F.C.G.) examined a Canadian species, *Taxus canadensis*. Numerous entries were observed through root hairs with a thickening of the mycelium at the tip of the hair, just as Prat depicted 70 years ago (compare Fig. 12 with Fig. 13). Branching within the root hair was also seen (Fig. 14). Prat (1926) concluded her work on the Taxaceae by saying that the mechanism she had described could be applied to all members of this family found in France. Her findings are thus extended to at least one member of the Canadian yews.

3.3.2 Dicotyledons

Jones (1924) studied the mycorrhizal associations found in roots of legumes. Although he mentioned fungal penetration, he did not elaborate on any mechanisms of entry. He made the distinction in response between annual, biennial and perennial legumes. In annuals, such as peas and sweet peas, the fungus colonizes the small lateral roots, which in general do not exhibit secondary thickening. In biennials and perennials, such as clover and alfalfa, considerable secondary growth occurs, and although no mycorrhizal fungi are observed in the large lateral roots, the small and young laterals are readily colonized. Jones (1924) correlated the size of the cortex to the degree of occupancy of the mycorrhizal fungus: whereas all the roots of the aforementioned species have thick cortices and exhibit extensive fungal colonization, those of beans and soybeans with their relatively smaller cortical regions exhibit reduced occupancy. Jones (1924) and Prat (1926) suggested that root characteristics such as thickness, tissue type, or presence of intercellular spaces have an effect on the extent of fungal colonization.

Root hair entry has been recently re-examined for two papilionoid legumes: pea and sweetclover (*Melilotus alba* Desr.). We have often observed fungal hyphae entering root hairs of *Pisum sativum* (F.C. Guinel unpublished results). The entry is not necessarily at the tip of the hair (Fig. 8), and no true appressorium is formed. Often the hypha seems to sense the root hair; it responds to contact by wrapping around the hair in a twisted coil, and grows along its surface for a while

Table 1. Occurrence of root hair entry and types of arbuscular mycorrhizal fungi in various plants

Order	Family	Subfamily (Tribe)	Species	Mycor. type	Root Hair Entry	Reference
Taxales	Taxaceae		*Taxus baccata*	Paris	+++[a]	Prat 1926
			Taxus canadensis	Paris	+++	F. Guinel unpublished results
Fabales	Leguminosae	Vicieae	*Pisum sativum*	Arum	++[b]	F. Guinel unpublished results
		Trifolieae	*Melilotus alba*	Arum	±[c]	Hirsch and Kapulnik 1998
		Trifolieae	*Medicago truncatula*	Arum	nd[d]	Harrison and Dixon 1993
		Phaseoleae	*Phaseolus vulgaris*	Arum	nd	Blee and Anderson 1996, Holley and Peterson 1979
Apiales	Araliaceae		*Panax quinquefolius*	Paris	++	Whitbread et al. 1995
Rosales	Rosaceae	Potentilleae	*Fragaria sp.*	Arum	nd	Hildebrand and Koch 1936
Solanales	Solanaceae		*Solanum tuberosum*	nd	nd	McArthur and Knowles 1992
			Nicotiana tabacum	Paris	nd	Hildebrand and Koch 1936
Gentianales	Gentianaceae		*Gentiana lutea*	Paris	nd	Jacquelinet-Jeanmougin and Gianinazzi-Pearson 1983
	Loganiaceae		*Gelsenium rankinii*	Paris	nd	Tiemann et al. 1994
	Apocynaceae		*Nerium*	Both	nd	Weber and Krämer 1994, Weber et al. 1995
Poales	Gramineae	Andropogoneae	*Zea* sp.	Arum	-[e]	Winter 1951
		Aveneae	*Avena* sp.	nd	+++	Winter 1951
		Triticeae	*Triticum* sp.	nd	±	Winter 1951
			Hordeum sp.	nd	±	Winter 1951
			Secale sp.	nd	±	Winter 1951
		Poeae	*Festuca rubra*	Paris?	+++	Nicolson 1959
Dioscorales	Trilliaceae		*Medeola virginiana*	Paris	-	Widden 1996
Asparagales	Alliaceae		*Allium porrum*	Arum	nd	Brundrett et al. 1985, Garriock et al. 1989
Liliales	Uvulariaceae		*Clintonia borealis*	Paris	++	Widden 1996
Liliales[f]			*Lilium formologo*	nd	+++	Leu and Chang 1995
	Amaryllidaceae		*Narcissus* sp.	Paris?	±	Chilvers and Daft 1981

[a]Main entry is through root hairs; [b]root hair entry occurs often; [c]occasional root hair entry; [d]not determined or not reported; [e]root hair entry has never been observed; [f]the authors did not give details about the species studied

before entering (Fig. 9). The topography of the internal hypha is not uniform and protuberances can be seen (Fig. 10). After having reached the base of the root hair, the hypha often branches and penetrates the cortical cells (Fig. 11). In sweetclover, root hair entry is more difficult to detect, suggesting that it occurs infrequently. We have not determined the culture or inoculum conditions that promote this type of entry. Preliminary results suggest that it is more common in roots that have copious root hairs. We are uncertain as to whether the fungal hypha that penetrates the hair is derived from a spore, from extraradical hyphae, or even from a branch off the appressorium.

Panax quinquefolius (ginseng), a member of the family Araliaceae, has a mycorrhizal association whereby the hypha enters through the root hair without the obvious assistance of an appressorium (Whitbread et al. 1995). The authors noted that the hyphae using the hair as a pathway were generally very straight. This could be explained by the constraint applied by the shape of the hair onto the hypha. As observed from their micrographs, all entries are made on the side of the hair, not at its tip (Whitbread et al. 1995). Although this mode of entry differs from what happens in the English yew (Prat 1926), it is similar to what is observed in pea.

3.3.3 Monocotyledons

Winter (1951) studied the importance of mycorrhizae in cultivated grasses: oat, rye, barley, wheat and corn, and made the astounding observation that in 99% of the cases, oat plants were colonized via their root hairs (Fig. 15) whereas corn was never colonized in this way. Wheat, barley and rye demonstrated intermediate behavior. In these species, entry usually takes place through the epidermis with occasional penetration through root hairs (Winter 1951). On the average, 2.2% of the total number of root hairs are infected in oat. Although all are Gramineae, these plants belong to different subfamilies. Corn (*Zea*) belongs to the subfamily Panicoideae, and oat (*Avena*), wheat (*Triticum*), rye (*Secale*) and barley (*Hordeum*) belong to the subfamily Pooideae. Within that subfamily, another distinction can be made between oat (tribe Aveneae) and the three other species (tribe Triticeae). Corn roots that had been infected by *Glomus aggregatum* were recently re-examined (F.C. Guinel unpublished results); the fungus was found to be very tightly appressed to the root and root hairs (Fig. 16), with the hypha wrapped around the length of the hair. No penetration was observed, however (Fig. 16). Thus, the process of root hair entry appears to be restricted to certain subfamilies.

Nicolson (1959), in his studies of sand-dune grasses, reported that entry is either through epidermal cells with an appressorium-like structure or via a root hair without the help of such a structure. When entry occurs through a root hair, the hypha was often broken off at the site of penetration. Nicolson (1959) found that pieces of root (4.5 mm long) of *Festuca rubra* var. *arenaria* contained up to 92% infected root hairs. That number is much higher than that associated with entry through epidermal cells.

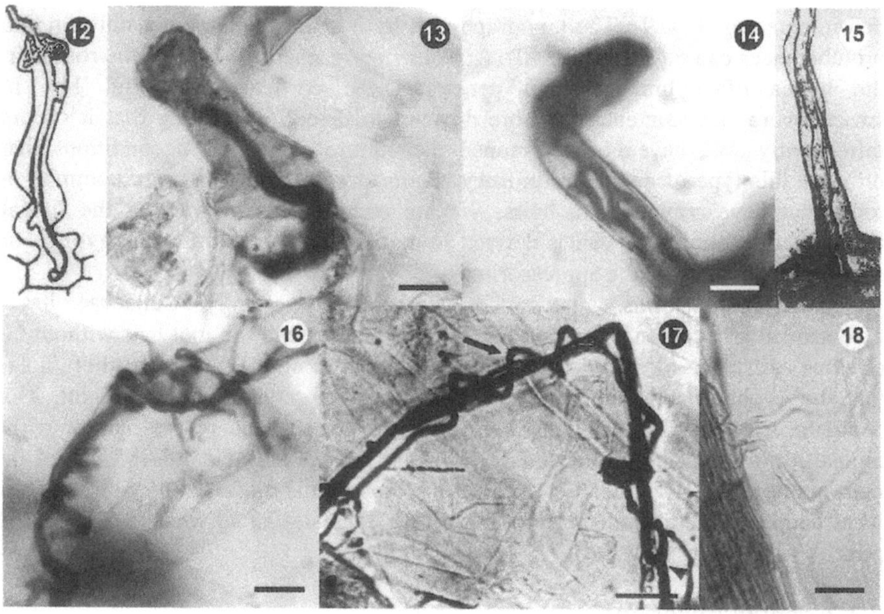

Fig. 12. Drawing of *Taxus baccata* infection by Prat (1926). The original legend was as follows: "Fungal penetration through absorbing root hair." Note outgrowths from main hyphae abutting the internal surface of the root hair wall. Note also the fungal cap at the tip of the hair. **Figs. 13-14.** Fungal hypha entering hairs of *Taxus canadensis* roots. *Fig. 13.* The fungus had grown downwards to the base of the root hair cell. Note the structure similar to the fungal cap seen in Fig. 12. Bar 140 μm. *Fig. 14.* The fungal hypha branches within the root hair, which it had infected at its tip. Bar 142 μm. **Fig. 15.** An oat root hair infected by a AM fungus. x500. Reprinted with permission from Winter (1951). **Fig. 16.** *Glomus aggregatum* hypha contacted a root hair of corn. The fungus had sensed the presence of the root hair and grew around it in a tight coil. This type of response was occasionally observed, but entry through the root hair was not observed. Bar 140 μm. **Fig. 17.** Entry of a root hair of *Lilium* sp. by *Gigaspora gigantea.* Note the coil (*arrow*) made by the fungal hypha around the root hair. Entry (*arrowhead*) does not appear to have occurred at the tip of the hair, but the hypha has progressed to the root hair base. Bar 80 μm. Reprinted with permission from Leu and Chang (1995). **Fig. 18.** Root surface of a non-mycotrophic *Carex* (*C. atherodes*) where the root hairs develop a characteristic bulbous-base. Bar 58 μm. Reprinted with permission from Miller at al. (1999)

Root hair penetration has also been observed in the order Liliales in more than one family. Widden (1996) surveyed the AM associations occurring among plants of deciduous forests. He focused on two liliaceous species, *Clintonia borealis* and *Medeola virginiana*. For the latter, which belongs to the order Dioscorales and to the family Trilliaceae, fungal penetration through root hairs was never observed. In contrast, in *Clintonia*, a member of the order Liliales and of the family Uvulariaceae, entry could be either through a root hair or through epidermal cells.

In another member of the Liliales, from the family Liliaceae, which was not identified by species, Leu and Chang (1995) noted that the penetration sites were mainly on root hairs. The hypha responds to the root hair contact in a manner similar to that on pea: the hypha coils around the length of the hair (Fig. 17). Once in the root hair, the hyphae branch and continue their growth towards the cortical cells. Leu and Chang (1995) even reported the formation of typical fungal-produced vesicles within the root hair.

3.4 Relationship Between Presence of Root Hairs and Absence/Presence of Mycorrhizae

3.4.1 Baylis' Hypothesis

According to Chilvers and Daft (1981), Pfeffer as early as 1877 had drawn a correlation between the frequency and length of root hairs and the ability of a plant to grow without mycorrhizal fungi. This interdependence whereby plant species with few root hairs are most consistently mycorrhizal was further supported by Metsävainio in 1931, who credited Höveler (1892) for making the correlation (cited in Baylis 1975). No matter who originated the hypothesis, it is still very much up-to-date. In the last 30 years, that correlation has been under great scrutiny.

Baylis (1970, 1975) proposed what is known today as "Baylis' hypothesis", suggesting that mycorrhizal fungi have a controlling influence on the evolution of roots. Studying the mycorrhizal colonization in woody Magnoliales, which are considered the most basal of the flowering plants, and comparing it to that found in more evolved plants, such as the grasses, Baylis noticed that magnolioid-like plants with their poor system of root hairs were strongly mycotrophic. In contrast, plants with a graminoid-like root system having long and thin root hairs became mycotrophic only when the availability of phosphorus was extremely low.

3.4.2 Dicotyledons

Since 1975, several authors have returned to Baylis' hypothesis and undertaken studies to prove or disprove it. Focusing on six tropical representatives of the Magnoliales, St. John (1980) confirmed Baylis' findings that magnolioid roots without root hairs were more likely to be infected by mycorrhizal fungi than plants with more graminoid-like root systems. Manjunath and Habte (1991) studied the aforementioned relationship in two genera of legumes: *Leucaena*, which belongs to the Mimosoideae and forms indeterminate nodules, and *Sesbania*, which belongs to the Papilionoideae and forms determinate nodules. Manjunath and Habte (1991) chose these two genera because they responded differently to fungal presence. *Leucaena* is a highly mycorrhizal-dependent genus and *Sesbania* is marginally to moderately-mycorrhizal. Among the several root characteristics that differ between the two species, of interest to us is root hair incidence: *Sesbania* has more root hairs than *Leucaena* (Manjunath and Habte 1991). Using a stepwise regression model with mycorrhizal dependency as the dependent variable and several root characteristics as independent ones, the authors concluded that root dry weight, root hair length, root density, root diameter and root hair incidence were factors important in determining mycorrhizal dependency. These results support the hypothesis that plant species having coarse roots with few or no root hairs are more dependent on VAM fungi than those having fine roots with numerous root hairs (Manjunath and Habte 1991).

Schweiger et al. (1995) tested the idea that root hairs and external AM hyphae might be alternative ways of getting phosphorus from beyond the root depletion zone. They studied five pasture species with distinct root hair length (four legumes with root hairs ranging in length from 125 μm to 500 μm and one grass with root hairs that were ~1 mm). The authors looked at the beneficial effect (i.e., the relative effectiveness of getting phosphorus) of being mycorrhizal, and found that their results were in good agreement with those of Baylis (1975). In other words, the shorter the root hairs of a species, the higher was the beneficial effect of being infected by *Glomus* sp. (Schweiger et al. 1995). Root characteristics such as root diameter, root length per plant or root/shoot ratio were not found as relevant as root hair length in the mycorrhizal equation.

3.4.3 Monocotyledons

Instead of considering species with different root hair characteristics, Baon et al. (1994) studied two cultivars of rye (*Secale cereale*), one possessing long hairs with diameters similar to those of hyphae, the other with shorter stubby hairs. However, the results were not clear-cut. Although the plants with long root hairs were less responsive to fungal infection, they had a much better P uptake than those with short hairs; this latter finding was true in the absence or presence of the fungus. This result could be explained by the fact that the long hairs, similar in morphology to fungal hyphae, are able to explore soil pores unreachable to short hairs. In the case of rye, the morphology of the root and the length of root hairs do not have an important effect on mycorrhizal response (Baon et al. 1994). The authors proposed that the degree of mycorrhizal infection could be under genetic control. Several studies support this proposal (see for example, Estaún et al. 1987; Mercy et al. 1990; Baon et al. 1993; Smith and Smith 1997).

Carex spp. are sand-dune sedges which, for a long time, were thought to be non-mycorrhizal. Recently, this view has been re-considered; the extent of mycotrophy in that species is indeed much greater than previously thought (Miller et al. 1999). These authors show that two factors, one environmental and one anatomical, influence the mycorrhizal status of the sedges. The first one relates to dryness, which has been known for a long time to facilitate fungal colonization.

Wetlands and flooded areas are not places where mycorrhizal associations usually thrive. Of interest to us is the second factor, the anatomical one. Certain species of *Carex,* which are not mycorrhizal, possess hairs with a unique morphology (Miller et al. 1999). These hairs have a large bulbous base (Fig. 18). The authors postulated that 1) such hairs have allowed the plants to live free of a mycorrhizal association, and 2) that these hairs could be the result of an adaptation to non-mycorrhizal growth. An evolutionary and phylogenetic study is required to support this hypothesis.

Chilvers and Daft (1981) studied three different genera of Liliales, some, such as *Narcissus*, with numerous root hairs and others, such as *Tulipa*, without any. Whereas *Narcissus* is a species that supports Baylis' hypothesis in that some cultivars with fewer root hairs are better mycotrophs than others with more abundant hairs, *Tulipa* is not. Numerous appressoria were present in the root surface of the tulip; however, only a few fungal hyphae were successful at penetrating the epidermis. Once inside, the fungus was slow-growing and the infection zone was restrained (Chilvers and Daft 1981).

4 Conclusions

4.1 A Species-Specific Characteristic?

As this review points out, there is increasing evidence that root hair entry by fungal hyphae appears to be, if not species-specific, then at least family-specific or order-specific. Moreover, root hair entry is probably not restricted to the families mentioned herein. It is practically certain that as more samples are studied, the list will become more complete. More details, as new (and old) mycorrhizae are described, are needed if we want to build upon this concept of a family-specific relationship. Researchers should not limit their research to the physiological aspects of the mycorrhizal symbiosis. Studies are still needed on the anatomy and morphology of the mycorrhizae, with special emphasis on the study at the ultrastructural level of how the fungus penetrates the epidermal surface of the root. Smith and Smith (1997) at the end of their review gave a checklist of items that they thought would be useful for standardizing mycorrhizae description. We propose that the frequency of root hairs infected by endomycorrhizal fungi be added to that list.

Although root hair entry is not common, the fact that it occurs most of the time in *Taxus baccata* (Prat 1926), in oat (Winter 1951), and in *Festuca rubra* var. *arenaria* (Nicolson 1959) leads us to conclude that this is not a chance type of entry. In these genera, fungal hyphae normally enter the root by penetrating root hairs. Moreover, the mechanism used by the fungus to cross the root hair cell wall appears different from that used to penetrate the outer periclinal wall of the epidermis. In the latter case, appressoria are formed whereas these are very rare, and maybe even nonexistent, on root hairs.

The host plant has been known to influence the type of mycorrhizae formed within the roots, i.e., the same fungus can develop into either *Arum*- or *Paris*- type depending on the species of the root it infects (Barrett 1958 cited in Smith and

Smith 1997). The entry of the fungal hyphae through root hairs does not seem restricted to either one of the two types (see Table 1). Thus, in *Taxus*, an extraradical mycelium can branch to produce fungal hyphae that enter through root hairs. Once in the base of the root hair cell, the hypha extends intracellularly and forms a *Paris*-type of mycorrhiza. In *Avena*, a fungal hypha utilizes the same type of entry and grows intracellularly through a few host cells. However, the hyphae become intercellular until they differentiate into arbuscules, in so doing, forming an *Arum*-type of mycorrhiza.

4.2 A Link Between Fungal Root Hair Entry, Mycotrophy and Plant Evolution?

The Magnoliales, the most basal of the flowering plants, have very few root hairs, but they are highly mycotrophic (Baylis 1975). It has been proposed that the fungal hyphae in species with no or few root hairs, or in species with short root hairs, substitute for these structures. Both are of a thin diameter and can grow in places that roots of a thicker diameter cannot reach (Schweiger et al. 1995). In less basal plant families, where root hairs are common, roots become mycotrophic only when phosphorous in the rhizosphere is in very low concentrations. The graminoid-like roots have long and thin root hairs, the presence of which is thought to reduce the mycorrhizal dependency of the plants (Baylis 1975). The families reported in the literature to possess fungal root hair entry for the most part have graminoid-like roots. These families are not closely related phylogenetically, suggesting that root hair entry evolved multiple times. Even a gymnosperm has evolved the root hair entry mode of colonization. *Taxus*, although a conifer, lies in a clade distinct from that containing pine, spruce, and larch based on 18S rRNA DNA (Chaw et al. 1997) and *rbc*L sequence analysis (http://www.herbaria.harvard.edu/~rice/treezilla/; Rice et al. 1997).

Other families or orders have evolved differently in response to fungal associations. One example would be that of the Cyperales, in the Commelinidae. In that order, some species are able to form mycorrhizae whereas others are not. In the latter, root hairs of some species actually acquire a different morphology (Miller et al., 1999), which is thought to be an adaptation to the non-mycorrhizal status of the plant.

4.3 Is the Root Hair Mode of Entry for Mycorrhizal Fungi an Old Trait or a Newly-Acquired One?

Although our focus has been primarily on mycorrhizae in angiosperms and gymnosperms, this important symbiosis is also found in bryophytes and lower tracheophytes. AM-associations are particularly ancient, having been found in fossils of *Aglaophyton major* dating from the Devonian (Remy et al. 1994). Plants, such as *Aglaophyton*, which were inhabited by AM fungi so many millions of years ago, had features of both bryophytes and lower vascular plants (Edwards 1986).

Unfortunately, the fossil record tells us nothing about how these fungi colonized the plant. For this, we have to look at present-day representatives. Roots are lacking in bryophytes and in the gametophytes of the lower tracheophytes, so the mode of entry for fungal hyphae assumedly must differ from the typical mycorrhizal pattern. Can we extrapolate patterns of evolution from a study of modern representatives?

Gallaud (1905) was probably the first to report that bryophytes were colonized by endomycorrhizal fungi. In his review, he mentioned that infection in *Hepatica* is through rhizoids and that the fungus remains always intracellular. Duckett et al. (1991) observed that a large number of British leafy liverworts are colonized by ascomycetes that establish mycorrhizal associations. The majority of these hepatics belong to the suborders Lepidoziineae and Cephaloziineae, two suborders of the 14 suborders in Jungermanniales. Rhizoids or fascicles of rhizoids were found to be infected by fungal hyphae. In some cases, the rhizoid apex became swollen before infection, whereas in others, the fungus penetrated the basal walls of rhizoids.

Duckett and Read (1995) resynthesized the interaction between the ericoid mycorrhizal fungus *Hymenoscyphus ericae* and rhizoids of a number of leafy liverworts to satisfy Koch's postulates. After inoculation, the fungal hyphae initially were seen to grow around the rhizoid surface, in close association with pockets of mucilage. The rhizoid tip became swollen and the hypha penetrated near the rhizoid apex. The fungal hyphae were tightly coiled within the bulbous tip but remained straight in the rest of the hair. This same fungus was also observed to colonize the hair roots of various ericaceous plants (Duckett and Read 1995). Hair roots are fine roots consisting of very few cell layers; the fungus completely colonizes the epidermal cell layer (Smith and Read 1997).

These same fungi, however, were incapable of infecting members of the Metzgeriales and Marchantiales, which if colonized, are normally associated with zygomycetes or basidiomycetes. Ligrone and Lopes (1989) determined that the thalloid liverwort *Conocephalum conicum* (Marchantiales) is colonized by an aseptate, arbuscular fungus that penetrated the smooth-walled versus tuberculate rhizoids that are characteristic of this group. Appressorium-like structures were rarely observed along the surface of the rhizoid.

Duckett et al. (1991) suggested for the bryophytes that the diverse mycotrophic relationships represent independent, and most likely, recent evolutionary events in the various genera. Based on a phylogenetic analysis using mitochondrial 19S rDNA sequences, the liverworts are monophyletic (Duff and Nickrent 1999). This suggests that the difference in mycorrhization patterns as well as in mycobionts occurred after the divergence of the liverwort orders. Because root hairs and rhizoids are analogous structures, this would suggest that the root hair/rhizoid mode of entry by mycorrhizal fungi is a newly acquired trait for the gymnosperms and angiosperms as well as for bryophytes.

A microscopic study of the gametophytes of *Lycopodium clavatum* demonstrated that an endophytic fungus of unknown identity formed hyphal coils and small vesicles in the cortical cells, but no arbuscules (Schmid and Oberwinkler 1993). Several other differences from the typical AM association were observed, leading the authors propose the term "Lycopodioid mycothallus interaction" to describe the relationship between the fungal endophyte and its host. As for bryophytes, fungal hyphae entered the host tissue through rhizoids that were swollen at their tips. The fungal hyphae also penetrated through or between degenerated epidermal cells (Schmid and Oberwinkler 1993).

4.4 What is the Advantage to the Plant or Fungus for Root Hair Entry?

Root hairs could represent a facilitating path for the growth of the fungal hypha. Specific exudates or specific receptors, absent from the epidermal surface, are likely to be present on the surface of the root hairs. The components of the cell wall of a still emerging root hair are also likely to be easier to dissolve than those of a differentiated epidermal cell. Prat (1926) proposed that for *Taxus*, entry through root hairs was possible because these cells have primary cell walls. In most cases, the fungal hypha penetrates the top third portion of the hair. Root hairs are known to have a strong polarity in their cytoplasmic arrangement (Miller et al. 1997), and this arrangement differs depending on the age of the hair. While the root hair is growing, a vesicle-rich region is present at its very tip, followed by an area with dense cytoplasm. In a mature root hair, a central vacuole that pushes the cytoplasm against the plasma membrane fills the tip (Miller et al. 1997). In *Taxus*, the fungus can infect both growing and mature root hairs. However, in the former, the root hair stops its growth upon infection (Prat 1926). Thus, the age of the hair and its cytological state could be important.

This type of root hair entry is analogous to the manner by which rhizobia enter legume root hairs. In legumes, root hairs respond to the presence of rhizobia by forming shepherd's crooks. Within these tight curls, the bacteria become encapsulated within the infection thread, which is a continuation of the root hair wall. When the rhizobia are released from the infection thread into recently divided nodule cells, they are separated from the host cell cytoplasm by the root hair plasma membrane. A similar situation exists in the fungal-root symbiosis except that the hypha, which enters the root hair and later differentiates into arbuscules within a root cortical cell, retains its cell wall. Thus, the fungal cell wall is included in the barrier between the root hair cytoplasm and the fungal plasma membrane. A detailed study of the interface formed as fungal hyphae enter root hairs is needed.

An additional reason for entry through root hairs could be the greater interface for increasing (signal or nutrient) exchange. Signal exchange may be limited in cases where the zone of colonization is restricted to a very small region of the root. An increase in surface area brought about by the differentiation of root hairs would allow an increase in signal exchange. As a counterpart to this proposal, roots that do not exhibit the root hair mode of entry should have a greater expanse of root surface devoid of root hairs that is available for fungal entry. However, to our knowledge, these types of correlations have not been made.

Bi-directional transfer of nutrients is thought to occur not only across the periarbuscular membrane, but also at the level of the intercellular hyphae and intracellular coils (Smith and Smith 1997). Does nutrient exchange take place as soon as the fungus has entered the root? Jones (1924) proposed a relationship between thick root cortex and extensive fungal colonization. In roots with a small cortex, exchange through the interface within the root hair would be beneficial because this would increase the surface area shared by the two symbionts, thereby improving the efficiency of the symbiosis. Although no observations have been made to support this hypothesis, it would be interesting to see if a correlation exists between cortex size and root hair entry, such that root hair entry occurs more frequently in plants having a smaller root cortex.

4.5 Prospects

We want to emphasize the need for a fuller description of fungal entry resulting in successful mycorrhizal associations to fill in gaps that exist. With a renewal of interest in microscopy brought about by the advent of confocal microscopy, as well as the use of molecular tools such as in-situ hybridization and immunogold localization, the few species mentioned here in which root hair entry is the norm could be used as models. They would be useful for the study of the mechanism(s) used by the fungus to penetrate the wall, for understanding the role(s) played by the cytoskeleton in that process, and for the search for root hair receptors able to recognize fungal signals.

Plant mutants with abnormalities in root hair development (either no root hairs or aberrant growth) could be useful too. A study of mycorrhizal formation in the wild-type and in the root hair mutants could help us understand the significance of root hair entry.

Nodulation mutants are also useful for study. Duc et al. (1989), who obtained Myc$^-$ mutants of pea and faba bean, were the first to show that the ability to form arbuscular-mycorrhizal associations and nodulate in response to rhizobia was genetically linked. However, to our knowledge, the non-nodulating Myc$^-$ mutants that have been described do not show any obvious defects in root hair development that would prevent the onset of the early stages of either nodulation or mycorrhizae. Nevertheless, the absence of a specific receptor in these Nod$^-$ Myc$^-$ mutants may be sufficient to eliminate root hair entry of either symbiont.

It would also be of interest to determine whether or not there are bryophyte model systems for studying root hair entry by mycorrhizal fungi. The haploid nature of the bryophyte genome makes it simpler to select for mutants that are incapable of establishing the symbiotic relationship.

Last but not least, are the parallels observed between the nitrogen-fixing and mycorrhizal symbioses significant or are they merely illusions brought about by piecemeal data? Only time and more research will tell.

ACKNOWLEDGMENTS

We gratefully acknowledge R. Larry Peterson (University of Guelph) and members of the Hirsch laboratory for helpful comments on the manuscript. We are also grateful to Michael Frohlich (University of Michigan) for insights on gymnosperm evolution. Special thanks to Ryan Geil (Wilfrid Laurier University, WLU) and Margaret Kowalczyk (UCLA) for their help with the illustrations. We thank Katharina Markmann (WLU) for taking some of the photographs and for her translation of Winter's article. We also acknowledge Joseph Chang (UCLA) who translated Leu and Chang's article. Additional thanks go out to Dorota Karlowska and Ed Ayala (UCLA) for their help with the library work.

Research on the molecular biology of mycorrhizal formation in A.M. Hirsch's lab was funded by National Science Foundation grants 96-30842 and 97-23982. F.C. Guinel's research was supported by a NSERC operating grant. Additional funding was provided by WLU.

References

Baon JB, Smith SS, Alston AM (1993) Mycorrhizal responses of barley cultivars differing in P efficiency. Plant Soil 157:97-105

Baon JB, Smith SS, Alston AM (1994) Growth response and phosphorus uptake of rye with long and short root hairs: Interactions with mycorrhizal infection. Plant Soil 167:247-254

Baylis GTS (1970) Root hairs and phycomycetous mycorrhizas in phosphorus-deficient soil. Plant Soil 33:713-716

Baylis GTS (1975) The magnolioid mycorrhiza and mycotrophy in root systems derived from it. In: Sanders FE, Mosse B, Tinker PB (Eds) Endomycorrhizae. Academic Press, New York London, pp 373-389

Béguiristain T, Lapeyrie F (1997) Host plant stimulates hypaphorine accumulation in *Pisolithus tinctorius* hyphae during ectomycorrhizal infection while excreted hypaphorine controls root hair development. New Phytol 136:525-532

Blee KA, Anderson AJ (1996) Defense-related transcript accumulation in *Phaseolus vulgaris* L. colonized by the arbuscular mycorrhizal fungus *Glomus intraradices* Schenk & Smith. Plant Physiol 110:675-688

Bonfante P, Perotto S (1995) Tansley Review No. 82. Strategies of arbuscular mycorrhizal fungi when infecting host plants. New Phytol 130:3-21

Bonfante-Fasolo P (1984) Anatomy and morphology of VA mycorrhizae. In: Powell CL, Bagyaraj DJ. (Eds) VA mycorrhiza; CRC Press Inc, Boca Raton pp 5-33

Bonfante-Fasolo P (1988) The role of the cell wall as a signal in mycorrhizal associations. NATO ISI Ser H17 pp. 219-235

Brundrett MC, Piché Y, Peterson RL (1984) A new method for observing the morphology of vesicular-arbuscular mycorrhizae. Can J Bot 62:2128-2134

Brundrett MC, Piché Y, Peterson RL (1985) A developmental study of the early stages in vesicular-arbuscular mycorrhiza formation. Can J Bot 63:184-194

Chaw, S-M, Zharkikh, A, Sung, H-M, Lau T-C, and Li W-H (1997) Molecular phylogeny of extent gymnosperms and seed plant evolution: analysis of nuclear 18S rRNA sequences. Mol Biol Evol 14:56-68

Chilvers MT, Daft MFJ (1981) Mycorrhizas of the Liliiflorae. II. Mycorrhiza formation and incidence of root hairs in field grown *Narcissus* L., *Tulipa* L., and *Crocus* L. cultivars. New Phytol 89:247-261

Duc G, Trouvelot A, Gianinazzi-Pearson V, Gianinazzi S (1989) First report of non-mycorrhizal plant mutants (Myc$^-$) obtained in pea (*Pisum sativum* L.) and fava bean (*Vicia faba* L.). Plant Sci 60:215–222

Duckett JG, Read DJ (1995) Ericoid mycorrhizas and rhizoid-ascomycete associations in liverworts share the same mycobiont: isolation of the partners and resynthesis of the associations in vitro. New Phytol 129:429-447

Duckett JG, Renzaglia KS, Pell K (1991) A light and electron microscope study of rhizoid-ascomycete associations and flagelliform axes in British hepatics with observations on the effects of the fungi on host morphology. New Phytol 118:233-258

Duff RJ, Nickrent DL (1999) Phylogenetic relationships of land plants using mitochondrial small-subunit rDNA sequences. Amer J Bot 86:372-386

Edwards, DS (1986) *Aglaophyton major*, a non-vascular land-plant from the Devonian Rhynie chert. Bot J Linn Soc 93:173-204

Estaún V, Calvet C, Hayman DS (1987) Influence of plant genotype on mycorrhizal infection: Response of three pea cultivars. Plant Soil 103:295-298

Fries N, Serck-Hannssen K, Häll Dimberg L, Theander O (1987) Abietic acid an activator of basidiospore germination in ectomycorrhizal species of the genus *Suillus* (Boletaceae). Exp Mycol 11:360-363

Gallaud I (1905) Etudes sur les mycorrhizes endotrophes, chapters 1 & 5, Revue Gén Bot 17:5-48 & 479-500

Garbaye J, Bowen GD (1989) Stimulation of ectomycorrhizal infection of *Pinus radiata* by some microorganisms associated with the mantle of ectomycorrhizas. New Phytol 112:383-388.

Garriock ML, Peterson RL, Ackerley CA (1989) Early stages in colonization of *Allium porrum* (leek) roots by the vesicular-arbuscular mycorrhizal fungus, *Glomus versiforme*. New Phytol 112:85-92.

Giollant M, Guillot J, Damez M, Dusser M, Didier P, Didier E (1993) Characterization of a lectin from *Lactarius deterrimus*. Research on the possible involvement of the fungal lectin in recognition between mushroom and spruce during the early stages of mycorrhizae formation. Plant Physiol 101:513-522

Gogala N (1991) Regulation of mycorrhizal infection by hormonal factors produced by hosts and fungi. Experentia 47:331-340

Harley JL (1969) The biology of mycorrhiza. Second edition. Plant Science Monograph, Polunin N (Ed) Leonard Hill Books Publishers, London. pp 242-269.

Harrison MJ, Dixon RA (1993) Isoflavonoid accumulation and expression of defense gene transcripts during the establishment of vesicular-arbuscular mycorrhizal associations in roots of *Medicago truncatula*. Mol-Plant Microbe Interact 6:643-654

Hebe G, Hager A, Salzer P (1999) Initial signalling processes induced by elicitors of ectomycorrhiza-forming fungi in spruce cells can also be triggered by G-protein-activating mastoparan and protein phosphatase-inhibiting cantharidin. Planta 207:418-425

Hildebrand AA, Koch LW (1936) A microscopical study of infection of the roots of strawberry and tobacco seedlings by micro-organisms of the soil. Can J. Res 14:11-25

Hirsch AM (1999) Role of lectins (and rhizobial exopolysaccharides) in legume nodulation. Curr Opin Plant Biol 2:320-326

Hirsch AM, Kapulnik Y (1998) Signal transduction pathways in mycorrhizal associations: Comparisons with the *Rhizobium*-legume symbiosis. Fungal Genet Biol 23:205-212

Holley JD, Peterson RL (1979) Development of a vesicular-arbuscular mycorrhiza in bean roots. Can J Bot 57:1960-1978

Horan DP, Chilvers GA (1990) Chemotropism: the key to ectomycorrhizal formation? New Phytol 116:297-301

Jacquelinet-Jeanmougin S, Gianinazzi-Pearson V (1983) Endomycorrhizas in the Gentianaceae. I. The fungi associated with *Gentiana lutea* L. New Phytol 95:663-666

Jacobs PF, Peterson RL, Massicotte HB (1989) Altered fungal morphogenesis during early stages of ectomycorrhiza formation in *Eucalyptus pilularis*. Scanning Microsc 3:249-255

Jones FR (1924) A mycorrhizal fungus in the roots of legumes and some other plants. J Agric Res 24:459-470

Lapeyrie F, Lei J, Malajczuk N, Dexheimer J (1989) Ultrastructural and biochemical changes at the pre-infection stage of mycorrhizal formation by two isolates of *Pisolithus tinctorius*. Ann Sci For 46s:754s-757s

Lei J, Wong KKY, Piché Y (1991) Extracellular concanavalin A-binding sites during early interactions between *Pinus banksiana* and two closely related genotypes of the ectomycorrhizal basidiomycete *Laccaria bicolor*. Mycol Res 95:357-363

Leu SW, Chang DCN (1995) Mycorrhizal infection and morphology of lily inoculated with vesicular-arbuscular mycorrhizal fungi, *Gigaspora gigantea* (Nic. & Gerd.) Gerd. & Trappe or *Glomus etunicatum* Becker & Gerdemann. Res Report Agric Inst Univ Taiwan 35:285-293 (In Chinese)

Ligrone R, Lopes C (1989) Cytology and development of a mycorrhiza-like infection in the gametophyte of *Conocephalum conicum* (L.) Dum. (Marchantiales, Hepatophyta). New Phytol 111:423-434.

Malajczuk N, Molina R, Trappe JM (1982) Ectomycorrhiza formation in *Eucalyptus*. I. Pure culture synthesis, host specificity and mycorrhizal compatibility with *Pinus radiata*. New Phytol 91:467-482

Manjunath A, Habte M (1991) Root morphological characteristics of host species having distinct mycorrhizal dependency. Can J Bot 69:671-676

Martin F, Tagu D (1995) Ectomycorrhiza development: a molecular perspective. In: Varma A, Hock B (Eds.) Mycorrhiza. Springer Verlag Berlin Heidelberg. pp. 29-58.

Massicotte HB, Melville LH, Peterson RL (1987) Scanning electron microscopy of ectomycorrhizae: potential and limitations. Scanning Microsc 1:1439-1454.

Massicotte HB, Peterson RL, Melville LH (1989) Ontogeny of *Alnus rubra-Alpova diplophloeus* ectomycorrhizae. I. Light and scanning electron microscopy. Can J Bot 67:191-200

Massicotte HB, Peterson RL, Ackerley CA, Melville LH (1990) Structure and ontogeny of *Betula alleghaniensis-Pisolithus tinctorius* ectomycorrhizae. Can J Bot 68:579-593

Mauseth JD (1995) Botany—An Introduction to Plant Biology. Second edition. Saunders College Publishing, Philadelphia

McArthur DAJ, Knowles NR (1992) Resistance responses of potato to vesicular-arbuscular mycorrhizal fungi under varying abiotic phosphorus levels. Plant Physiol 100:341-351

Mercy MA, Shivashankar G, Bagyaraj DJ (1990) Mycorrhizal colonization in cowpea is host dependent and heritable. Plant Soil 121:292-294

Miller DD, de Ruijter NCA, Emons AMC (1997) From signal to form: aspects of the cytoskeleton-plasma membrane-cell wall continuum in root hair tips. J Exp Bot 48:1881-1896

Miller RM, Smith CI, Jastrow JD, Bever JD (1999) Mycorrhizal status of the genus *Carex* (Cyperaceae). Amer J Bot 86:547-553

Nehls U, Béguiristain T, Ditengou F, Lapeyrie F, Martin F (1998) The expression of a symbiosis-regulated gene in eucalypt roots is regulated by auxins and hypaphorine, the tryptophan betaine of the ectomycorrhizal basidiomycete *Pisolithus tinctorius*. Planta 207:296-302

Nicolson TH (1959) Mycorrhiza in the Gramineae. I. Vesicular-arbuscular endophytes, with special reference to the external phase. Trans Brit Mycol Soc 42:421-438

Nylund JE, Unestam T (1982) Structure and physiology of ectomycorrhiza. I. The process of mycorrhiza formation in Norway spruce in vitro. New Phytol 91:65-79

Peterson RL, Farquhar ML (1996) Root hairs: Specialized tubular cells extending root surfaces. Bot Rev 62:1-40

Piché Y, Peterson RL, Massicotte HB (1988) Host-fungus interactions in ectomycorrhizae. Cell to cell signals in plant, animal and microbial symbiosis. NATO ISI Ser H17:55-71

Pingret J-I, Journet E-P, Barker DG (1998) *Rhizobium* Nod factor signalling: Evidence for a G protein-mediated transduction mechanism. Plant Cell 10:659-671

Prat H (1926) Etude des mycorrhizes du "*Taxus baccata*". Ann Sci Nat (Bot Biol Vég) 8:141-163

Remy W, Taylor TN, Haas H, Kerp H (1994) Four hundred-million-year-old vesicular-arbuscular mycorrhizae. Proc Natl Acad Sci USA 91:11841-11843

Rice KA, Donoghue MJ, Olmstead RG (1997) Analyzing large data sets: *rbc*L 500 revisited. System Biol 46:554-563

Salzer P, Hebe G, Reigh A, Zitterell-Haid B, Stransky H, Gaschler K, Hager A (1996) Rapid reactions of spruce cells to elicitors released from the ectomycorrhizal fungus *Hebeloma crustuliniforme*, and inactivation of these elicitors by extracellular spruce cell enzymes. Planta 198:118-126

Schmid E, Oberwinkler F (1993) Mycorrhiza-like interaction between the achlorophyllous gametophyte of *Lycopodium clavatum* L. and its fungal endophyte studied by light and electron microscopy. New Phytol 124:69-81

Schweiger PF, Robson AD, Barrow NJ (1995) Root hair length determines beneficial effect of a *Glomus* species on shoot growth of some pasture species. New Phytol 131:247-254

Smith FA, Smith SE (1997) Tansley Review No. 96. Structural diversity in (vesicular)-arbuscular mycorrhizal symbioses. New Phytol 137:373-388

Smith SE, Read DJ (1997) Mycorrhizal symbiosis. Second edition. Academic Press, Harcourt Brace & Company, Publishers, San Diego

Sprent JI (1989) Tansley Review No. 15. Which steps are essential for the formation of functional legume nodules? New Phytol 111:129-153

St. John TV (1980) Root size, root hairs and mycorrhizal infection: A re-examination of Baylis' hypothesis with tropical trees. New Phytol 84:483-487

Tagu D, Martin D (1996) Molecular analysis of cell wall proteins expressed during the early steps of ectomycorrhiza development. New Phytol 133:73-85

Taiz L, Zeiger E (1998) Plant Physiology. Second edition. Sinauer Associates, Inc., Publishers, Sunderland, Massachusetts

Thomson J, Melville LH, Peterson RL (1989) Interaction between the ectomycorrhizal fungus *Pisolithus tinctorius* and root hairs of *Picea mariana* (Pinaceae). Amer J Bot 76:632-636

Tiemann C, Demuth K, Weber HC (1994) Zur VA-Mycorrhiza von *Gelsenium rankinii* and *G. sempervirens* (Loganaceae). Beiträge Biol Pflanz 68:311-321

Vierheilig H, Alt-Hug M, Engel-Streitwolf R, Mäder P, Wiemken A (1998) Studies on the attractional effect of root exudates on hyphal growth of an arbuscular mycorrhizal fungus in a soil compartment-membrane system. Plant Soil 203:137-144

Weber HC, Krämer M (1994) VA-Mycorrhiza bei Menyanthaceae. Beiträge Biol Pflanz 68: 351-362

Weber HC, Klahr A, Marron-Heimbuch M (1995) Anatomical structures of the VA mycorrhiza in the Apocynaceae (Gentianales). Bot Acta 108:525-534

Weiss M, Schmidt J, Neumann D, Wray V, Christ R, Strack D (1999) Phenylpropanoids in mycorrhizas of the Pinaceae. Planta 208:491-502

Wessels JG (1993) Tansley Review No. 45. Wall growth, protein excretion and morphogenesis in fungi. New Phytol 123:397-413

Whitbread F, McGonigle TP, Peterson RL (1995) Vesicular-arbuscular mycorrhizal associations of American ginseng (*Panax quinquefolius*) in commercial production. Can J Bot 74:1104-1112

Widden P (1996) The morphology of vesicular-arbuscular mycorrhizae in *Clintonia borealis* and *Medeola virginiana*. Can J Bot 74:679-685

Winter AG (1951) Untersuchungen über die Verbreitung und Bedeutung der Mykorrhiza bei kultivierten Gramineen und einigen anderen landwirtschaftlichen Nutzpflanzen. Phytopath Zeit 17:421-432

Chapter 18
Root Hair-*Frankia* Interactions in Actinorhizal Symbioses

T.V. Bhuvaneswari and Bjørn Solheim

Department of Biology, Faculty of Science, University of Tromsø, N-9037 Tromsø, Norway

1 Introduction

The term actinorhizae refers to the nitrogen-fixing nodular structures induced on roots of higher plants by the actinomycete *Frankia* (Tjepkema and Torrey 1979). Occurrence of actinorhizal nodules is reported for about 200 different plant species from 23 genera distributed over eight dicotyledonous families (Table 1; van Ghelue 1994). Except for the genus *Datisca,* all other actinorhizal hosts are woody shrubs and their geographical distribution is primarily temperate. Only some hosts of the families Casuarinaceae and Myricaceae are tropical. Actinorhizal plants possess great economic value as timber and fuel wood. They are also used extensively in land reclamation and as ornamental shrubs. Ecologically, actinorhizal plants are pioneer invaders of nitrogen poor sites (Baker and Schwintzer 1990).

The presence of actinorhizae was first reported for alder (*Alnus glutinosa)* in 1829 by Meyen and these were shown to provide nitrogen for plant growth by Nobbe and Hiltner in 1904 (see Quispel 1990). Morphologically and developmentally, actinorhizal nodules are modified lateral roots. Nodules are coralloid clusters with multiple lobes, which can be either well defined or densely packed (Berry and Sunell 1990). They possess a central vascular tissue, which is surrounded by a cortex containing cells infected with *Frankia.* In the infected cortical cells, *Frankia* can be found differentiated into three different cell types: septate hyphae, terminally born vesicles and multicellular sporangia with numerous immotile spores. Sporulation occurs only in nodules of some hosts. In all the actinorhizal genera except *Casuarina* and *Allocasuarina,* biological nitrogen fixation occurs in the differentiated vesicles. In these two latter hosts nitrogen fixation occurs in the absence of vesicles (Baker and Schwintzer 1990; Berry and Sunell 1990).

Key words. actinorhiza, deformation factors, *Frankia*, nodulation, nitrogen fixation, root hair, symbiosis, signal molecules

Table 1. List of actinorhizal plant genera (adapted from van Ghelue 1994)

Family	Genus	No. of nodulated species
Betulaceae	*Alnus*	42*
Casuarinaceae	*Allocasuarina*	58*
	Ceuthostoma	2#
	Gymnostoma	18*
Coriariaceae	*Coriaria*	16#
Datisacaceae	*Datisca*	2#
Elaeagnaceae	*Elaeagnus*	35#
	Hippophae	2#
	Shepherdia	2#
Myricaceae	*Comptonia*	1*
	Myrica	28*
Rhamnaceae	*Ceanothus*	31#
	Colletia	3#
	Discaria	5#
	Kentrothamnus	1#
	Retanilla	1#
	Talguenea	1#
	Trevoa	2#
Rosaceae	*Cercocarpus*	4#
	Chammaebatia	1#
	Cowania	1#
	Dryas	3#
	Purshia	2#

* Infection through root hairs; # Infection through intercellular penetration

Frankia is a filamentous actinomycetous Gram-variable soil bacterium and belongs to the family *Frankiaceae*. It is a facultative symbiont and is the only actinomycete genus that is known to form root nodules on higher plants. It was first cultured in vitro by Callaham et al. (1978). The three differentiated cell types seen in nodules also occur in cultures. Growth of *Frankia* is usually quite slow in static cultures. Exponential growth rates of some isolates can be prolonged by gradual addition of certain complex fatty acids or by culturing in a fermentor (Ringø et al. 1995). Presence of vesicles in cultures is restricted to nitrogen limiting conditions in the growth medium. *Frankia* cultures also fix atmospheric nitrogen under these conditions (Baker and Schwintzer 1990).

Frankia can infect its host plants in two different ways. In some hosts, it penetrates the root through a deformed root hair, while in others it gains entry between root epidermal cells. In the hosts where infection occurs via root hairs, cortical cells near the infected root hairs are induced to divide and these divisions result in the formation of a prenodular structure in the cortex. *Frankia* penetrates the cells in the prenodule and branch and ramifies within these cells (Miller and Baker 1986).

Formation of the prenodule is restricted to hosts that permit root hair infection. In hosts where *Frankia* entry is intercellular, the bacterium passes through the root cortex intercellularly by digestion of the middle lamella. Regardless of the mode of entry, one to several nodule lobe primordia are initiated in the root pericycle in close proximity (Carmona 1974) which eventually develop into mature nitrogen-fixing nodules

Some of the early infection events in actinorhizal symbioses, root hair interactions in particular, bear striking resemblance to similar events in the legume-*Rhizobium* symbioses (Pawlowski and Bisseling 1996). The latter being a better-studied model system we have chosen to refer to some of these events very briefly (for comparison) in this review. The readers are advised to consult the other chapters in this volume for detailed information on the root hair interactions in legume model systems.

2 Infection

2.1 Infectivity Groups and Evolution

Based on cross-inoculation tests performed, *Frankia* isolates are grouped into three host infectivity groups (Cournoyer et al. 1993).

1) Strains infective on *Alnus* species and *Myrica* species
2) Strains infective on *Casuarina* species and *Myrica* species
3) Strains infective only on members of Elaeagnaceae or strains infective on Elaeagnaceae and *Myrica*

Actinorhizal hosts, likewise fall into two groups based on the mode of *Frankia* entry. In hosts *Gymnostoma, Casuarina, Allocasuarina, Alnus, Myrica* and *Comptonia* symbiont entry is always via root hairs while in all other actinorhizal hosts investigated the entry is only through intercellular penetration (Table1; Swensen and Mullin 1997). With *Frankia* strains that can infect both host groups (majority of the strains in the infectivity group 3), the host determines the mode of entry. For example, *Frankia* strain HFPGpI1 infects the roots of *Gymnostoma papuanum* (Casuarinaceae) via root hairs, but its entry is intercellular in *Shepherdia argentea* (Elaeagnaceae) (Racette and Torrey 1989). Phylogenetic relationships between *Frankia* strains based on 16S rRNA gene sequences and nitrogenase genes *nif HD* sequences match the mode of host-plant infection. Strains that infect hosts that possess root hair entry form a single monophyletic clade. Strains that nodulate hosts with intercellular penetration form another cohesive group with some "flexible strains" capable of both modes of entry (Cournoyer et al. 1993). Actinorhizal plant hosts, which permit root hair entry likewise cluster within a single lineage in the phylogenetic tree constructed, based on the analysis of their *rbcL* sequences. Hosts which permit only intracellular entry cluster in three separate lineages. These results suggest a close phylogenetic relationship among the hosts which support root hair infection and a co-evolution

Table 2. Promiscuity of actinorhizal hosts toward *Frankia*. (Adapted from Maggia and Bousquet 1994)

Frankia Strains		Host promiscuity in nodulation				
Host of origin	Strain	*Allo-casuarina*	*Casuarina*	*Gymno-stoma*	*Alnus*	*Myrica*
Allocasuarina	AlI1	+	+	+	-	+
Casuarina	CcI3	+	+	+	-	+
	CeI5	-	+	+	ND	+
Gymnostoma	Gp1	-	-	+	-	+
Alnus	ArI3	ND	ND	+	+	ND
	AvsI3	ND	-	ND	+	+
Comptonia	CpI1	ND	-	+	+	+
	CpI3	ND	-	ND	+	+
Myrica	MpI3	ND	-	ND	+	+
	M+gI5	ND	-	+	+	+
Elaeagnaceae	Ea118	ND	-	ND	-	+
	EUN1f				+	+
Colletia	WgCc1.17	ND	-	ND	-	+

+ Nodulation in at least one species tested within the genus; - no nodulation
ND Data not available

between these hosts and *Frankia* that nodulate them (Cournoyer et al. 1993; Swensen and Mullin 1997). Based on these analyses, Swensen and Mullin (1997) postulate that actinorhizal symbioses may have four independent origins in the evolution of land plants and that root hair mode of entry has originated only once in the evolutionary history of actinorhizal symbioses.

A phylogenetic study based on the *rbcL* sequences also indicated a strong correlation between the evolutionary history of the hosts and their promiscuity towards *Frankia*. Among the actinorhizal hosts tested, *Myrica* is the most promiscuous and *Casuarina* the least promiscuous (Table 2; Maggia and Bousquet 1994).

Swensen and Mullin (1997) suggest that evolution in actinorhizal symbioses has progressed from promiscuity to selectivity, perhaps even leading to loss of susceptibility altogether. For example, among the hosts that support only root hair entry, *Myrica* and *Gymnostoma* occupy an earlier position in the lineage and are nodulated by genetically diverse strains of *Frankia*. *Casuarina* and *Allocasuarina*, which are more selective with regard to *Frankia* strains, are in derived positions in this evolutionary lineage (Fig.1). Among the families that possess root hair mode of entry, Myricaceae and Betulaceae appear earliest in the fossil records followed by Casuarinaceae (Swensen and Mullin 1997; Maggia and Bousquet 1994).

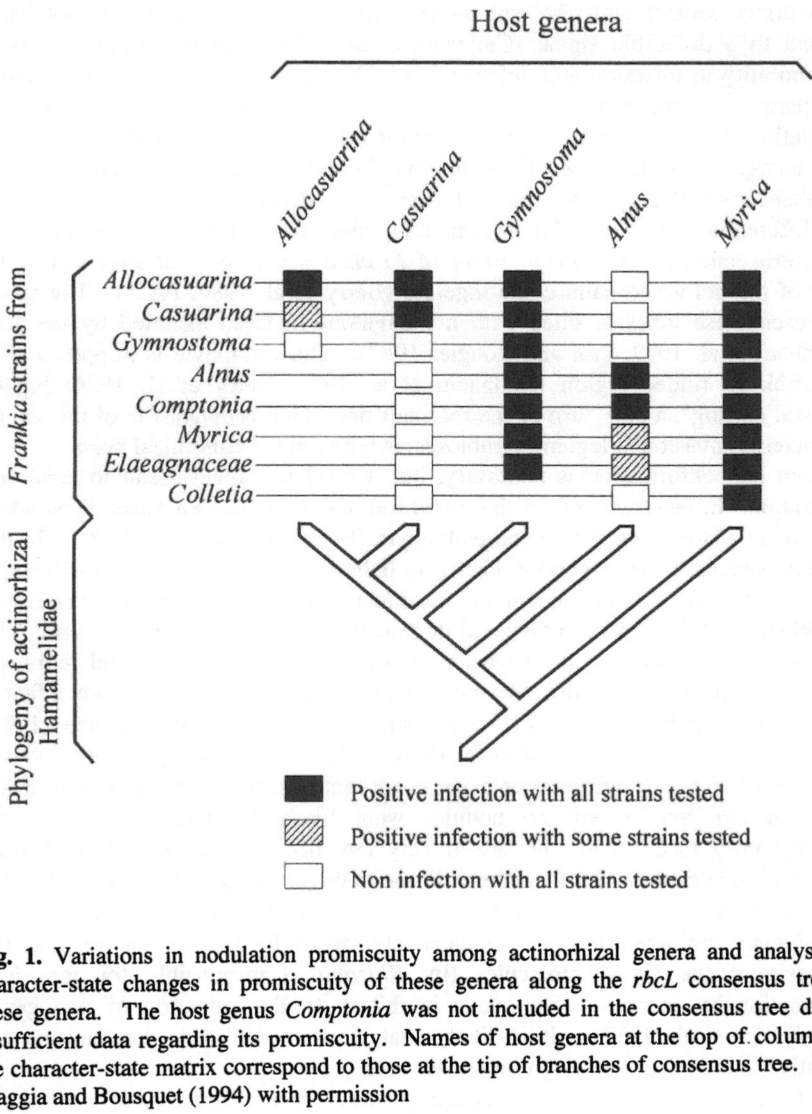

Fig. 1. Variations in nodulation promiscuity among actinorhizal genera and analysis of character-state changes in promiscuity of these genera along the *rbcL* consensus tree of these genera. The host genus *Comptonia* was not included in the consensus tree due to insufficient data regarding its promiscuity. Names of host genera at the top of columns of the character-state matrix correspond to those at the tip of branches of consensus tree. From Maggia and Bousquet (1994) with permission

2.2 Root Hair Infections

The earliest observable event in root hair infections is deformation of root hairs which begins as a swelling at the root hair tip, and occurs as early as 1 h after inoculation (Prin and Rougier 1987). Fully deformed hairs with an extensively branched or lobed appearance have been reported within 6 h after inoculation (van Ghelue et al. 1995). The branching is thought to result from the initiation of new growing points in the root hairs (Fig. 2). Each branch point or root hair need not be in direct contact with the endophyte filament indicating that the root hairs respond to a diffusible signal (Callaham et al. 1979; Prin and Rougier 1987). Susceptibility to infection and deformation is developmentally restricted to young root hairs just emerging and root hairs that are still growing at the time of inoculation (Burggraaf et al. 1983; Prin and Rougier 1987). Old root hairs that are fully elongated at the time of inoculation do not deform and remain straight. Inoculated roots thus typically possess a "slope" of deformed hairs, with root hairs that differentiate after inoculation remaining short and branched (Callaham et al. 1979; Prin and Rougier 1987). Entry of *Frankia* occurs only at sites where the lobes of the deformed hairs come together (Berry et al. 1986; Fig.3). The space inbetween these lobes is filled with mucilaginous material excreted by the host (Callaham et al. 1979; Prin and Rougier 1987). The endophyte is trapped within the crook or folded region (Callaham et al. 1979; Berry et al. 1986; Fig.4). "Marked curling", a 360° turn of the root hair tip, which is diagnostic of the site of the bacterial invasion in legume symbioses, is not seen in actinorhizal hosts.

Root hair deformation is necessary, but it need not always lead to symbiont penetration. In fact, most of the deformed root hairs are not penetrated even when they are in close contact with the endophyte (Berry and McCully 1990). In the legume systems, there is good evidence to believe that frequency of infections in the deformed hairs is dependent on the inoculum concentration (Bhuvaneswari unpublished 1985), host genotype and its sensitivity to the hormone ethylene. The supernodulating mutants of soybean (Gresshoff et al. 1988) and ethylene insensitive mutants of alfalfa (Penmetsa and Cook 1997) possess more infected root hairs compared to their wild type parents. Wall and Huss Danell (1997) suggested that symbiotic infections in *Alnus* could be regulated by the host even at the stage of root hair deformation. These authors observed that the region of the *Alnus incana* root, where no nodules were likely to develop due to host autoregulatory mechanisms, possessed very crowded short root hairs with high degree of deformation. Further, down in the younger parts of these roots the root hairs were no longer deformed even in the presence of *Frankia*. In legume-*Rhizobium* symbioses it has been conclusively shown that the host can regulate the synthesis of the signal molecules (by *Rhizobium*) responsible for root hair deformation by secreting compounds inhibitory to the induction of *nod* genes (Djordjevic et al. 1987). It is likely that similar regulatory mechanisms are operative in the actinorhizal symbioses as well.

Infected root hairs in *Alnus rubra* possess some characteristic cytological features. Deposition of secondary cell wall material is a characteristic feature in actinorhizal infections. The primary cell wall material of an axenically grown and uninfected root hair of *Alnus* is composed of a coarsely fibrillar wall material deposited in at least two layers with an outer mucilaginous covering. The mucilage layer has a coarsely textured, non-fibrillar appearance and it is secreted outside the cell wall (Prin and Rougier 1986). Presence of a cuticular layer between the mucilage and the outer wall was noted by Dawes and Bowler (1959), but was not observed by Berry et al. (1986).

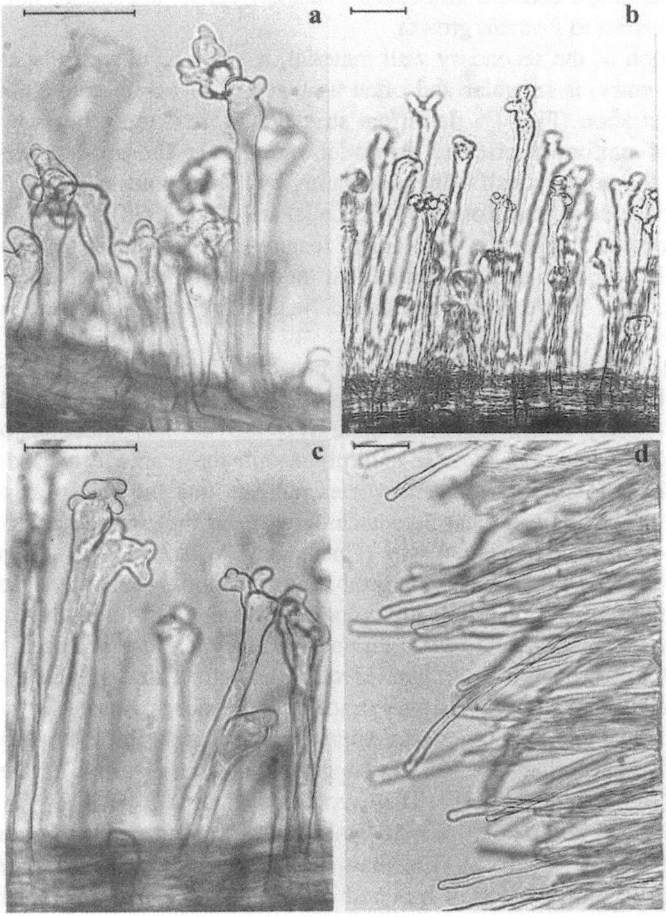

Fig. 2. Root hair deformations in *Alnus glutinosa-Frankia* interactions. Bars 20μm. *a* Root hairs of *A. glutinosa* seedlings exposed to 10^{-5} dilution of *Frankia* culture in stationary phase. Culture was squeezed through a syringe to break up the bacterial filaments and diluted with sterile water. Photograph taken 24 h after exposure. *b* and *c* Deformed root hairs of *A. glutinosa* seedlings exposed for 24 h to cell free filtrate from a stationary phase culture of *Frankia*. *b* Root exposed to 10^{-5} dilution of the filtrate. *c* Root exposed to 10^{-4} dilution. Note that most root hairs are deformed. *d* Root hairs of *A. glutinosa* seedlings exposed for 24 h to sterile *Frankia* growth medium diluted to 10^{-4} with sterile water

The cell walls of the infected root hairs become thicker and exhibit altered histochemical properties (Berry et al. 1986). The mucilage surrounding the infected root hairs also appears to be different from the mucilage excreted by the uninfected and axenic root hairs. In both *A. rubra* and *A. glutinosa*, the infected root hairs become surrounded by numerous host-excreted droplets of mucilaginous material (Fig. 4; Berry et al. 1986; Lalonde 1977). This mucilage layer of host origin contains embedded *Frankia* hyphae, but it does not constitute a physical barrier to *Frankia* growth.

Deposition of the secondary wall material, which occurs typically at the point of *Frankia* entry, is irregular and often arranged in elaborate convolutions at the site of penetration (Fig. 4). It differs structurally from the primary wall and is made up of uniformly oriented arrays of fine fibrils and appears less electron dense than the primary wall. The penetrating hypha becomes encapsulated by this wall material (Berry and Torrey 1983; Callaham et al. 1979). Berry and Torrey (1983) compared the general cytological features of the infected root hairs to the transfer cells in plants and postulated that these cytological details might indicate a metabolically active state associated with new cell wall and membrane synthesis.

A single infective *Frankia* hypha that penetrates the root hair stays continuous from the rhizosphere through the hair cell wall layers into the infected root hair (Berry and Torrey 1983). This is in contrast to the root hair infections of legumes by *Rhizobium* where the bacteria multiply within the markedly curled region of the root hair before entry. Recent studies indicate that the bacterium's ability to synthesize the exo-polysaccharide, succinoglycan, in the right amounts is critical for the root hair colonization step (Cheng and Walker 1998). Some of the *Rhizobium meliloti* mutants overproducing succinoglycans induced marked curling of the root hairs but failed to colonize this region or infect these root hairs. Since extensive growth of *Frankia* around the entry site has not been noted, this step may not be critical for root hair infections in actinorhizal symbioses. It is not known whether *Frankia* synthesizes extracellular polysaccharides.

At the actual site where *Frankia* enters the root hair cell, the primary wall has an altered microfibrillar organization suggesting a local digestion of the root hair cell wall (Berry et al. 1986). They reported an abrupt break in the root hair primary cell wall at the point of transition from outside to the wall interior. As in the case of legume infections (Callaham and Torrey 1981) this was postulated to involve localized dissolution of root hair cell wall by enzymes. The host cell wall degrading enzymes, particularly cellulases, were implicated in this process (Berry and Torrey 1983). The source(s) of enzymes for dissolution of root hair cell walls seems to be poorly documented and understood in both symbiotic systems. van Spronsen et al. (1994) suggested that cell wall degradation during infection thread formation in legumes might be a two step process. The first step is a local cell wall modification by plant enzymes, induced by rhizobial Nod factors. Munoz et al. (1998) have identified a polygalacturonase gene in *Medicago sativa* that is induced only in response to infection by nodulating strains of *Sinorhizobium meliloti*. These authors implicate a role for this enzyme in the degradation of host cell walls during root hair infection and nodule development.

The second step is complete cell wall degradation in the presence of rhizobia. Mateos et al. (1992) detected cell bound pectinolytic and cellulolytic enzymes in *Rhizobium leguminosarum* biovar *trifolii*. Krishnan and Pueppke (1994) overproduced cellulase in *Rhizobium fredii* with a cloned cellulase from *Erwinia*, but found that cellulase overproduction had no effect on the symbiotic phenotype of this genetically modified strain. Although there is convincing microscopic

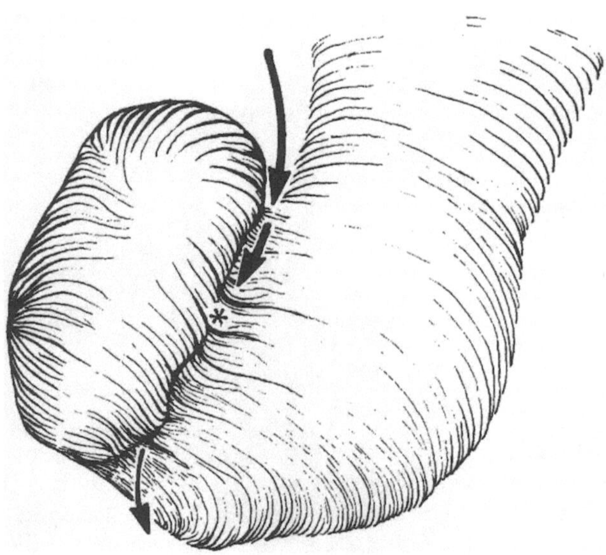

Fig. 3. Artist's representation of the pathway of *Frankia* infection and branching in an infected root hair of *A. rubra* (*arrows*). (*) The site of entry into the root hair. From Berry et al. (1986) with permission

documentation for cell wall dissolution in the root hair infections in both systems (Berry et al. 1986; Callaham and Torrey 1981) equally convincing genetic evidence to implicate the enzymes is lacking. van Spronsen et al. (1994) reported that they never observed complete cell wall degradation in the absence of rhizobia. However, to date no mutants of rhizobia (or legumes), which lack cellulolytic or pectinolytic enzymes have been isolated and tested for infection related phenotypes. DNA sequences homologous to pectinases and cellulases have not been detected or cloned from rhizobia but sequences homologous to pectate lyase and cellulase have been detected in *Frankia* (Seguin et al. 1989). Culture fluids of *Frankia* possess cell wall degrading enzyme activities (Safo Sampah and Torrey 1988). Whether the production of these enzymes by the *Frankia* is required for the dissolution of the root hair cell wall at the point of entry is yet unresolved.

In successful infections, the hypha entering the root hair becomes completely encapsulated by the newly deposited cell wall material and this structure is analogous to the infection threads formed during intracellular invasion of legumes by rhizobia. The infection threads in legumes are filled with a matrix substance, which is composed of both bacterial and host-secreted products. Production of exo-polysaccharide by *Rhizobium* is necessary for induction and further growth of infection threads (Cheng and Walker 1998). Mutants defective in exo-polysaccharide production and mutants that produce exo-polysaccharide with altered chemical composition fail to induce infection thread formation (Cheng and Walker 1998). Exo- polysaccharide production by rhizobia is thought to aid in the

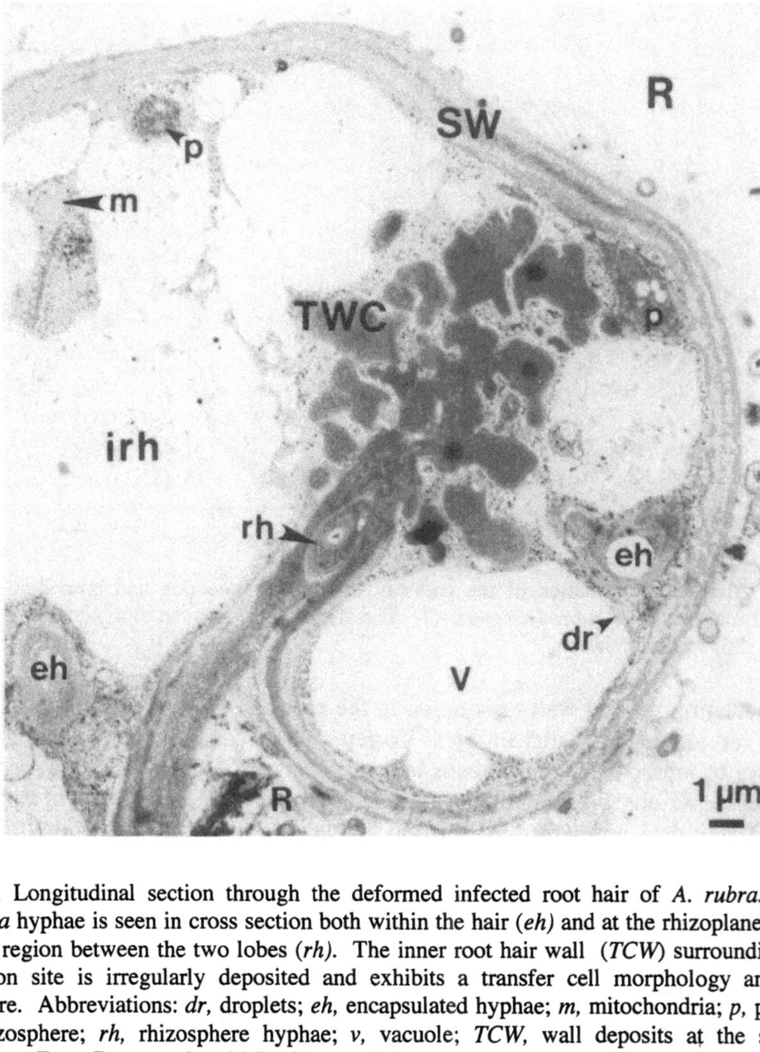

Fig. 4. Longitudinal section through the deformed infected root hair of *A. rubra.* The *Frankia* hyphae is seen in cross section both within the hair *(eh)* and at the rhizoplane in the folded region between the two lobes *(rh)*. The inner root hair wall *(TCW)* surrounding the infection site is irregularly deposited and exhibits a transfer cell morphology and sub structure. Abbreviations: *dr*, droplets; *eh*, encapsulated hyphae; *m*, mitochondria; *p*, plastid; *R*, rhizosphere; *rh*, rhizosphere hyphae; *v*, vacuole; *TCW*, wall deposits at the site of infection. From Berry et al. (1986) with permission

acceleration of root hair curling so that bacterial penetration of the root precedes the induction of the host's defense responses (van Workum et al. 1998)

Infection threads in actinorhizal symbioses lack matrix substances (Pawlowski 1997). *Frankia* hyphae are thus in direct contact with the host encapsulating material. The endophyte remains encapsulated by the secondary wall material even in the infected nodule cells (Pawlowski 1997). This situation is analogous to those found in nodules of many woody legumes and *Parasponia* where rhizobia remain inside the persistent infection threads (Hirsch and LaRue 1997). Persistence of infection threads in nodules is thought to be an evolutionarily less specialized symbiotic state than bacteroid differentiation and release (Sprent et al.

1989). The encapsulated *Frankia,* the infection threads, as well as the differentiated rhizobial bacteroids remain external to the host cell plasma membrane at all times.

2.3 Intercellular Infections

Frankia infections occur by intercellular penetration in all hosts that belong to the families Coriariaceae, Datiscaceae, Elaeagnaceae, Rhamnaceae, and Rosaceae. In these hosts, not only the entry is intercellular, but further proliferation of *Frankia* in the root cortex and nodule primordia is also intercellular. In these hosts, *Frankia* becomes intracellular and encapsulated only in the nodules (Liu and Berry 1991; Miller and Baker 1985; Racette and Torrey 1989). This mode of infection involves extensive digestion of the host middle lamella. Cell wall degrading enzymes of *Frankia* could play a very significant role in the infection process. No obvious interactions between the root hairs and *Frankia* are evident during this mode of entry.

2.4 Unsuccessful Infections

Frankia infections are initiated in many root hairs but only a fraction of these leads to the successful penetration of root hairs and initiation of prenodule formation. In root hairs where infections are arrested, callose deposits surround the *Frankia* hyphae where they penetrate the root hair wall. Cytoplasm of these root hairs, which possess arrested infections, show varying degrees of disintegration and contain few or no recognizable organelles. This may be a reaction induced in the host to regulate the numbers of successful infections (Berry and McCully 1990). Presence of callose-containing cell wall appositions were also noticed by Berry and McCully (1990) in several deformed root hairs that lie near successfully infected root hairs. These papillae were localized along the sidewalls of the root-hair lobes or branches, but never at the tips of these root hairs and they may indicate unsuccessful attempts by *Frankia* to enter root hairs. The secondary wall material elaborated to encapsulate *Frankia* at the site of penetration does not contain callose (Berry and McCully 1990). Whether callose deposition by the host occurs also in the unsuccessful intercellular infections is not yet known.

3 Root Hair Deformation

3.1 Biology

Deformation of root hairs is the first morphological indication of molecular signal exchange between the host plants and *Frankia.* Interactions between root hairs and a diffusible molecule from *Frankia* were first inferred as the signal for deformation by Callaham et al. (1979). It is now well established that direct cell to cell contact between *Frankia* and host root hairs is not necessary for root hair deformation (Burggraaf et al. 1983). Deformations can be induced in root hairs of actinorhizal hosts by exposing the roots to sterile cell free filtrates of *Frankia*. Besides *Frankia*, other soil bacteria like *Bradyrhizobium japonicum*, *Rhizobium* sp. 32H1, and *Pseudomonas cepacia* also deform root hairs of *Alnus* sp. (Berry et al. 1983; Knowlton et al. 1980; Prin and Rougier 1987).

Significance of root hair deformation, particularly its role as an indicator of host specificity in the legume-*Rhizobium* symbioses, was debated for at least three decades (Bhuvaneswari and Solheim 1985). It is now well established that the root hair deformation indeed is the first visible event indicative of the biologically specific interactions between the legume hosts and their symbionts (Heidstra et al. 1994). The molecules responsible for root hair deformation have been purified from several rhizobia and chemically characterized as lipochito-oligosaccharides. Elegant studies on their structure-function relationships have elucidated how host specificity is determined by these molecules (Roche et al. 1991; Spaink and Lugtenberg 1994; Dénarié et al. 1992). Purified lipochito-oligosaccharides or "Nod factors" induce deformation of root hairs, preinfection thread formation, as well as nodule organogenesis in a concentration-dependent manner (Spaink et al. 1991; Truchet et al. 1991; van Brussel et al. 1992). It is therefore reasonable to accept the root hair deformation induced by the cell free extracts of *Frankia* as early indicators of host specific symbiotic interactions. Biological significance of root hair deformations induced by the other soil bacteria mentioned above or their cell free filtrates, on the other hand, is more doubtful.

Deformations induced by cell free filtrates of *Frankia* on *A. glutinosa* root hairs are morphologically identical to that of deformation seen in the presence of *Frankia* (Fig. 2a, b, and c) and it follows a similar time course. Beginning of the deformation process is noticeable as tip swelling occurring one hour after exposure to the filtrates. By 6 h, root hairs are fully deformed and possess many branches at the tips (Fig. 2b; Prin and Rougier 1987; van Ghelue et al.1995). There is thus a short lag period before morphological changes are obvious. Whether this is required for the synthesis of new gene products is currently unknown. The deformation response is restricted to root hairs that are still actively growing. Root hairs, which are fully elongated at the time of exposure to *Frankia* or its cell free filtrates, remain undeformed even after prolonged exposure (Prin and Rougier 1987).

The degree of deformation caused by the cell free extracts is dependent on the concentration of the deformation factor(s) (DF) present in the filtrates. This was examined by studying the extent of deformation seen in root hairs that were exposed for 6 h to increasing dilutions of the cell free filtrates. At 10^{-3} dilution most root hairs were fully deformed with branches while above 10^{-6} dilutions the majority of the deformed root hairs showed no more than tip swelling. A period of 2 min contact between the culture filtrate and the root hairs is sufficient to start the process (van Ghelue et al. 1995). This time course of deformation observed for *Alnus* is remarkably similar to the deformation time course observed for the legumes *Vicia sativa* (Heidstra et al. 1994) and *Trifolium repens* (Bhuvaneswari and Solheim 1985). Deformation induced on *Alnus* root hairs by *P. cepacia* also follows a similar time course (Berry and Torrey 1983).

It is still not clear whether *Frankia* needs to be cultured in the presence of its host roots in order to induce the synthesis of DF. Prin and Rougier (1987) and Schwencke and Selim (1995) reported that the production of DF in vitro needed exudates from seeds and/or roots in the medium. van Ghelue et al. (1995) on the other hand, found this requirement critical for only some strains of *Frankia* (Fig. 2). Synthesis of lipochito-oligosaccharide molecules (Nod factors) by rhizobia occurs only upon exposure to the specific flavanoid inducers secreted by their host plants (Long 1989). Flavanone-like compounds have been detected in the seed and root exudates of *A. rubra* and some of them promote nodulation but others inhibit the process (Benoit and Berry 1997). Flavones have been shown to act as both inducers and anti-inducers of nod factor synthesis in the *Rhizobium*-legume symbiosis as well (Djordjevic et al. 1987)

3.2 Deformation Factor(s)

The chemical nature of the DF in the cell free filtrates of *Frankia* is still unknown. There have been relatively few attempts to purify this molecule due to technical difficulties associated with slow rate of growth and biomass production in *Frankia* cultures and our inability to genetically manipulate *Frankia*. A preliminary characterization of DF has shown that:

a) Its biological activity is stable to heating up to 100°C (Prin and Rougier 1987), but is lost gradually when stored at 4°C (van Ghelue 1994).
b) Periodate oxidation of the cell free filtrates destroys its biological activity (van Ghelue 1994).
c) Incubation of the culture filtrates with endochitinase or a mixture of N-acetyl glucosaminidase + chitobiosidase does not affect its biological activity (van Ghelue 1994).
d) Biological activity is also retained after treatment with enzymes cyclohelicase, ProteinaseK, β1-3 and β1-6 glucanases (van Ghelue 1994).
e) Biological actvity is lost after treatment with Pronase (Solheim unpublished 1997).

The available data are too tenuous to draw any conclusions about DF's chemical identity or about its similarity to the Nod factors of rhizobia. The striking similarities in the biology of the root hair deformation and nodule development, has led to the view that the molecule(s) that influence the developmental responses in actinorhizal hosts may also be lipochito-oligosaccharides. Phylogenetic evidence, which suggests that predisposition to develop nodules in association with prokaryotes, may have originated only once in angiosperm evolution (Doyle 1998), lends support to this view. Attempts to determine whether *Frankia* harbors structural and/or functional homologs to *Rhizobium nodA* and *nodD* genes, however, have failed to provide unequivocal evidence for their presence in *Frankia* (Chen et al. 1991; Reddy et al. 1992). DF from *Frankia* does not purify in the same manner as the Nod factors of rhizobia and on a sizing column the DF activity from *Frankia* elutes at an apparent molecular weight of 5000 d (van Ghelue 1994). Therefore, the possibility that the deformation factor(s) of *Frankia* may be chemically distinct from the Nod factors of rhizobia cannot be ruled out.

4 Conclusions

The 200 or more actinorhizal hosts are traditionally assigned to eight families. These families belong to different evolutionary lineages within the angiosperms, which implies that symbioses evolved independently in these lineages. Recent results from molecular phylogenetic analysis of *Frankia* and its hosts suggest that actinorhizal symbioses may have originated at least four times during angiosperm evolution and that the root hair mode of infection arose only in one of these lineages. Phylogenetic evidence for co-evolution between the two symbiotic partners is stronger in the actinorhizal symbioses than it is in the legume-*Rhizobium* system (Sprent 1994). Due to the striking similarities in the infection and nodulation events between the actinorhizal and plant-rhizobia symbioses, we can expect root hair deformation in actinorhizal hosts to be indicative of the host specific interactions in nodulation. Early signal exchange events between the interacting partners seem to be highly conserved in all well-investigated plant-prokaryote associations (Rasmussen et al. 1996). Thus, it is likely that the regulation of the early events in actinorhizal symbioses may follow a similar theme. However, evidence for this is still tenuous in the actinorhizal system. The chemical nature of signal molecule(s) from either *Frankia* or the plant hosts is yet not known. Our inability to genetically manipulate *Frankia* is the major obstacle to progress in this area. The need to direct our future research efforts towards overcoming this single obstacle cannot be overemphasized.

ACKNOWLEDGMENTS

The authors thank Prof. Ann Hirsch, University of California, Los Angeles, USA, and Dr. John Beck Jensen, University of Tromsø, for critically reading the manuscript.

References

Baker DD and Schwintzer CR (1990) Introduction. In: Schwintzer CR and Tjepkema JD (Eds) The biology of *Frankia* and Actinorhizal Plants. Academic press, New York. pp 3-11

Benoit LF and Berry AM (1997) Flavonoid-like compounds from seeds of red alder (*Alnus rubra*) influence host nodulation by *Frankia* (Actinomycetales). Physiol Plant 99:588-593

Berry AM and McCully ME (1990) Callose-containing deposits in relation to root-hair infections of *Alnus rubra* by *Frankia*. Can J Bot 68:798-802

Berry AM, McIntyre L and McCully ME (1986) Fine structure of root hair infection leading to nodulation in the *Frankia Alnus* symbiosis. Can J Bot 64:292-305

Berry AM and Sunell LA (1990) The infection process and nodule development. In: Schwintzer CR and Tjepkema JD (Eds) The Biology of *Frankia* and Actinorhizal plants. Academic Press Inc., New York.pp 61-81

Berry AM and Torrey JG (1983) Root hair deformation in the infection process of *Alnus rubra*. Can J Bot 61:2863-2876

Berry AM, Torrey JG and McCully ME (1983) The fine structure of the root hair wall and surface mucilage in the actinorhizal host *Alnus rubra*. In: Goldberg R (Ed) Plant Molecular Biology Alan R. Liss inc, New York. pp 319-327

Bhuvaneswari TV and Solheim B (1985) Root hair deformation in the white clover (*Trifolium repens*) and *Rhizobium trifolii* symbiosis. Physiol Plant 63:25-34

Burggraaf AJP, van der Linden J and Tak T (1983) Studies on the localization of infectable cells on *Alnus glutinosa* roots. Plant Soil 74:175-188

Callaham D, Newcomb W, Torrey JG and Peterson RL (1979) Root hair infection in actinomycete-induced root nodule initiation in *Casuarina, Myrica, a*ND *Comptonia*. Bot Gaz 140:S1- S9

Callaham D, Tredici PD and Torrey JG (1978) Isolation and cultivation *in vitro* of the actinomycete causing rot nodulation in *Comptonia*. Science (Washington) 199:899-902

Callaham DA and Torrey JG (1981) The structural basis for infection of root hairs of *Trifolium repens* by *Rhizobium trifolii*. Can J Bot 59:1647-1664

Carmona AF (1974) La formation des nodule fixateurs d´azote chez *Alnus glutinosa (L.)* Vill. Acta Bot Ned 23:257-303

Chen L-M, Cui Y-H, Qin M, Wang Y-l, Bai X-l and Ma Q-S (1991) Identification of a *nod*D -like gene in *Frankia* by direct complementation of a *Rhizobium nodD*- mutant. Mol Gen Genet 233:311-314

Cheng HP and Walker GC (1998) Succinoglycan is required for initiation and elongation of infection threads during nodulation of alfalfa by *Rhizobium meliloti*. J Bacteriol 180:5183-91

Cournoyer B, Gouy M and Normand P (1993) Molecular phylogeny of the symbiotic actinomycetes of the genus *Frankia* matches host-plant infection processes. Mol Biol Evol 10:1303-1316

Dawes CJ and Bowler E (1959) Light and electron microscope studies of the cell wall structure of the root hairs of *Rhaphanus sativus*. Am. J. Bot. 46:561-565

Dénarié J, Debellé F and Rosenberg C (1992) Signaling and host range in nodulation. Annu Rev Microbiol 46:497-531

Djordjevic MA, Redmond JW, Bately M and Rolfe BG (1987) Clovers secrete specific compounds which stimulate or repress nod gene expression in *Rhizobium trifolii*. EMBO J 6:1173-1179

Doyle JJ (1998) Phylogenetic perspectives on nodulation: evolving views of plants and symbiotic bacteria. Trends Plant Sci 3:473- 478

Gresshoff PM, Mathews A, Krotsky A, Olsson JE, Carroll BJ, Delvis AC, Kosslak R, Applebaum ER and Day DA (1988) Supernodulation and non-nodulation mutants of soybean. In: Palacios R and Verma DPS (Eds) Molecular genetics of Plant-microbe Interactions. APS press St. Paul, MN pp 364-369

Heidstra R, Geurts R, Franssen H, Spaink HP, van Kammen A and Bisseling T (1994) Root hair deformation activity of nodulation factors and their fate on *Vicia sativa*. Plant Physiol 105:787-797

Hirsch AM and LaRue TA (1997 Is the legume nodule a modified root or stem organ or an organ sui generis? Critical Reviews in Plant Science 16:361-392

Knowlton S, Berry A and Torrey JG (1980) Evidence that associated soil bacteria may influence root hair infection of actinorhizal plants by *Frankia*. Can J Microbiol 26:971-977

Krishnan HB and Pueppke SG (1994) A cloned cellulase gene from *Erwinia carotovora* subsp. carotovora is expressed in *Rhizobium fredii* but does not influence nodulation of cowpea. FEMS Microbiol Lett 119:289-294

Lalonde M (1977) The infection process of *Alnus* root nodule symbiosis. In Recent Developments in Nitrogen fixation. Newton W, Postgate JR and .Rodriguez-Barrueco C (Eds) Academic Press New York pp 569-589

Liu Q and Berry AM (1991) Localization and characterization of pectic polysaccharides in roots and root nodules of *Ceanothus* spp. during intercellular infection by *Frankia*. Protoplasma 163:93-101

Long SR (1989) *Rhizobium*- legume nodulation: Life together in the underground. Cell 56:203-214

Maggia L and Bousquet J (1994) Molecular phylogeny of the actinorhizal *Hamamelidae* and relationships with host promiscuity towards *Frankia*. Mol Ecol 3:459-467

Mateos FM, Jimenez-Zuro JI, Chen J, Squartini AS, Haack SK, Martinez-Molina E, Hubbel DH and Dazzo FB (1992) Cell-associated pectinolytic and cellulolytic enzymes in *Rhizobium leguminosarum* biovar *trifolii*. Appl. Environ. Microbiol. 58:1816-1822

Miller IM and Baker DD (1985) The initiation, development and structure of root nodules in *Elaeagnus angustifolia* (Elaeagnaceae). Protoplasma 128:107-119

Miller IM and Baker DD (1986) Nodulation of actinorhizal plants by *Frankia* strains capable of both root hair infection and intercellular penetration. Protoplasma 131:

Munoz, JA, Coronado, C, Perez-Hormaeche, J, Kondorosi, A, Ratet, P, and Palomares,AJ (1998) MsPG3, a *Medicago sativa* polygalacturonase gene expressed during the alfalfa-*Rhizobium meliloti* interaction PNAS 95:9687-9692

Pawlowski K (1997) Nodule-specific gene expression. Physiol Plant 99:617-631

Pawlowski K and Bisseling T (1996) Rhizobial and actinorhizal symbioses: What are the shared features? Plant Cell 8:1899-1913

Penmetsa RV and Cook DR (1997) A legume ethylene-insensitive mutant hyperinfected by its rhizobial symbiont. Science (Washington) 275:527-530

Prin Y and Rougier M (1986) Cytological and histochemical characteristics of the axenic root surface of *Alnus glutinosa*. Can J Bot 64:2216-2226

Prin Y and Rougier M (1987) Preinfection events in the establishment of *Alnus- Frankia* Symbiosis: Study of root hair deformation step. Plant Physiology (Life SciAdv) 6:99-106

Quispel A (1990) Discoveries, discussions and trends in research on actinorhizal root nodule symbioses before 1978. In: Schwintzer CR and Tjepkema JD (Eds) The Biology of *Frankia* and Actinorhizal plants Academic Press Inc New York. pp 17-28

Racette S and Torrey JG (1989) Root nodule initiation in *Gymnostoma* (*Casuarinaceae*) and *Shepherdia* (*Elaeagnaceae*) induced by *Frankia* strain HFPGpI1. Can J Bot 67:2873-2879

Rasmussen U, Johansson C, Renglin A, Petersson C and Bergman B (1996) A molecular characterization of the *Gunnera-Nostoc* symbiosis: comparison with *Rhizobium* and *Agrobacterium* plant interactions. New Phytol 133:391-398

Reddy A, Bochenek B and Hirsch AM (1992) A new *Rhizobium meliloti* symbiotic mutant isolated after introducing *Frankia* DNA sequence into a *nodA*::Tn5 strain. Mol Plant Micro Inter 5:62-71

Ringø E, Clausen E, Løvaas E, van Ghelue M and Solheim B (1995) Effects of extracts of *Alnus glutinosa* seeds on growth of *Frankia* strain ArI3 under static and fermentor culture conditions. Plant Soil 176:283-288

Roche P, Lerouge P, Ponthus C and Prome JC (1991) Structural determination of bacterial nodulation factors involved in the *Rhizobium meliloti* alfalfa symbiosis. J Biol Chem 266:10933-10940

Safo Sampah S and Torrey JG (1988) Polysaccharide-hydrolyzing enzymes of *Frankia* (Actinomycetales). Plant Soil 112:89-98

Schwencke J and Selim S (1995) Deformation factors of *Casuarina* root hairs (dnF) are induced in *Frankia* by flavonoids from *Casuarina* seeds. In: Berry AM (Ed) 10th international conference on *Frankia* and actinorhizal plants. University of California, Davis.

Seguin A, Lalonde M, Winship LJ and Benson DR (1989) Detection of pectolytic activity and *pel* homologous sequences in *Frankia*. Plant Soil 118:1-2

Spaink HP and Lugtenberg BJJ (1994) Role of rhizobial lipo-chitin oligosaccharide signal molecules in root nodule organogenesis. Plant Mol Biol 26:1413-1422

Spaink, HP, Sheeley, D.M, van Brussel, AA, Glushka, J, York W S, Tak, T, Geiger O, Kennedy EP, Reinhold V, and Lugtenberg BJ (1991) A novel highly unsaturated fatty acid moiety of lipo-ologosaccharide signal determines host specificity of *Rhizobium*. Nature 328:337-340

Sprent JI (1994) Evolution and diversity in the legume-*Rhizobium* symbioses: Chaos theory? Plant Soil 161:1-10

Sprent JI, Sutherland JM, F, De Faria SM, Stirton CH and Zarucchi JL (1989) Structure and function of root nodules from woody legumes. Advances in legume biology Monographs in Systematic Botany from the Missouri Botanical Garden 29:559-578

Swensen SM and Mullin BC (1997) Phylogenetic relationships among actinorhizal plants: The impact of molecular systematics and implications for the evolution of actinorhizal symbiosis. Physiol Plant 99:565-573

Tjepkema JD and Torrey JG (1979) Symbiotic nitrogen fixation in actinomycete-nodulated plants. Bot Gaz (supplement) 140:i-ii

Truchet G, Roche P, Lerouge P, Vasse J, Camut S, de Billy F, Prome JC and Denarie J (1991) Sulfated lipo-oligosaccharide signals of *Rhizobium meliloti* elicit root nodule organogenesis in alfalfa. Nature 351:670-673

van Brussel AAN, Bakhuizen R, van Spronsen PC, Spaink HP, Tak T, sLugtenberg BJJ and Kijne JW (1992) Induction of pre-infection thread structures in the leguminous host plant by mitogenic lipo-oligosaccharides of *Rhizobium*. Science (Washington) 257:70-72

van Ghelue M (1994) Interactions in Actinorhizal symbiosis. Thesis, University of Tromsø.

van Ghelue M, Løvaas E, Ringø E and Solheim B, (1995) Early interactions between *Alnus* glutinosa and *Frankia* strain ArI3. Production and specificity of root hair deformation factor(s). Physiol Plant 99:579-587

van Spronsen PC, Bakhuizen R, van Brussel AAN and Kijne JW (1994) Cell wall degradation during infection thread formation by the root nodule bacterium *Rhizobium leguminosarum* is a two-step process. Euro J Cell Biol 64:88-94

van Workum W. A.T, van Slageren Sophie, van Brussel AAN and Kijne JW (1998) Role of exopolysaccharides of *Rhizobium leguminosarum bv. viciae* as host plant -specific molecules required for infection thread formation during nodulation of *Vicia sativa*. Mol Plant Micro Inter 11:1233-1241

Wall LG and Huss Danell K (1997) Regulation of nodulation in *Alnus incana-Frankia* symbiosis. Physiol Plant 99:594-600

Index

c

d

g

h

i

k

l

MIX
Papier aus verantwortungsvollen Quellen
Paper from responsible sources
FSC® C105338

If you have any concerns about our products,
you can contact us on
ProductSafety@springernature.com

In case Publisher is established outside the EU,
the EU authorized representative is:
Springer Nature Customer Service Center GmbH
Europaplatz 3, 69115 Heidelberg, Germany

Printed by Libri Plureos GmbH
in Hamburg, Germany